Tissue Engineering Strategies for Organ Regeneration

Editors

Naznin Sultana

Medical Academy, Prairie View A&M University, Texas, USA

Sanchita Bandyopadhyay-Ghosh

Department of Mechanical Engineering
Manipal University Jaipur
Rajasthan, India

Chin Fhong Soon

Biosensor and Bioengineering Laboratory
Microelectronics and Nanotechnology-Shamsuddin
Research Centre (MiNT-SRC)
AND
Faculty of Electrical and Electronic Engineering
Universiti Tun Hussein Onn Malaysia
Batu Pahat, Johor, Malaysia

CRC Press
Taylor & Francis Group
Boca Raton London New York

CRC Press is an imprint of the
Taylor & Francis Group, an **informa** business
A SCIENCE PUBLISHERS BOOK

Cover credit: Cover illustrations reproduced by kind courtesy of Dr. Mohd. Izzat Hassan and Mr. Kapender Phogat.

CRC Press
Taylor & Francis Group
6000 Broken Sound Parkway NW, Suite 300
Boca Raton, FL 33487-2742

First issued in paperback 2021

CRC Press is an imprint of Taylor & Francis Group, an Informa business

ISBN-13: 978-1-138-39154-3 (hbk)
ISBN-13: 978-1-03-217526-3 (pbk)
DOI: 10.1201/9780429422652

Publisher's Note
The publisher has gone to great lengths to ensure the quality of this repribut points out that some imperfections in the original copies may be apparent.

Library of Congress Cataloging-in-Publication Data

Names: Sultana, Naznin, editor. | Bandyopadhyay-Ghosh, Sanchita, 1971- editor. | Soon, Chin Fhong, 1974- editor.
Title: Tissue engineering strategies for organ regeneration / editors, Naznin Sultana, Sanchita Bandyopadhyay-Ghosh, Chin Fhong Soon.
Description: Boca Raton : CRC Press, Taylor & Francis Group, [2020] | Includes bibliographical references and index.
Identifiers: LCCN 2019036051 | ISBN 9781138391543 (hardback)
Subjects: LCSH: Tissue scaffolds. | Guided tissue regeneration. | Tissue engineering. | Biomedical materials. | Regeneration (Biology)
Classification: LCC R857.T55 T574 2020 | DDC 610.28--dc23
LC record available at https://lccn.loc.gov/2019036051

Visit the Taylor & Francis Web site at
http://www.taylorandfrancis.com

and the CRC Press Web site at

Preface

Tissue Engineering Strategies for Organ Regeneration addresses the multidisciplinary tissue engineering approaches in regenerating different types of tissues and organs. This book provides a comprehensive summary of the recent improvements of biomaterials used in scaffold-based tissue engineering and describes the different protocols for the manufacture of scaffolds. In addition, it describes the mechanisms behind cell-biomaterials interactions. In addition, this book focuses on advanced technologies such as microfabrication techniques for tissue engineering approaches. Several *in vitro* and *in vivo* functions of tissue engineering scaffolds for several applications are provided. Tissue engineering scaffolds with multifunctional properties such as biocompatibility, conductivity and antibacterial characteristics are addressed in the book. This book not only addresses the present constraints in tissue engineering applications but also highlights the future directions of tissue engineering applications.

This book is written by experts from around the world in tissue engineering field. Chapter 1 focuses on designing biomaterials for regenerative medicine: state-of-the-art and future perspectives. Chapter 2 deals with new generation materials for applications in bone tissue engineering and regenerative medicine. Chapter 3 discusses enhanced scaffold fabrication techniques for optimal characterization. Chapter 4 describes next generation tissue engineering strategies by combination of organoid formation and 3D bioprinting. Chapter 5 provides a strategy for regeneration of three-dimensional (3D) microtissues in microcapsules: aerosol atomization technique. Chapter 6 highlights BioMEMS devices for tissue engineering. Chapter 7 focuses on injectable scaffolds for bone tissue repair and augmentation. Chapter 8 presents the details associated with bio-ceramics for tissue engineering. Chapter 9 reviews stimulus-receptive conductive polymers for tissue engineering. Chapter 10 delivers PCL/Chitosan/Nanohydroxyapatite/Tetracycline composite scaffolds for bone tissue engineering.

All the chapters of this book are self-contained and focused on current tissue engineering strategies for organ restoration. It is expected that the book will be a great resource and reference for the multidisciplinary societies such as the Researchers, Advanced Undergraduate and Postgraduate students in Biomedical Engineering, Materials Engineering, Chemical Engineering, and Clinical investigators.

Naznin Sultana
Sanchita Bandyopadhyay-Ghosh
Chin Fhong Soon

Preface

Contents

1

Designing Biomaterials for Regenerative Medicine: State-of-the-Art and Future Perspectives

Zohreh Arabpour[1], Mansour Youseffi[2], Chin Fhong Soon[3], Naznin Sultana[4], Mohammad Reza Bazgeir[5], Masoud Mozafari[6,7] and Farshid Sefat[2,8*]

1.1 INTRODUCTION

The complexity of the human body can be simplified when stating the matter from which it is composed. To expand, the body is known to be made up of four tissue types, including epithelial tissue, neural tissue, muscle tissue and connective tissue. Each tissue type is created from a varying physiology, which contributes to the functionality of the matter. For example, the muscle tissue is rich in mitochondria due to the excessive need for oxygen in order for it to function with great exertion of energy. Table 1.1 demonstrates the four tissue types, and clarifies both the functionality of the matter alongside the cells within the tissue which allows the tissue type to work as it should.

[1] Department of Tissue Engineering and Applied Cell Sciences, School of Advanced Technologies in Medicine, Tehran University of Medical Sciences, Tehran, Iran.

[2] Department of Biomedical and Electronics Engineering, School of Engineering, University of Bradford, Bradford, UK.

[3] Biosensor and Bioengineering Lab, MiNT-SRC, Faculty of Electrical and Electronic Engineering, Universiti Tun Hussein Onn Malaysia, 86400 Parit Raja, Batu Pahat, Johor, Malaysia.

[4] Medical Academy, Prairie View A&M University, TX 77446, USA.

[5] Royal National Orthopaedic Hospital, Brockley Hill, London, UK.

[6] Bioengineering Research Group, Nanotechnology and Advanced Materials Department, Materials and Energy Research Center (MERC), Tehran, Iran.

[7] Department of Tissue Engineering & Regenerative Medicine, Faculty of Advanced Technologies in Medicine, Iran University of Medical Sciences, Tehran, Iran.

[8] Interdisciplinary Research Centre in Polymer Science & Technology (IRC Polymer), University of Bradford, Bradford, UK.

* Corresponding author: F.Sefat1@Bradford.ac.uk

TABLE 1.1 The four main tissue types in the human body

Tissue Type	Location	Functions	Subtype
Epithelial	Cover inner and outer organ and body surfaces	Protection, secretion, absorption	Squamous, columnar, cuboidal, simple, pseudostratified, stratified
Connective	Between other tissues	Support and protect body	Loose and dense
Muscle	Attached to skeletal system, digestive system	Movement	Skeletal or striated, cardiac and smooth
Nervous	Distributed in the body	Regulates and controls physical functions	Central and peripheral system

1.2 ORGAN SYSTEMS

There are 11 main organ systems in the human body, composed of different variations of the main tissue types. These organ systems are vital to quality of life, and if one organ within a system fails to carry out its purpose, fatality could occur, hence the need of tissue engineering intervention. Trauma is one of the main causes for organ failure, and the body responds through expressing genes, growth factors and activating cells as a healing process. Unfortunately, humans do not possess the capability to regrow limbs, such as the salamander; however in terms of natural tissue growth, the extracellular matrix for some tissues (such as simple connective) can be rebuilt to a certain extent (Krafts 2010, Zadpoor 2015).

As technology has advanced, novel methods have been developed and tested, portraying that synthetic tissues can exponentially increase the duration of a life cycle, by allowing continuity of functioning organ systems. Examples of beneficial tissue regeneration include creation of blood vessels for cardiac patients, bone scaffolds for amputees, skin grafts for burn patients and many more, to be discussed further in the chapter.

1.3 ESSENTIAL REQUIREMENTS IN DESIGNING BIOMATERIALS FOR TISSUE ENGINEERING AND REGENERATIVE MEDICINE APPLICATIONS

1.3.1 Mechanical Requirements

The chemical and physical optimization of new biomaterials in order to interact with living cells are being studied by many research groups (Khan and Tanaka 2017). Synthetic or hybrid biomaterials should be developed to adapt for living systems or live cells *in vitro* and *in vivo*. The selection and design of an appropriate biomaterial is determined by specific application of scaffold. Some of the mechanical properties that are of utmost importance are hardness, plasticity, elasticity, tensile strength and compressibility. For example, ceramics such as hydroxyl apatite (HAp), and tricalcium phosphate (TCP) are appropriate for bone regeneration (Khan and Tanaka 2017). The scaffold of the bioceramic should mimic mechanical properties of the anatomical location that will be planted and the degradation rate should be consistent with bioactive surface for suitable tissue regeneration. Since the regeneration rates of bone are different for different age groups, this must be taken into consideration when designing scaffolds because the rate of regeneration in older adults is slower than young individuals (O'Brien 2011).

Maintaining the mechanical behavior of implanted scaffolds structure and tolerance of stress and loads during the reconstruction is very important. The stability of scaffolds in biological systems depends on some factors such as stress, strength, elasticity, temperature and absorption of the material associated with chemical degradation. Therefore, in order to select an appropriate biomaterial, it is important to assess some of the following properties: (1) Elastic behavior—measurement of pressure in response to tensile or compressed stress during the force. This reversible behavior could be assessed by linear relation between stress and strain. Stress is a measure of load and strain is a measure of displacement; (2) Plastic behavior—when (or compression) uniaxial tensile stress reaches yield strength, permanent deformation occurs; (3) Tensile strength—the highest stress that material can endure before breakdown; (4) Ductility—the plastic strain at failure. Plasticity before breaking; (5) Toughness—the energy needed to break a unit volume of material; (6) Flexural behavior—the relationship between a flexural stress and strain in response to a tensile or compressive stress perpendicular to the bar. The mechanical behavior of materials can be equated by some factors. Swelling, porosity pore size, shape, orientation, and connectivity are some of these factors that directly impacts mechanical properties of the biomaterial (Olson et al. 2011).

The balance between mechanical behaviors and porous pattern allowing cell penetration and vascularization is necessary to ensure success of scaffolds in tissue engineering. The mechanical stiffness as well as the roughness of materials and the physical stimulation of the three-dimensional microstructure of the scaffold significantly influence the cellular regeneration, cellular polarization and balanced intracellular signaling (Olson et al. 2011).

1.3.2 Biological Requirements

Production of appropriate scaffolds to support the proliferation and differentiation of cells to mimic biological function of extracellular matrix proteins is another essential step to generate appropriate 3D biomimetic scaffolds in tissue engineering (Chiono et al. 2009). Biocompatibility of scaffolds must be ensured, to avoid undesirable immune responses to the implant and ectopic calcifications *in vivo*. The surface of biomaterials should have excellent chemical properties to improve attachment, migration, proliferation, and differentiation of cells (Mandal et al. 2009).

Biomaterials used as scaffold in tissue engineering should be non-toxic to eliminate inflammatory or allergic reactions in the human body (Moztarzadeh et al. 2018). The human body's response to the implant determines the success of the implanted biomaterial, and assesses the degree of biocompatibility of a substance. The tissue response to the materials and materials' degradation in the body system are two major factors that affect the biocompatibility of biomaterial (Sefat et al. 2018). In this context, biodegradability should be controllable to support the formation of new tissue (Grayson et al. 2003). After a biomaterial implant is exposed to the body, tissues start to react to the implant surface. The body responses to implants are: (1) Thrombosis or coagulation of blood after platelets are attached to the surface of implant and, (2) Formation of fibrous capsule around the surface of implant (Chiono et al. 2009). The kind of reactions depends on type of biomaterial that is used in the implants. Biomaterials based on the body responses can be classified into three main groups: bioinert, bioactive and bioresorbable. The bioresponses and examples of each classified biomaterials are as shown in Table 1.2 (Geetha et al. 2009).

TABLE 1.2 Biomaterials classification and interaction with tissue

Classification	Response	Examples
Bioinert materials	Minimal interaction with tissue. Formation of connective tissue capsules (0.1-10 lm) around the implant, without any attachment to the implant surface	Zirconia, polymethyl metha acrylate (PMMA), alumina, titanium, etc
Bioactive materials	Interaction with tissue. Formation of new tissue around the implant and strongly merges with the implant surface	Bioglass, glass ceramic, synthetic, hydroxyl apatite (HAP)
Bioresorbable materials	Dissolved and replaced by the advanced tissue	Polyglycolic acid and Polylactic, tricalcium phosphate, composites of proteins

Bioinert materials have minimal interaction with its surrounding tissue and the connective tissue capsules. Usually, very few attachments to the implant surface would be formed around the implant. Examples of bioinert materials are zirconia, polymethyl metha acralyte (PMMA), alumina, titanium and stainless steel. Bioactive materials can interact and form adhesions with the surrounding bone and soft tissue. New tissues around the implant would be formed and strongly integrated with the implant surface. Some examples of bioactive materials are bioglass, glass ceramic, synthetic and hydroxyl apatite (HAP). Bioresorbable materials are dissolvable and can be replaced by new tissue after implanted in the body. These materials are polyglycolic acid, polylactic acid, composites of proteins, and tricalcium phosphate (Heness and Ben-Nissan 2004). For production of implants, bioactive materials are highly preferred than bioinert materials because of the high integrability with the surrounding bone and soft tissue (Geetha et al. 2009).

Recently, much attention has been paid to bioresorbable materials (Kaur et al. 2017). These scaffolds appear as more suitable technologies for the treatment of patients. These materials will promote restoration of both tissue function and anatomic formation through regeneration. Penetration of new tissue in reabsorbed site leads to decrease risk of thrombosis and fibrosis (Garcia-Garci et al. 2014).

1.4 POLYMERIC-BASED BIOMATERIALS

1.4.1 Natural Polymeric-based Biomaterials

Natural and synthetic polymer-based biomaterials were widely used as cell supporting matrices in tissue regeneration (See Fig. 1.1) (Jafari et al. 2017). These kinds of biomaterials have been considered as attractive approaches, because of their chemical and biological similarities to extracellular matrix (ECM) (Phu et al. 2011). The ideal material for tissue engineering should have suitable properties such as desired biocompatibility, microstructure, mechanical property, degradation rate and potency for cell supporting to retrieve metabolic functions (Sun and Tan 2013).

Naturally derived materials can prepare the biological environment to recognize and connect with metabolic system. Diminishing inflammation or immunological reactions and cytotoxicity, as well as excellent biocompatibility are some advantages of natural biomaterial. A large class of polymers can be extracted from living organisms according to their chemical structures. They can be classified into three groups: (i) polysaccharides, (ii) proteins, and (iii) nucleic acids. Hyaluronic acid (HA), chondroitin sulfate, chitin and chitosan, alginates, and cellulose belong to polysaccharides group. Proteins have essential role in biological activities. Elastin, silk, collagen and fibrinogen are classified in the protein group. Both polysaccharides and proteins are widely used as tissue engineering scaffold or in drug delivery.

Biotechnology by the micro-organism's fermentation or manufactured by enzymatic processes offers new materials to produce natural polymers (Widner et al. 2005). Animal, plants and algae are still the major sources for polymer extraction (Mano et al. 2007). The manipulating of these materials into porous scaffold usually needs more complicated techniques than synthetic polymers. Several methods like using appropriate cross linker are necessary for resolving mechanical and physical defects of natural polymers. The cell toxicity caused by cross-linkers is another challenge for natural polymer utilization (Wu et al. 2016). There are also clinical requirements for processing of natural biomaterials into other forms such as nano/micro particles to release control of application, or performance into two or three-dimensional structures.

As mentioned before, this group of materials has excellent biocompatibility but poor mechanical and thermal properties while synthetic polymers can be optimized for desirable properties. Combination of synthetic and natural polymers and selecting an appropriate cross-linker helps to overcome these challenges.

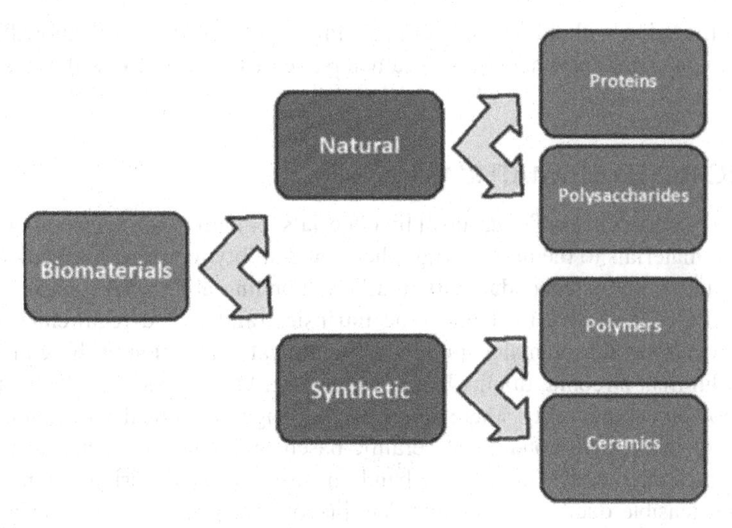

FIGURE 1.1 Classification of biomaterials.

1.4.2 Synthetic Polymeric-based Biomaterials

The synthetic biodegradable polymers are major groups of polymers that are popular in tissue engineering. These biodegradable and noncytotoxic materials can support cell attachment, proliferation and differentiation to reconstruct tissue defect (Guelcher 2008). A major difference between natural biopolymers and synthetic polymers is in their structures. Most synthetic polymers have much simpler than natural polymer structures. The degradation rate of these groups of materials are adjustable by changing the mixing ratio, molecular weight of components and other parameters to match with the regeneration rate of the tissue. Polyanhydrides, polyesters, polyphosphazenes, poly (glycerol sebacate) and polyurethanes are classified in this group of materials. This group can be manipulated to the desirable characteristics. Since polyanhydrides possess high hydrophobicity and favorable degradation pattern (degradation from the surface to the inside), they are appropriate for drug-delivery applications (Jain et al. 2008).

Polyesters are polymers formed from a dicarboxylic acid and a diol. Polyesters can be degraded by hydrolysis of the ester groups. The degradation rates can be controlled by the polymer concentration, structure and molecular weight. Polylactide (PLA), polyglycolide (PGA), their copolymer poly (lactide-co-glycolide) (PLGA) and Poly (ε-caprolactone) (PCL) are popular synthetic biopolymers. These polymers have suitable property for bone and cartilage regeneration (Liu and Ma 2010, Deshpande et al. 2013).

Polyphosphazenes usually have high molecular weight with an inorganic structure of alternating phosphorous and nitrogen atoms where each phosphorous atom connects to two groups of organic substituents. Different properties of polymer can be prepared by changing these attached groups. Polyphosphazenes are suitable for tissue-engineering scaffolds and drug delivery (Guo and Ma 2014).

Poly (glycerol sebacate) (PGS) is a content of glycerol and sebacic acid, which requires reasonable procedure to be produced with thermo set elastomeric properties. Also, the mechanical properties and degradation rates of PGS can be controlled by optimizing the concentrations, curing time, curing temperature and the degree of acrylation in acrylated form of PGS for specific performance. The most common use of this polymer is for soft tissue regeneration, such as retina, cardiac muscle, nerve and cartilage. Due to its elastomeric nature of this polymer, it is applied in drug controlled-release systems and hard-tissue regeneration (Masoumi et al. 2013).

Polyurethanes (PUs) are popular because of their hardness, durability, biocompatibility, and stability in biological systems. In addition, Polyurethanes were demonstrated to improve cell adhesion and proliferation without any side effects (Phan et al. 2005). Conventionally, PUs have been used as permanent and inert materials in catheters, heart valves and vascular grafts (Santerre et al. 2005).

1.5 CERAMIC-BASED BIOMATERIALS

Ceramic-based biomaterials are major group of biomaterials for hard tissue regeneration. Because of similarity of these materials to the mineral ingredients and ability to mimic mechanical properties of natural bones and teeth, osteoconductivity and bone bonding ability have made them suitable for bone and teeth's regenerative medicine. The intrinsic fragility and requirements of tedious manufacturing processing are the major limitation for clinical application of these materials (Wei and Ma 2004). Alumina, zirconia, titania, hydroxyapatite (HA), and calcium silicate are common ceramics with good biocompatibility. There are many challenges to clinical applications of ceramic biomaterials due to inadequate bonding of ceramic-based bioimplants to adjacent bones during long-term use in the human body. On the other hand, the use of these materials to repair large bone defects is still not feasible due to the inherent fragility of ceramic. Therefore, designing of new ceramic materials that are linked to the bones with osteoinductive and osteoconductive potency can be revolutionized for orthopedic implants (Liu et al. 2010).

1.6 NANO-BIOMATERIALS

The communication between surface of biomaterials and living tissues plays essential role in reaction of the body to implanted devices. Microstructure, chemical composition and morphology of surface determined the surface property that leads to mediate the cells' behavior on the surfaces. Many studies (Sefat 2017) have been performed to understand the process of these interactions to improve the success of transplantation of biomaterial in the body systems. Microscale and nanoscale of surfaces have been shown with considerable effects on cellular and sub-cellular behaviors. In fact, ions, DNA, proteins, and many factors are constituents of nanostructures in the human body.

Previous studies (Nuffer and Siegel 2009) showed that nanoscale materials can interact effectively with some proteins in the live tissues. Nanostructured surface of biomaterials promotes attachment, migration and amplification of cells (Urbanska et al. 2018). For example, in bone regeneration, nanoscale materials can mimic components and structures of natural bone such as proteins and minerals (Tariverdian et al. 2018). Connection between hydroxyapatite and nanofibers can induce osteoblast performance and some adhesive protein such as vitronectin and fibronectin. In addition, the nanoscale structure can help to enhance deposition of calcium on it (Liu et al. 2010). In bone regeneration, the degree of nanoscale of implants surface is more effective than materials chemistry in bone cells function (Manjubala et al. 2006). Therefore, a nanoscale of surface can improve biological properties of the implants. There are various techniques for surface modification of biomaterials such as the nanocoatings of ceramics on polymers and metals implants to improve clinical outcomes. Nanoscale scaffolds are more stable than microscaled scaffolds and they can provide large surface-to-volume ratio of material that is favorable for cell attachment and proliferation. Various methods were designed for producing nanoparticles and nanofibers. Phase separation, electrospinning, electrospray and self-assembly are becoming popular in tissue engineering (Barnes et al. 2007).

1.7 COMPOSITE BIOMATERIALS

Recently, composite of biomaterials have been considered in the construction of scaffolds for tissue engineering (Nejatian et al. 2017). As previously discussed, natural polymers are more biocompatible than synthetic polymers in living tissue, but they have a lower mechanical strength compared with synthetic polymers (Syed et al. 2018). Several methods were used to overcome the mechanical strength defects of natural polymers, using appropriate cross-linkers or combining these polymers with synthetic polymers. These strategies can considerably improve the mechanical property and chemical stability of natural polymers. In addition, using natural polymers in composite leads to the improvement of biocompatibility of synthetic polymers. On the other hand, optimizing the degradation rate of biomaterials in accordance with target tissue regeneration rate is one of the most important points in scaffolds designing for tissue engineering. Designing predictable degradation rate of scaffold for different applications can be achieved by using composite of biomaterials with different degradation rate (Mohamadi et al. 2017). For example, in order for the regeneration of human hard tissues, multiple hydroxyapatite composites with ceramic, metal, or polymer have been used widely to improve mechanical and chemical properties of implants (Suchanek and Yoshimura 1998). Moreover, combination of polyesters/ceramic and calcium phosphates as an osteoconductive have been used to improve biodegradability of scaffold for bone tissue engineering (Wei and Ma 2004).

1.8 DISCOVERY OF NOVEL BIOMATERIALS

Stimulant or environmentally sensitive polymers are a class of novel polymers. Environmental stimulation such as temperature, pH and ion changes, and also magnetism, light and ultrasound, can induce physical changes in the polymer. Thermo sensitive polymers exhibit physical changes where temperatures are more popular than other stimulant sensitive polymers. The temperature can change the phase of polymers (Tcr) and hence, induce changes in the physical properties of the polymer. The polymer at temperature below the phase changing temperature behaves as hydrophilic and switches to a hydrophobic behavior when stimulated at a higher temperature. Consequently, the cells show affinity to attach on the polymer at a temperature higher than the phase changing temperature and the cell layers can be detached from the polymer at a lower temperature. Hence, the cell layer can be separated without using enzymes such as crude trypsin or mechanical dissociation. Poly (N-isopropylacrylamide) (PNIPAAm) is one of the most commonly used thermoresponsive polymers in tissue engineering (Ravichandran et al. 2012).

1.9 CONCLUDING REMARKS AND FUTURE PERSPECTIVES

Although the transplantation of organs from one person to another seemed feasible for organ regeneration, it presented some limitations. The greatest limitation is the accessibility to adequate donor tissues or organs for all the people who are waiting for them (Meyer et al. 2009). According to statistics in 2018, more than 120,000 people in the United States are waiting for organ transplants. In addition, immune problems after transplantation cause chronic rejection and destruction in implanted organ. The usage of immunosuppressive drugs can lead to formation of new tumor. These limitations require new solutions to provide for highly demanding regenerative tissue.

Tissue engineering is a progressive technology, opening up new ways to fathom the dream of regenerative medicine (Olson et al. 2011, Mahjour et al. 2015). Nanotechnology and tissue engineering have the potential to revolutionize medical treatments in the near future (Olson et al. 2011). The basis of tissue engineering is cell culture on scaffold that is synthesized from natural membranes or artificial membranes made of natural or synthetic polymers. Natural and synthetic biopolymers have shown very different advantages. Although natural membranes contain growth factors and cytokines that promote cell viability, these compounds may induce immune protective responses (Sun and Tan 2013). Combination of synthetic and natural polymers has the potential to overcome poor mechanical property of natural polymers and low biocompatibility of synthetic polymers. Scaffolds in tissue engineering should support the proliferation and differentiation of cells to mimic biological function of extracellular matrix. Biocompatible and non-cytotoxic materials can support cell viability, proliferation and differentiation for tissue regeneration. The degradation rate of these materials is controllable by changing the ratio and the molecular weight of the composites. Accordingly, the rate of destruction of the scaffolds must be proportional to the reconstruction rate of the target tissue. Additionally, a biomaterial must imitate the mechanical properties of the natural tissue. The surface of materials should present suitable chemistry to support attachment, migration, proliferation and differentiation of cells. Many improvements can be made through bionanotechnology. The nanostructure of a surface can be modified to improve the material interactions with cells and tissues. Nano-functionalization of surfaces by various techniques can enhance cellular and subcellular functions on the scaffolds. Bionanotechnology can provide great prospect for future biomaterials' design in tissue engineering.

REFERENCES

Barnes, C.P., S.A. Sell, E.D. Boland, D.G. Simpson and G.L. Bowlin. 2007. Nanofiber technology: designing the next generation of tissue engineering scaffolds. Advanced Drug Delivery Reviews 59(14): 1413-1433.

Chiono, V., C. Tonda-Turo and G. Ciardelli. 2009. Artificial scaffolds for peripheral nerve reconstruction. International Review of Neurobiology, Academic Press 87: 173-198.

Deshpande, P., C. Ramachandran, F. Sefat, I. Mariappan, C. Johnson, R. McKean, M. Hannah, V.S. Sangwan, F. Claeyssens and A.J. Ryan. 2013. Simplifying corneal surface regeneration using a biodegradable synthetic membrane and limbal tissue explants. Biomaterials 34(21): 5088-5106.

Garcia-Garcia, H.M., P.W. Serruys, C.M. Campos, T. Muramatsu, S. Nakatani, Y.-J. Zhang, Y. Onuma and G.W. Stone. 2014. Assessing bioresorbable coronary devices: methods and parameters. JACC: Cardiovascular Imaging 7(11): 1130-1148.

Geetha, M., A.K. Singh, R. Asokamani and A.K. Gogia. 2009. Ti based biomaterials, the ultimate choice for orthopaedic implants: a review. Progress in Materials Science 54(3): 397-425.

Grayson, A.C.R., I.S. Choi, B.M. Tyler, P.P. Wang, H. Brem, M.J. Cima and R. Langer. 2003. Multi-pulse drug delivery from a resorbable polymeric microchip device. Nature Materials 2: 767.

Guelcher, S.A. 2008. Biodegradable polyurethanes: synthesis and applications in regenerative medicine. Tissue Engineering Part B: Reviews 14(1): 3-17.

Guo, B. and P.X. Ma. 2014. Synthetic biodegradable functional polymers for tissue engineering: a brief review. Science China Chemistry 57(4): 490-500.

Heness, G. and B. Ben-Nissan. 2004. Innovative Bioceramics. Materials Forum, Institute of Materials Engineering Australasia Ltd.

Jafari, M., P. Paknejad, M. Rezai Rad, S.R. Motamedian, M.J. Eghbal, N. Nadjmi and A. Khojasteh. 2017. Polymeric scaffolds in tissue engineering: a literature review. Journal of Biomedical Materials Research 105(2): 431-459.

Jain, J.P., D. Chitkara and N. Kumar. 2008. Polyanhydrides as localized drug delivery carrier: an update. Expert Opinion on Drug Delivery 5(8): 889-907.

Kaur P., S.A. Khaghani, A. Oluwadamilola, Z. Khurshid, M. Sohail Zafar, M. Mozafari, M. Youseffi and F. Sefat. 2017. Fabrication and characterizations of hydrogels for cartilage repair. Advances in Tissue Engineering & Regenerative Medicine 2(6): 283-288.

Khan, F. and M. Tanaka. 2017. Designing smart biomaterials for tissue engineering. International Journal of Molecular Sciences 19(1): 17.

Krafts, K.P. 2010. Tissue repair; the hidden drama. Organogenesis 6(4): 225-233.

Liu, X. and P.X. Ma. 2010. The nanofibrous architecture of poly (L-lactic acid)-based functional copolymers. Biomaterials 31(2): 259-269.

Liu, X., P.K. Chu and C. Ding. 2010. Surface nano-functionalization of biomaterials. Materials Science and Engineering: R: Reports 70(3-6): 275-302.

Mahjour, S.B., X. Fu, X. Yang, J. Fong, F. Sefat and H. Wang. 2015. Rapid creation of skin substitutes from human skin cells and biomimetic nanofibers for acute full-thickness wound repair. Burns 41(8): 1764-1774.

Mandal, B.B., T. Das and S.C. Kundu. 2009. Non-bioengineered silk gland fibroin micromolded matrices to study cell-surface interactions. Biomedical Microdevices 11(2): 467-476.

Manjubala, I., S. Scheler, J. Bössert and K.D. Jandt. 2006. Mineralisation of chitosan scaffolds with nano-apatite formation by double diffusion technique. Acta biomaterialia 2(1): 75-84.

Mano, J., G. Silva, H.S. Azevedo, P. Malafaya, R. Sousa, S.S. Silva, L. Boesel, J.M. Oliveira, T. Santos and A. Marques. 2007. Natural origin biodegradable systems in tissue engineering and regenerative medicine: present status and some moving trends. Journal of the Royal Society Interface 4(17): 999-1030.

Masoumi, N., K.L. Johnson, M.C. Howell and G.C. Engelmayr Jr. 2013. Valvular interstitial cell seeded poly (glycerol sebacate) scaffolds: toward a biomimetic *in vitro* model for heart valve tissue engineering. Acta biomaterialia 9(4): 5974-5988.

Meyer, U., T. Meyer, J. Handschel and H.P. Wiesmann. 2009. Fundamentals of Tissue Engineering and Regenerative Medicine. Springer Nature Switzerland.

Mohamadi, F., S. Ebrahimi-Barough, M.R. Nourani, K. Mansoori, M. Salehi, A.A. Alizadeh, S.M. Tavangar, F. Sefat, S. Sharifi and J. Ai. 2017. Enhanced sciatic nerve regeneration by human endometrial stem cells in an electrospun poly (ε-caprolactone)/collagen/NBG nerve conduit in rat. Artificial cells, Nanomedicine and Biotechnology 46(8): 1731-1743.

Moztarzadeh, S., K. Mottaghy, F. Sefat, A. Samadikuchaksaraei and M. Mozafari. 2018. Nanoengineered biomaterials for lung regeneration. pp. 305-323. *In*: M. Mozafari, J. Rajadas and D. Kaplan [eds.]. Nanoengineered Biomaterials for Regenerative Medicine. Elsevier, USA.

Nejatian, T., Z. Khurshid, M. Zafar, S. Najeeb, S. Zohaib, M. Mazafari, L. Hopkinson and F. Sefat. 2017. Dental biocomposite. pp. 65-84. *In*: L. Tayebi and K. Moharamzade [eds.]. Biomaterials for Oral and Dental Tissue Engineering. Woodhead Publishing, USA.

Nuffer, J.H. and R.W. Siegel. 2009. Nanostructure–biomolecule interactions: implications for tissue regeneration and nanomedicine. Tissue Engineering Part A 16(2): 423-430.

Olson, J.L., A. Atala and J.J. Yoo. 2011. Tissue engineering: current strategies and future directions. Chonnam Medical Journal 47(1): 1-13.

O'Brien, F.J. 2011. Biomaterials & scaffolds for tissue engineering. Materials Today 14(3): 88-95.

Phan, T.T., I.J. Lim, E.K. Tan, B.H. Bay and S.T. Lee. 2005. Evaluation of cell culture on the polyurethane-based membrane (Tegaderm): implication for tissue engineering of skin. Cell Tissue Bank 6(2): 91-97.

Phu, D., L.S. Wray, R.V. Warren, R.C. Haskell and E.J. Orwin. 2011. Effect of substrate composition and alignment on corneal cell phenotype. Tissue Eng Part A 17(5-6): 799-807.

Ravichandran, R., S. Sundarrajan, J.R. Venugopal, S. Mukherjee and S. Ramakrishna. 2012. Advances in polymeric systems for tissue engineering and biomedical applications. Macromolecular Bioscience 12(3): 286-311.

Santerre, J., K. Woodhouse, G. Laroche and R. Labow. 2005. Understanding the biodegradation of polyurethanes: from classical implants to tissue engineering materials. Biomaterials 26(35): 7457-7470.

Sefat, F., T. Raja, M. Zafar, Z. Khurshid, S. Najeeb, S. Zohaib, E. Ahmadi, M. Rahmati and M. Mozafari. 2018. Nanoengineered biomaterials for cartilage repair. pp. 39-71. *In*: M. Mozafari, J. Rajadas and D. Kaplan [eds.]. Nanoengineered Biomaterials for Regenerative Medicine. Elsevier, USA.

Suchanek, W. and M. Yoshimura. 1998. Processing and properties of hydroxyapatite-based biomaterials for use as hard tissue replacement implants. Journal of Materials Research 13(1): 94-117.

Sun, J. and H. Tan. 2013. Alginate-based biomaterials for regenerative medicine applications. Materials 6(4): 1285-1309.

Syed, M., M. Khan, F. Sefat, Z. Khurshid, M. Zafar and A. Khan. 2018. Bioactive glass and glass fiber composite: biomedical/dental applications. pp. 467-495. *In*: G. Kaur [ed.]. Biomedical, Therapeutic and Clinical Applications of Bioactive Glasses. Woodhead Publishing, USA.

Tariverdian, T., P. Zarintaj, P. Milan, M. Saeb, S. Kargozar, F. Sefat, A. Samadikuchaksaraei and M. Mozafari. 2018. Nanoengineered biomaterials for kidney regeneration. pp. 325-344. *In*: M. Mozafari, J. Rajadas and D. Kaplan [eds.]. Nanoengineered Biomaterials for Regenerative Medicine. Elsevier, USA.

Urbanska, A., F. Sefat, S. Yousaf, S. Kargozar, P. Milan and M. Mozafari. 2018. Nanoengineered biomaterials for intestine regeneration. pp. 363-378. *In*: M. Mozafari, J. Rajadas and D. Kaplan [eds.]. Nanoengineered Biomaterials for Regenerative Medicine. Elsevier, USA.

Wei, G. and P.X. Ma. 2004. Structure and properties of nano-hydroxyapatite/polymer composite scaffolds for bone tissue engineering. Biomaterials 25(19): 4749-4757.

Widner, B., R. Behr, S. Von Dollen, M. Tang, T. Heu, A. Sloma, D. Sternberg, P.L. DeAngelis, P.H. Weigel and S. Brown. 2005. Hyaluronic acid production in Bacillus subtilis. Applied and Environmental Microbiology 71(7): 3747-3752.

Wu, Z., X. Su, Y. Xu, B. Kong, W. Sun and S. Mi. 2016. Bioprinting three-dimensional cell-laden tissue constructs with controllable degradation. Scientific Reports 6: 24474.

Zadpoor, A.A. 2015. Mechanics of Biological Tissues and Biomaterials: Current Trends, Multidisciplinary Digital Publishing Institute, Materials, Basel.

2

New Generation Materials for Applications in Bone Tissue Engineering and Regenerative Medicine

Ravikumar K[1], Ashutosh Kumar Dubey[2] and Bikramjit Basu[1,*]

2.1 INTRODUCTION

Research and development in the biomedical materials sector has grown exponentially in the last few decades due to the advent of new materials, technological expertise and novel applications in the affordable healthcare sector. According to a report by global market research companies, the biomedical materials industry is estimated to be worth $130 billion worldwide by the year 2020 with emphasis on growth and development in the Asia-Pacific region, especially in India and China (MarketsandMarkets.com 2015). The materials used in the biomedical industry can be classified based on their properties and applications. Based on their characteristics, biomedical materials can be categorized into metallic materials such as stainless steel, titanium and its alloys, cobalt-chrome alloys, etc. which have well known applications. This is followed by ceramic materials and the common examples are calcium phosphates, apatites, calcium sulphates, alumina (Al_2O_3), zirconia (ZrO_2), glass ceramics and graphite. The other class is the polymeric materials, which include polyethylenes, polyesters, polymethylmethacrylate (PMMA), nylon, and silicone rubber which are few of the well known materials. Also, some of the naturally occurring materials derived from plants or animals such as cellulose, fibrin, collagen, gelatin, hyaluronic acid, alginates, chitin and silk are the excellent materials for biomedical applications. However, the materials currently being used in biomedical industry have longstanding issues that make it more difficult to use such materials in conjuction with the human body. For example, stress shielding owing to high elastic modulus as well as corrosion during longer service are the common concerns, associated with the widely used metallic implants such as stainless steel, CoCrMo, and Ti as well as Ti-based alloys. In addition, fibrous tissue formation, which may lead to local and systemic effects, is also a concern associated with these metallic implants (Geetha et al. 2009, Basu and Ghosh 2017). Other classes of materials such as ceramics and polymers bring in their own advantages and disadvantages. Generally, the merits of ceramic materials are resistance to corrosion, bioactivity/bioinertness

[1] Materials Research Centre, Indian Institute of Science, Bangalore-560012.
[2] Department of Ceramic Engineering, Indian Institute of Technology (BHU), Varanasi-221005.
[*] Corresponding author: bikram@iisc.ac.in

(depending on the chemical composition) and good wear resistance whereas the downsides are most of the ceramic materials are not bioresorbable (except certain bioglasses) resulting in a prolonged foreign body response in addition to their inherent poor mechanical properties (fracture toughness) (Basu et al. 2010). On the other hand, commonly used polymeric materials are mainly bioinert and bioresorbable. Bioinert polymers such as teflon (PTFE), ultra-high molecular weight polyethylene (UHMWPE), and high density polyethylene (HDPE) have been shown to have applications in hard tissue replacement and some of the well-known biodegradable polymers such as polylactic acid, poly (latic-co-glycolic acid), polyesters, polyanhydrides, and polyglycolic acid are usually used as drug delivery vehicles, sutures and biodegradable prosthetics and scaffolds (Mano et al. 2004). A major issue with these biodegradable materials is the nature of foreign body response of not only the parent material but also of the degradation products and a significant amount of effort is still being spent in studying this response of the host tissue to a variety of different materials in order to reduce, if not overcome, the undesirable effects of the host response (Anderson et al. 2008, Morais et al. 2010, Chandorkar et al. 2019). The recent advances in the field are undergoing a paradigm shift from bioactive and bioresorbable materials to a new generation of materials that can elicit specific cellular and sub-cellular responses to control cell fate processes to facilitate tissue repair and regeneration. This has given way to what are now being termed as Third Generation Biomaterials (Hench and Polak 2002). The techniques for tissue repair and regeneration are currently focused on alternative strategies for tissue engineering and *in situ* tissue regeneration. While the underlying principles of biomaterials, biocompatibility and host tissue response remain unchanged, the trend is to realize alternate and equally (or more) efficient routes for repair and regeneration of tissues. The chapter elucidates this concept with some of the prime examples of materials influencing cell response that can potentially aid in cell and tissue regeneration.

2.2 ELECTRICAL PROPERTIES OF BONE

Inspite of the significant advancement in materials for orthopedic applications during the last few decades, the development of number of bioengineering approaches such as biochemical, mechanical, physical, electrical and magnetic stimuli etc. in combination with the prosthetic implants are continuously being pursued in order to achieve effective integration and long-lasting service of the implant with the host bone tissue (Basu 2017). In addition to the biomechanical compatibility, the functional property resemblence of prosthetic implants with the natural living bone has been the major concern in recent years. The natural bone is a hybrid nano-composite material, primarily consisiting of crystalline hydroxyapatite nanoplates and collagenous proteins. In fact, the living bone is also regarded as the quintessence of biologically controlled self-assembly instead of a composite because of its inherent repairing and remodeling ability. This is mainly due to the unique hierarchical structure (shown in Fig. 2.1) which is responsible for its mechanical and electrical properties (Taylor et al. 2007, Pethig 1985). The relationship between the mechanical and electrical nature of the bone is generally described using its piezoelectric properties. The mechanical loading results in the generation of electrical charges.

The existance of piezoelectric phenomenon in bone was reported by the pioneer work of Yasuda (Yasuda 1954). Thereafter, Fukada and Yasuda (Fukada and Yasuda 1957) suggested that the bone demonstrates both direct as well as inverse piezoelectric effects. The non-centrosymmetric structure of the collagen molecule in bone has been suggested as one of the possible reasons for such phenomenon (Hastings and Mahmud 1988). It is known that collagen fibers slip past over one another under the application of internal or external mechanical stimulation, which generates piezoelectricity (Fukada and Yasuda 1964).

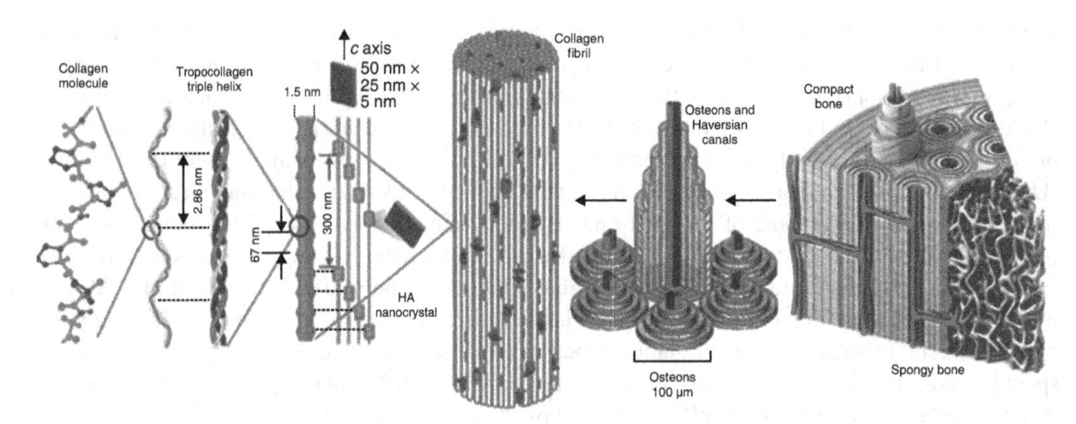

FIGURE 2.1 Hierarchical structure of natural bone, representing the sub-nanostructure of collagen molecules and tropocollagen helix, submicrostructure (collagen fibrils) and its macrostructure. Image adapted with permission from Pethig (1985).

Color version at the end of the book

The compressive and tensile stresses on the bone tissue generate the negative and positive charges, respectively, in the closed vicinity of physiologically stressed sections. The negatively and positively charged regions promote the osteoblastic and osteoclastic activities, respectively, and consequently, the bone deposition or bone desorption takes place from these polarized regions (Bassett and Becker 1962, Friedenberg et al. 1971). It has been reported that the bone piezoelectricity also helps in healing of fractured bone. Owing to such electro-mechanical coupling response, natural bone is suggested to act as a transducer (Isaacson and Bloebaum 2010). The bone piezoelectricity also helps in bone regeneration by means of mechanotransduction (Moss 1997).

In addition to piezoelectric phenomenon, the natural bone also exhibits pyroelectric behavior, i.e. the bone can generate momentary potential in response to its heating or cooling (El Messiery et al. 1979, Lang 1966, Lang 1981). The natural bone has also been reported to exhibit ferroelectricity. The microscopic arrangement of natural bone can be correlated with the domain organization in a typical ferroelectric material, where these domains align in response to the externally applied mechanical stimulation (Hastings et al. 1981). The mechanical stimulation induced ferroelectric domain alignment dissipates energy which has important consequence as far as the toughening of any ferroelectric implant is concerned.

Overall, the inherent electrcial properties of the tissue play an important role in assisting the number of metabolic activities in bone. However, the biocompatibility as well as the mechanical performance of material has become the primary concern during design and development of orthopedic implant materials. In addition to these aspects, the development of materials mimicking the electrical properties of bone can be an appealing choice over the existing implants by not only supporting bone ingrowth but also healing. Novel materials with electrical properties similar to that of bone have recently gained much attention for tissue engineering and regenerative medicine applications. This chapter will discuss some of the well-known systems.

2.2.1 Materials for Bone Tissue Engineering

With the importance of electrical properties of materials for biomedical applications coming to forefront, many of the known electrically active materials have been considered for such applications. However, the necessary condition that remains is the biocompatibility of the material which rules out several material systems leaving behind a few materials with potential multifunctional properties than can mimic the electrical and mechanical response of bone tissue. In particular, materials with conducting and piezoelectric properties have been chosen as examples in this chapter.

2.2.2 Electroactive Ceramics for Bone Tissue Engineering

Among the class of electroactive ceramics, materials with perovskite structure such as $CaTiO_3$, $BaTiO_3$, $Na_{0.5}K_{0.5}NbO_3$ etc. have been investigated as materials for potential electroactive implants. These well-known perovskites have good electrical properties such as conductivity and piezoelectricity that can be tuned to match that of natural bone. In addition, the suitability of perovskites as implant materials in terms of *in vitro* and *in vivo* biocompatibility has been extensively studied with several materials (Tandon et al. 2018). Owing to fact that the bone possesses significant electrical activities, the influence of externally applied electrical stimulation and electrical poling induced surface charges on cellular response is another issue that is worth considering. In this scenario, the following section focuses mainly on perovskite-based materials with conducting or/ and piezoelectric properties that are a promising alternative for prosthetic orthopedic implants. As far as the biomedical application of the perovskite material is concerned, the following discussion with examples attempts to highlight the advantages of such multi-functional implants which are designed to support cell and tissue growth.

2.2.2.1 $BaTiO_3$-based Materials

$BaTiO_3$ (BT) is a well-known piezoelectric ceramic that exists in different crystal structures-hexagonal, cubic, tetragonal, orthorhombic and rhombohedral (order of appearance from high temperature to low temperature). All of them, except cubic structure, exhibit ferroelectric effect (hence pyro- and piezoelectric effects too) due to non-centrosymmetric crystal structure and the displacement of Ti^{4+} ion in the lattice below the Curie temperature, creating a dipole. It is a high permittivity material used in capacitors and its piezoelectric property allows it to be used in transducers and microphones (Jaffe 2012). While it has good electrical properties, it does not enhance or support cell growth to a significant extent. In order to compensate for this, BT is usually used in conjunction with a bioactive phase such as Hydroxyapatite (HA) in a ceramic composite. While pure HA is generally used as a coating on implant materials to enhance the integration of the prosthetic material with the host, it cannot be used as a load bearing material due to its poor mechanical and electrical properties (Cook et al. 1988, Akao et al. 1981). Hence, additives are selected so that it can contribute to many/all of the above mentioned properties of HA for advanced biomedical applications. HA-ceramic composites are the most common among the HA-based composites, since HA being a ceramic does not encounter rejection and debonding from the matrix such as in metals and polymer matrix composites. Due to their inherent chemical inertness, ceramics are also the most common choices for implantable materials. Ceramics, though suitable implant materials in terms of bioactivity, do not possess bone-like strength, toughness and electrical properties.

The driving force behind the idea of HA-BT ceramic composites is to give implant materials (HA-BT) bone-like properties, mainly piezoelectricity. The composites have been processed by several routes but advanced sintering techniques like Spark Plasma Sintering (SPS) are usually preferred in order to obtain a dense pellet (Dubey et al. 2013a). In the study by Bowen et al. (Bowen et al. 2006), pressureless sintering was carried out at 1300°C for 2 h and corona poled at 130°C at 28 kV (with potential source 70 mm away) to measure the piezoelectric characteristics. The dielectric behavior of HA-BT composites, examined at 100 Hz, showed that they exhibited a trend close to the series Wiener law, implying that composites with lesser vol.% of BT exhibit very low dielectric constant and composites with at least 80 vol.% of BT showed good dielectric behavior. Similarly, it was observed that composites with less than 80 vol.% BT did not exhibit piezoelectric effect and the highest piezoelectric strain co-efficient d_{33} was seen in 95 vol.% BT composites ($\approx$ 20 pC/N) whereas for pure BT, it is around 80 pC/N. But the study failed to examine (or report) the phase distribution of HA-BT composites after sintering. The poor dielectric and piezoelectric behavior of these composites might be due to the formation of sintering products as is usually the case with pressureless sintering techniques. At 1300°C, it is known that HA can undergo dehydroxylation and form β-TCP, α-TCP and even reactions with BT might have occurred, reducing the performance of the composite. The lack of phase characterization does not allow one to come to a firm conclusion as to the reason behind the low dielectric constant and absence of piezoelectric properties in less vol.% BT composites. A more recent study by Dubey et al. (Dubey et al. 2013a) used multi-stage SPS to obtain better properties in HA-BT composites. The sintering was carried out in multiple steps (at 850°C for 5 min and at 950°C without holding) to achieve the maximum density (>95% of theoretical density) without the dissociation of HA. The phase analysis with X-ray diffraction indicated the presence of small amounts of β, α-TCP and even small amounts of $CaTiO_3$ and $Ba_{10}(PO_4)_6(OH)_2$. Investigation of electrical properties of SPSed HA-BT composites showed a similar behavior (series Wiener law), in that the dielectric constant and piezoelectric co-efficient were very low for composites below 80 wt. % BT. The summary of electrical properties is given in Table 2.1. It was observed that the composite HA-40BT possesses piezoelectric properties close to that of bone (Fukada and Yasuda 1957).

TABLE 2.1 Electrical properties of SPSed HA-BT composites, adapted with permission from Dubey et al. 2013a

Composite	Dielectric Constant (@ 10 kHz)	Conductivity (AC@ 10 kHz) (S.cm^{-1})	Piezoelectric Strain Co-efficient (pC/N)
HA	11.7	5×10^{-10}	–
HA-20BT	15.5	2.5×10^{-10}	0.6
HA-40BT	20.8	1.3×10^{-9}	0.9
HA-60BT	36.7	4×10^{-9}	1.0
HA-80BT	223.7	2×10^{-6}	1.2

The ferroelectric property of the BT was taken into account and the effect of surface charges and electric field stimulation on the osteoblast cell response on HA-BT composites were examined in a study with HA-BT composites (Dubey and Basu 2014). It was found that osteoblasts proliferation was highest, when DC electric field of 1 V/cm with pulse width of 400 μs was used. As far as the surface charge is concerned, negatively charged surfaces were found to support cell adhesion better than the positively charged and neutral surfaces. *In vivo* implantation in dog femoral bone with pure BT ceramics have reported a reduction in mechanical properties of bone-implant interface due to ageing and found no difference in the tissue response of poled and unpoled BT implant. Nevertheless, the poled BT samples in the study showed a higher co-efficient after 16 and 86 days of implantation, indicating that load is being transferred efficiently to the composites (Jianqing et al. 1997, Park et al. 1981, Park et al. 1977). However, recent studies with HA-BT composites indicate that poling controls the bone formation and cell response both *in vitro* and *in vivo,* due to the adsorbed Ca^{2+} ions on a negatively charged surface. Also, bone growth has been reported to be direction dependent and aligned in the direction of poling as observed through histological studies (Baxter et al. 2010). It is clear that HA-BT composites are excellent bone-replacement materials with multi-functional properties. Therefore, further improvement in their mechanical properties either by microstructural modification or by additives will move these composites higher in the list of promising bone-replacement materials.

2.2.2.2 *$CaTiO_3$-based Ceramics*

$CaTiO_3$ (CT) is a polymorphic with Cubic → Tetragonal → Orthorhombic transitions as it is cooled from high temperature. At room temperature, it is orthorhombic with a perovskite structure where

Ti^{4+} occupies octahedral voids and Ca^{2+} has a co-ordination number 12. Though it has a structure similar to BT, it does not possess piezoelectric properties but has better conductivity than BT due to its relatively lower band gap (3.46 eV) (Balachandran et al. 1982). $CaTiO_3$ is also known to nucleate apatite layer (mineralization) *in vitro* by adsorbing both Ca^{2+} and PO_4^{3-} ions in SBF; it can also leach out Ca^{2+} ions according to the reaction (Coreno and Coreno 2005)–

$$CaTiO_3 + H_2O \rightarrow Ca_{2+}(aq) + TiO_2(s) + H_2O\,(aq)$$

The reaction implies that CT surface would more likely be a negatively charged surface at pH 7.4 and the general consensus is that negatively charged surfaces support the nucleation and growth of apatite layer better than a neutral or positively charged surface (Viitala et al. 2002). This can be explained by the imbalance of Ca^{2+}, OH^- and PO_4^{3-} ions on the surface of CT. In a study by Dubey et al. (Dubey et al. 2013b), HA-CT composites were sintered using multi-stage SPS technique with holding at three different temperatures (950, 1100 and 1200°C) for 5 min at a pressure of 50 MPa to achieve good densification and to limit the formation of sintering reaction products. The dielectric properties indicate that beyond 60 wt. % of CT, significant increase in dielectric constant and conductivity were observed. Most importantly, HA-80CT displayed a dielectric constant of 155 and AC conductivity of 10^{-5} S.cm^{-1} at 1 MHz frequency. Phase analysis by XRD shows the absence of sintering reaction products apart from a small amount of TiO_2. A summary of the dielectric properties of HA-CT composites is presented here in Table 2.2.

TABLE 2.2 Electrical properties of SPSed HA-CT composites, adapted with permission from Dubey et al. 2013b

Composite	Densification (%)	Dielectric Constant	AC Conductivity (S.cm^{-1})
HA	99.4	8.9	–
HA20CT	99.3	–	–
HA40CT	99.0	24.9	1.2×10^{-7}
HA60CT	99.3	22.4	3.2×10^{-7}
HA80CT	98.6	155	4×10^{-5}
CT	98.2	490	10^{-3}

The conductivity and the high dielectric constant in HA80CT have been attributed to the presence of a fraction of Ti^{3+} state in CT and oxygen vacancies generated during sintering. After preliminary investigation of *in vitro* cellular response with L929 fibroblast cells and SaOS2 osteoblast like cells, the cell viability and imaging results show good proliferation and adhesion on HA-CT substrates (Dubey et al. 2010). *In vivo* implantation studies in rabbit animal model with HA and HA80CT composite (chosen for its high conductivity and is similar to that of natural bone) to evaluate bone regeneration in cylindrical bone defects were carried out and histological assessment of the efficacy of HA80CT as a bone implant material was made at different time points over the course of 3 months (1, 4 and 12 weeks). It was found that the amount of neobone formation was about 150% higher in HA80CT as compared to monolithic HA at the end of the duration of experiment (Mallik and Basu 2014). While the tissue response and the formation of neobone was found to be higher on conducting composites, the cell response, however, has a more subtle dependence on the conductivity of the substrate. This was demonstrated in another study with myoblasts, grown on HA-CT composites (Thrivikraman et al. 2013). It was observed that the alignment of the myoblasts and maturation into myotubes was enhanced on more conducting HA80CT compared to HA and HA20CT (see Fig. 2.2).

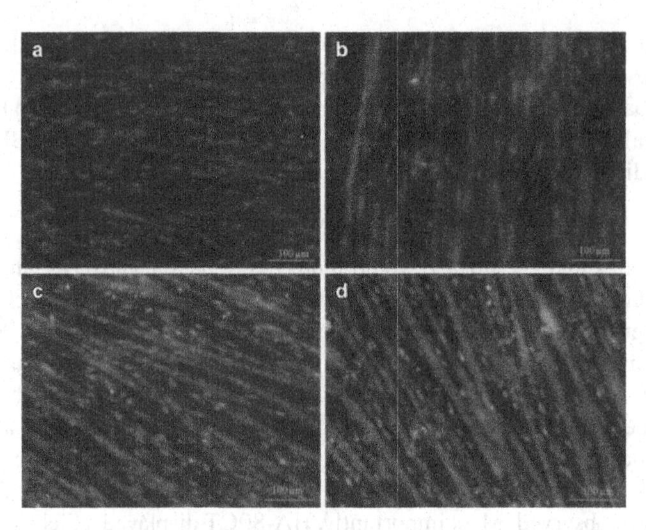

FIGURE 2.2 Myogenin immunostaining of C2C12 myoblasts cultured on (a) HA (b) HA40CT (c) HA80CT and (d) CT substrates showing a marked increase of expression on substrates with higher conductivity. Adapted with permission from Thrivikraman et al. 2013.

Further studies on these HA-CT composites focused on determining the bone healing/ regeneration potential of the substrates by examining the adult human mesenchymal stem cell (hMSC) response on these substrates with a range of conductivities (Ravikumar et al. 2017). In addition to the substrate conductivity, external electric field stimulation sequence used in the study showed that a synergy exists between these two parameters (substrate conductivity and electric field stimulation) that can direct stem cell differentiation towards bone-like cells. The transformation of hMSCs to bone-like cells was tracked with imaging and biochemical techniques. The initial sign of differentiation was observed in the form of change in the morphology of the cells on conducting substrates with electric field exposure. The cells, cultured on HA40CT, HA60CT and HA80CT in the presence of electric field showed a reduction in the aspect ratio as compared to other substrates (See Fig. 2.3).

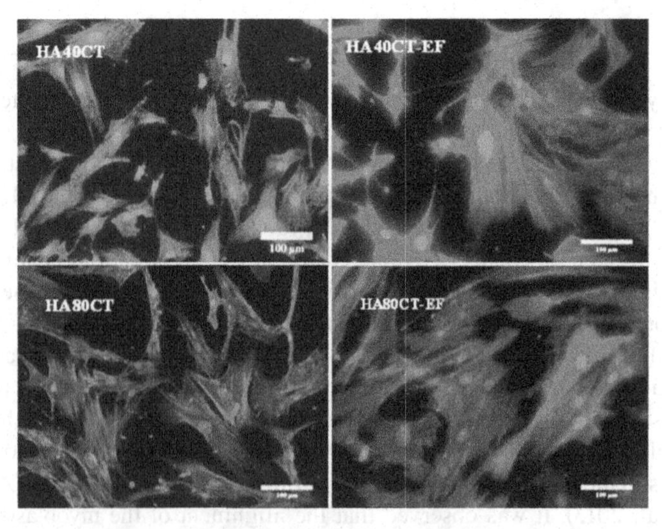

FIGURE 2.3 Change in hMSC morphology observed on conducting ceramics with external electric field stimulation. Adapted with permission from Ravikumar et al. 2017.

Color version at the end of the book

The differentiation behavior was further observed by measuring the activity of Alakline Phosphatase (ALP) and the amount of collagen in the cells to find that substrates with higher conductivity showed better results in the presence of electric field. The osteogenesis of hMSCs was confirmed by gene expression analysis and calcium deposition assay at the end of 4 weeks. This study indicates that with appropriate selection of substrate conductivity and electric field parameters, hMSC differentiation towards bone-like cells can be achieved on HA-CT composites which have better mechanical properties (Ravikumar et al. 2016a), showing a tremendous potential as a material for not only implant prosthesis but also for bone tissue engineering and regeneration.

2.2.3 Electroactive Polymers for Bone Tissue Engineering

Polymers are large molecules made of a number of repeating units called 'monomers'. The chemical bonding in polymers is predominantly covalent in nature along with weaker hydrogen bonding in the presence of O, N and F atoms and Van der Waal's interaction between the polymeric chains. Polymers and their derivatives are the largest class of materials used in the biomedical industry mainly due to the ease of processing and the versatile nature of the applications of polymers ranging from drug delivery vehicles to implant prosthesis. This is possible in part due to a wide variation in properties (such as stiffness and elastic modulus) that can be obtained in polymeric materials depending on the chemical composition and processing route. While the commonly known polymers are generally insulators and have been used in the biomedical industry for a long time, in the recent times the attention of the scientific community has shifted towards polymers with unique electrical properties commonly referred to as "electroactive polymers". These polymers show electrical activity generally due to a permanent dipole in the structure or due to presence of conjugated double bonds (delocalized electrons) in the structure which can be made conducting by the addition of a dopant. These polymers have found applications as electrodes, supercapacitors, photovoltaics and organic LEDs but their potential applications in tissue engineering and regenerative medicine has been explored to a lesser degree. Some of the common examples of conducting polymers that have been studied extensively in the biomedical context include polypyrrole, polythiophene, polyacetylene, poly (vinylidene difluoride), poly (l-lactic acid) and polyaninline. These polymers are generally prepared by oxidative coupling (in the presence of a dopant) of monomers that usually involves dehydrogenation that results in the generation of electrons that are responsible for electrical conduction. On the other hand, polymers that exhibit piezoelectric properties require the presence of a strong net dipole moment in the structure. The most commonly studied example is that of poly (vinyldene difluoride) (PVDF) which has a permanent C-F dipole moment in some of its polymorphs which is responsible for its piezoelectric nature.

2.2.3.1 PVDF-based Piezoelectric Polymers

PVDF is a semi-crystalline polymer which exists in several polymorphs. The most common phase is the α-PVDF which has a zero net dipole moment due to the TGTG' arrangement of C-F bonds and the β-PVDF with a TTTT arrangement of C-F bonds, giving it a net dipole moment and the piezoelectric property. PVDF is already in use as an inert membrane for filtration and protein detection due to its chemical stability. On the other hand, electroactive β-phase has received more attention for applications such as tactile and bio-sensors, transducers and energy harvesting devices which are particularly effective due to the low dielectric constant and elastic stiffness of the polymer (Bauer 2000, Jo et al. 2013, In'acio et al. 2003). In addition to PVDF which has a modest piezoelectric co-efficient ($d_{33} \approx -32$ pC/N) (Ueberschlag 2001), its co-polymer with trifluoroethylene P (VDF-TrFE) shows much better properties with $d_{33} = -38$ pC/N (Legrand 1989). In context of biomedical applications, the β-PVDF is of more interest and the focus during processing stage is to maximize the yield of the electroactive β-phase in the polymer by methods such as poling, stretching, shear and with the use of nanofillers (Sharma et al. 2014). The piezoelectric character of PVDF based polymers is important because the natural bone itself has piezoelectric properties ($d_{15} \approx 0.7$ pC/N) (Minary-Jolandan and Yu 2010) and as discussed before, it helps in maintaining the balance between bone forming and bone resorbing cells by altering the surface charge density of bone as a response to mechanical stress. The *in vitro* studies reported in the literature show a strong correlation between the electroactive PVDF and cell response. In the study by Zhou et al. (Zhou et al. 2016), Ti surface was modified with a layer of PVDF film coating using solvent casting method, followed by corona poling. It was observed that the hMSCs, grown on these substrates showed increased ALP activity and upregulated bone specific gene markers indicating a differentiation response of stem cells towards bone-like cells. Similar study using mechanically drawn PVDF films with MC3T3-E1 cells and with the presence of dynamic mechanical stimuli showed increased proliferation (Ribeiro et al. 2012). While PVDF has reasonable ferroelectric properties, it is a weak dielectric. In order to impart conductivity to the polymer; a recent study used multiwall carbon nanotubes as nanofillers in the PVDF matrix (Ravikumar et al. 2016b). C2C12 mouse muscle cells cultured on these conducting and piezoelectric substrates under the influence of external electric field (1 V/cm) showed enhanced proliferation as compared to neat PVDF substrates. One of the important advantages piezoelectric polymers offer (over metals and ceramics) is in the preparation of scaffolds for tissue engineering applications. Electrospun fibers and 3D scaffolds of PVDF and P(VDF-TrFE) used as substrates for hMSC with dynamic loading conditions have shown enhanced differentiation towards osteogenic cells (Damaraju et al. 2013, Damaraju et al. 2017). These examples show that PVDF-based materials show a significant potential as biomedical materials, especially in the area of bone tissue engineering and healing, owing to its mechanical and electrical properties which can play a major role in shaping the biomedical implant scenario in the near future.

2.2.3.2 Conducting Polymers

The concept of conducting polymers is well explored in the areas of sensors, actuators, organic-LEDs and battery applications (Chandrasekhar 1999). Though they were discovered as early as 1960s (Polypyrrole), their applications in the field of biomedical engineering has been growing steadily mainly due to the recent development in the understanding of the electrical nature of living cells and tissues (bone, muscle and nerve) and their response to electric field (Guimard et al. 2007). The major advantages of conducting polymers is their ease of processing, wide range of conductivities and the ability to process structures that are favorable for biomedical applications (sutures, patches and scaffolds) and more importantly the biocompatibility of the polymers. However, these polymers are usually non-biodegradable and are hydrophobic. Some of the commonly known conducting polymers have been discussed earlier and their structures are shown in Fig. 2.4.

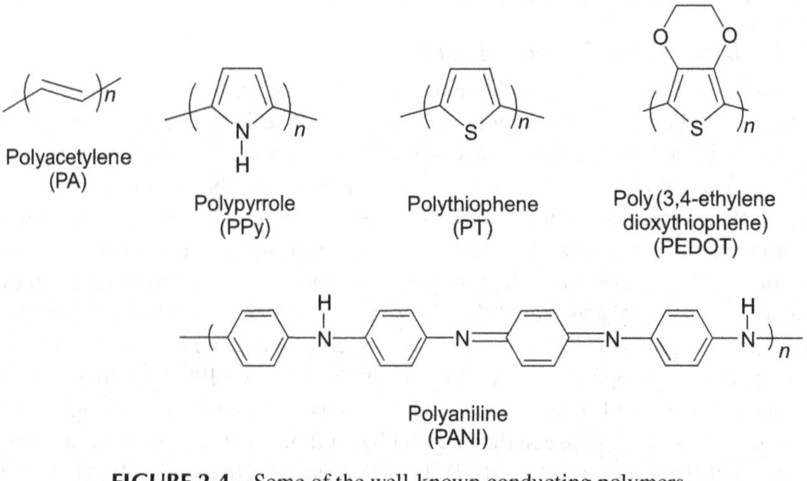

Polyacetylene (PA)

Polypyrrole (PPy)

Polythiophene (PT)

Poly(3,4-ethylene dioxythiophene) (PEDOT)

Polyaniline (PANI)

FIGURE 2.4 Some of the well-known conducting polymers.

Polypyrrole (PPy) is the earliest material that was explored for applications in tissue engineering. Collier et al. (Collier et al. 2000) showed the *in vitro* cytocompatibilty of PPy-Hyaluronic acid composite substrates and enhanced vascularization in *in vivo* rat model. The importance of effect of electrical conductive polymers and their effect on cell morphological changes was shown with the help of transparent PPy thin films, cultured with mammalian cells (Wong et al. 1994). It was then observed that such conducting polymers can also enhance neurite outgrowth *in vitro* (Schmidt et al. 1997). The differentiation response on PPy substrates, formed by admicellar polymerization showed an inclination towards osteogenic cells, which essentially helps in bone healing and regeneration (Castano et al. 2004). Apart from PPy, Polyaniline (PANI) has been an excellent conducting polymer for biomedical applications. The *in vivo* tissue response of PANI, as early as late 1990s, showed limited inflammation even after 50 weeks, a favorable aspect of any biomaterial for tissue engineering applications (Wang et al. 1999). More recently, HCl doped PANI films used to culture hMSCs under the influence of external electric field show neurite outgrowth and upregulation of neural gene markers (Thrivikraman et al. 2014). Modification of PANI with collagen and sulfated hyaluronan and pulse electric field stimulation of hMSCs result in a significant change in differentiation response towards osteogenic cell line. Further studies with doped PANI with a coating of Gold nanoparticles indicated that with appropriate selection of electric field stimulation parameters (pulsed field and DC), stem cell differentiation can be directed towards either neural like cells or cardiomyogenic cells, respectively (Thrivikraman et al. 2016). The above set of studies indicate that in addition to the substrate conductivity, electric field and chemical modification of the surface can be used to modulate the cell differentiation on conducting polymers. Other conducting polymers that have known applications in the biomedical and tissue engineering field are polythiphenes (Waugaman et al. 2003), biodegradable copolymers of pyrrole and thiophene with ester linkages (Rivers et al. 2002), controlled drug delivery and release vehicles and implantable electrodes such as PEDOT (Abidian et al. 2006, Ludwig et al. 2006) carbon nanotube based bioactuators (Tahhan et al. 2003, Spinks et al. 2005, Otero and Sansinena 1997) for muscle/nerve stimulation and biosensors based on PPy and PANI for glucose, cholesterol and urea sensing (Foulds and Lowe 1986, Singh et al. 2006, Adeloju et al. 1996). The above discussion, though not focused on bone tissue engineering applications, is intended to showcase the wide range of applications of conducting polymers and their importance to the field of not only tissue engineering but also drug delivery, bio-sensing and actuation.

2.3 CONCLUDING REMARKS

Recent studies in tissue engineering have indicated that bioactive and bioresorbable materials cannot cater to all the necessary applications that have given way to new generation of biomedical materials. The field of biomaterials has witnessed an emergence of advanced multi-functional materials for various applications such as bone replacement and healing, biodegradable scaffolds infused with drugs that aid healing, drug delivery devices/vehicles, conducting extracellular matrices that direct stem cell growth are among the few selected examples. Hence, understanding the mechanism of material interaction with cells and tissues is vital in designing application-specific materials. Materials that are biocompatible but also possess other multi-functional properties such as conductivity, ferroelectricity, strength, toughness, bioactivity, and bioresorbability are in high demand. For example, in bone replacement applications, piezoelectricity and conductivity of the substrate have been shown to enhance the performance of the material and tissue engineering scaffolds. The use of stem cells in regenerative therapies remains of utmost importance and this new generation of materials have the ability to directly affect stem cell fate process to elicit a particular cell response. The recent popularity of the use of external electric fields to alter cell functionality *in vitro* has been brought to notice mainly by establishing the synergy of cell response with substrate conductivity and electric field stimulation. Taken together, piezoelectric and conducting materials in addition to external stimulation (mechanical and electrical) form the basis of new, interesting and efficient strategies for bone tissue engineering and regenerative medicine applications in the foreseeable future that involves interdisciplinary and exploratory research in the field of materials, biology and medicine.

REFERENCES

Abidian, M.R., D.H. Kim and D.C. Martin. 2006. Conducting-polymer nanotubes for controlled drug release. Advanced Materials 18: 405-409.

Adeloju, S.B., S.J. Shaw and G.G. Wallace. 1996. Polypyrrole-based amperometric flow injection biosensor for urea. Analytica Chimica Acta 323: 107-113.

Akao, M., H. Aoki and K. Kato. 1981. Mechanical properties of sintered hydroxyapatite for prosthetic applications. Journal of Materials Science 16: 809-812.

Anderson, J.M., A. Rodriguez and D.T. Chang. 2008. Foreign body reaction to biomaterials. Seminars in Immunology 20: 86-100.

Balachandran, U., B. Odekirk and N.G. Eror. 1982. Electrical conductivity in calcium titanate. Journal of Solid State Chemistry 41: 185-194.

Bassett, C.A.L. and R.O. Becker. 1962. Generation of electric potentials by bone in response to mechanical stress. Science 137: 1063-1064.

Basu, B., D.S. Kattia and A. Kumar. 2010. Advanced Biomaterials: Fundamentals, Processing, and Applications. John Wiley & Sons, American Ceramic Society, New Jersey, USA.

Basu, B. 2017. Biomaterials Science and Tissue Engineering: Principles and Methods: Cambridge University Press, Cambrigde, UK.

Basu, B. and S. Ghosh. 2017. Biomaterials for Musculoskeletal Regeneration. Springer-Nature, Singapore.

Bauer, F. 2000. PVDF shock sensors: applications to polar materials and high explosives. IEEE Transactions on Ultrasonics, Ferroelectrics, and Frequency Control 47: 1448-1454.

Baxter, F.R., C.R. Bowen, I.G. Turner and A.C.E. Dent. 2010. Electrically active bioceramics: a review of interfacial responses. Annals of Biomedical Engineering 38: 2079-2092.

Bowen, C.R., J. Gittings, I.G. Turner, F. Baxter and J.B. Chaudhuri. 2006. Dielectric and piezoelectric properties of hydroxyapatite-$BaTiO_3$ composites. Applied Physics Letters 89: 132906 (1-3).

Castano, H., E.A. O'Rear, P.S. McFetridge and V.I. Sikavitsas 2004. Polypyrrole thin films formed by admicellar polymerization support the osteogenic differentiation of mesenchymal stem cells. Macromolecular Bioscience 4: 785-794.

Chandorkar, Y., K. Ravimumar and B. Basu. 2019. The foreign body response demystified. ACS Biomaterials Science & Engineering 5(1): 19-44.

Chandrasekhar, P. 1999. Conducting Polymers, Fundamentas and Applications. Springers Massachusetts, USA.

Collier, J.H., J.P. Camp, T.W. Hudson and C.E. Schmidt. 2000. Synthesis and characterization of polypyrrole–hyaluronic acid composite biomaterials for tissue engineering applications. Journal of Biomedical Materials Research 50(4): 574-584.

Cook, S.D., K.A. Thomas, J.F. Kay and M. Jarcho. 1988. Hydroxyapatite-coated titanium for orthopedic implant applications. Clinical Related Research 232: 225-243.

Coreno, J. and O. Coreno. 2005. Evaluation of calcium titanate as apatite growth promoter. Journal of Biomedical Materials Research Part A: An Official Journal of The Society for Biomaterials, The Japanese Society for Biomaterials, and The Australian Society for Biomaterials and the Korean Society for Biomaterials 75(2): 478-484.

Damaraju, S.M., S. Wu, M. Jaffe and T.L. Arinzeh. 2013. Structural changes in PVDF fibers due to electrospinning and its effect on biological function. Biomedical Materials 8: 045007.

Damaraju, S.M., Y. Shen, E. Elele, B. Khusid, A. Eshghinejad, J. Li, M. Jaffe and T.L Arinzeh. 2017. Three-dimensional piezoelectric fibrous scaffolds selectively promote mesenchymal stem cell differentiation. Biomaterials 149: 51-62.

Dubey, A.K., G. Tripathi and B. Basu. 2010. Characterization of hydroxyapatite-perovskite ($CaTiO_3$) composites: Phase evaluation and cellular response. Journal of Biomedical Materials Research Part B: Applied Biomaterials 95: 320-329.

Dubey, A.K., K. Balani and B. Basu. 2013a. Multifunctional properties of multistage spark plasma sintered HA–BaTiO3-based piezobiocomposites for bone replacement applications. Journal of the American Ceramic Society 96: 3753-3759.

Dubey, A.K., P.K. Mallik, S. Kundu and B. Basu. 2013b. Dielectric and electrical conductivity properties of multi-stage spark plasma sintered HA–$CaTiO_3$ composites and comparison with conventionally sintered materials. Journal of the European Ceramic Society 33: 3445-3453.

Dubey, A.K. and B. Basu. 2014. Pulsed electrical stimulation and surface charge induced cell growth on multistage spark plasma sintered hydroxyapatite-barium titanate piezobiocomposite. Journal of the American Ceramic Society 97: 481-489.

El Messiery, M.A., G.W. Hastings and S. Rakowski. 1979. Ferro-electricity of dry cortical bone. Journal of Biomedical Engineering 1: 63-65.

Foulds, N.C. and C.R. Lowe. 1986. Enzyme entrapment in electrically conducting polymers. Immobilisation of glucose oxidase in polypyrrole and its application in amperometric glucose sensors. Journal of the Chemical Society, Faraday Transactions 1: Physical Chemistry in Condensed Phases 82: 1259-1264.

Friedenberg, Z.B., R.H. Dyer jr and C.T. Brighton. 1971. Electro-osteograms of long bones of immature rabbits. Journal of Dental Research 50: 635-639.

Fukada, E. and I. Yasuda. 1957. On the piezoelectric effect of bone. Journal of the Physical Society of Japan 12: 1158-1162.

Fukada, E. and I. Yasuda. 1964. Piezoelectric effects in collagen. Japanese Journal of Applied Physics 3: 117-122.

Geetha, M., A.K., Singh, R. Asokamani and A.K. Gogia. 2009. Ti based biomaterials, the ultimate choice for orthopaedic implants–a review. Progress in Materials Science 54: 397-425.

Guimard, N.K. N. Gomez and C.E. Schmidt. 2007. Conducting polymers in biomedical engineering. Progress in Polymer Science 32: 876-921.

Hastings, G.W., M.A. El Messiery and S. Rakowski. 1981. Mechano-electrical properties of bone. Biomaterials 2: 225-233.

Hastings, G.W. and F.A. Mahmud. 1988. Electrical effects in bone. Journal of Biomedical Engineering 10: 515-521.

Hench, L.L. and J.M. Polak. 2002. Third-generation biomedical materials. Science 295: 1014-1017.

In'acio, P., J.N. Marat Mendes and C.J. Dias. 2003. Development of a biosensor based on a piezoelectric film. Ferroelectrics 293: 351-356.

Isaacson, B.M. and R.D. Bloebaum. 2010. Bone bioelectricity: what have we learned in the past 160 years? Journal of Biomedical Materials Research Part A 95: 1270-1279.

Jaffe, B. 2012. Piezoelectric Ceramics. Academic Press, Berkeley Square, London.

Jianqing, F., Y. Huipin and Z. Xingdong. 1997. Promotion of osteogenesis by a piezoelectric biological ceramic. Biomaterials 18: 1531-1534.

Jo, C., D. Pugal, I.-K. Oh, K.J. Kim and K. Asaka. 2013. Recent advances in ionic polymer–metal composite actuators and their modeling and applications. Progress in Polymer Science 38: 1037-1066.

Lang, S.B. 1966. Pyroelectric effect in bone and tendon. Nature 212: 704-705.

Lang, S.B. 1981. Pyroelectricity: occurrence in biological materials and ossible physiological implications. Ferroelectrics 34: 3-9.

Legrand, J.F. 1989. Structure and ferroelectric properties of P(VDF-TrFE) copolymers. Ferroelectrics 91: 303-317.

Ludwig, K.A., J.D. Uram, J., Yang, D.C. Martin and D.R. Kipke. 2006. Chronic neural recordings using silicon microelectrode arrays electrochemically deposited with a poly (3, 4-ethylenedioxythiophene) (PEDOT) film. Journal of Neural Engineering 3: 59-70.

Mallik, P.K. and B Basu. 2014. Better early osteogenesis of electroconductive hydroxyapatite–calcium titanate composites in a rabbit animal model. Journal of Biomedical Materials Research Part A 102: 842-851.

Mano, J.F., R.A. Sousa, L.F. Boesel, N.M. Neves and R.L. Reis. 2004. Bioinert, biodegradable and injectable polymeric matrix composites for hard tissue replacement: state of the art and recent developments. Composites Science and Technology 64: 789-817.

MarketsandMarkets.com. 2019. Biomaterials market by type of material (metallic, ceramic, polymers, natural biomaterials) application (cardiovascular, orthopedic, dental, plasticsurgery, wound healing, neurology, tissue engineering, ophthalmology) - Global Forecast to 2024.

Minary-Jolandan, M. and M.-F. Yu. 2010. Shear piezoelectricity in bone at the nanoscale. Applied Physics Letters 97: 153127.

Morais, J.M., F. Papadimitrakopoulos and D.J. Burgess. 2010. Biomaterials/tissue interactions: possible solutions to overcome foreign body response. The AAPS Journal 12: 188-196.

Moss, M.L. 1997. The functional matrix hypothesis revisited. 1. The role of mechanotransduction. American Journal of Orthodontics and Dentofacial Orthopedics 112: 8-11.

Otero, T.F. and J.M. Sansinena. 1997. Bilayer dimensions and movement in artificial muscles. Bioelectrochemistry and Bioenergetics 42: 117-122.

Park, J.B., G.H. Kenner, S.D. Brown and J.K. Scott. 1977. Mechanical property changes of barium titanate (ceramic) after *in vivo* and *in vitro* aging. Biomaterials, Medical Devices, and Artificial Organs 5: 267-276.

Park, J.B., B.J. Kelly, G.H. Kenner, A.F. von Recum, M.F. Grether and W.W. Coffeen. 1981. Piezoelectric ceramic implants: *in vivo* results. Journal of Biomedical Materials Research 15: 103-110.

Pethig, R. 1985. Dielectric and electrical properties of biological materials. Journal of Bioelectricity 4: vii-ix.

Ravikumar, K., P.K. Mallik and B. Basu. 2016a. Twinning induced enhancement of fracture toughness in ultrafine grained hydroxyapatite–calcium titanate composites. Journal of the European Ceramic Society 36: 805-815.

Ravikumar, K., G.P. Kar, S. Bose and B. Basu. 2016b. Synergistic effect of polymorphism, substrate conductivity and electric field stimulation towards enhancing muscle cell growth *in vitro*. RSC Advances 6: 10837-10845.

Ravikumar, K., S.K. Boda and B. Basu. 2017. Synergy of substrate conductivity and intermittent electrical stimulation towards osteogenic differentiation of human mesenchymal stem cells. Bioelectrochemistry 116: 52-64.

Ribeiro, C., S. Moreira, V. Correia, V. Sencadas, J.G. Rocha, F.M. Gama, J.L. Gomez Ribelles and S.L. Mendez. 2012. Enhanced proliferation of pre-osteoblastic cells by dynamic piezoelectric stimulation. Rsc Advances 2: 11504-11509.

Rivers, T.J., T.W. Hudson and C.E. Schmidt. 2002. Synthesis of a novel, biodegradable electrically conducting polymer for biomedical applications. Advanced Functional Materials 12: 33-37.

Schmidt, C.E., V.R. Shastri, J.P. Vacanti and R. Langer. 1997. Stimulation of neurite outgrowth using an electrically conducting polymer. Proceedings of the National Academy of Sciences 94: 8948-8953.

Sharma, M., G. Madras and S. Bose. 2014. Process induced electroactive β-polymorph in PVDF: effect on dielectric and ferroelectric properties. Physical Chemistry Chemical Physics 16: 14792-14799.

Singh, S., P.R. Solanki, M.K. Pandey and B.D. Malhotra. 2006. Cholesterol biosensor based on cholesterol esterase, cholesterol oxidase and peroxidase immobilized onto conducting polyaniline films. Sensors and Actuators B: Chemical 115: 534-541.

Spinks, G.M., B. Xi, V.-T. Truong and G.G. Wallace. 2005. Actuation behaviour of layered composites of polyaniline, carbon nanotubes and polypyrrole. Synthetic Metals 151: 85-91.

Tahhan, M., V.-T. Truong, G.M. Spinks and G.G. Wallace. 2003. Carbon nanotube and polyaniline composite actuators. Smart Materials and Structures 12: 626.

Tandon, B., J.J. Blaker and S.H. Cartmell. 2018. Piezoelectric materials as stimulatory biomedical materials and scaffolds for bone repair. Acta Biomaterialia 73: 1-20.

Taylor, D., J.G. Hazenberg and T.C. Lee. 2007. Living with cracks: damage and repair in human bone. Nature Materials 6: 263-268.

Thrivikraman, G., P.K. Mallik and B. Basu. 2013. Substrate conductivity dependent modulation of cell proliferation and differentiation *in vitro*. Biomaterials 34: 7073-7085.

Thrivikraman, G., G. Madras and B. Basu. 2014. Intermittent electrical stimuli for guidance of human mesenchymal stem cell lineage commitment towards neural-like cells on electroconductive substrates. Biomaterials 35: 6219-6235.

Thrivikraman, G., G. Madras and B. Basu. 2016. Electrically driven intracellular and extracellular nanomanipulators evoke neurogenic/cardiomyogenic differentiation in human mesenchymal stem cells. Biomaterials 77: 26-43.

Ueberschlag, P. 2001. PVDF piezoelectric polymer. Sensor Review 21: 118-126.

Viitala, R., M. Jokinen, T. Peltola, K. Gunnelius and J.B. Rosenholm. 2002. Surface properties of *in vitro* bioactive and non-bioactive sol-gel derived materials. Biomaterials 23: 3073-3086.

Wang, C.H., Y.Q. Dong, K. Sengothi, K.L. Tan and E.T. Kang. 1999. *In vivo* tissue response to polyaniline. Synthetic Metals 102: 1313-1314.

Waugaman, M., B. Sannigrahi, P. McGeady and I.M. Khan. 2003. Synthesis, characterization and biocompatibility studies of oligosiloxane modified polythiophenes. European Polymer Journal 39: 1405-1412.

Wong, J.Y., R. Langer and D.E. Ingber. 1994. Electrically conducting polymers can noninvasively control the shape and growth of mammalian cells. Proceedings of the National Academy of Sciences 91: 3201-3204.

Yasuda, I. 1954. On the piezoelectric activity of bone. The Japanese Society of Orthopaedic Surgery 28: 267.

Zhou, Z., W. Li, T. He, L. Qian, G. Tan and C. Ning 2016. Polarization of an electroactive functional film on titanium for inducing osteogenic differentiation. Scientific Reports 6: 35512.

3

Enhanced Scaffold Fabrication Techniques for Optimal Characterization

Tshai Kim Yeow[1,*], Lim Siew Shee[2], Yong Leng Chuan[3] and Chou Pui May[3]

3.1 INTRODUCTION

Today, the utilization of three-dimensional (3D) biomimetic scaffolds plays a crucial role in tissue engineering application due to their unique nanoscale structure. In conventional practice, single or mixed type of freshly isolated functional cells can be formed via two-dimensional (2D) culture technique from human tissue or organs' cell (Nelson and Bissell 2006), where fresh nutrients and metabolic waste are controlled by the culture media. Nevertheless, human tissues or organs' cells are arranged in 3D structure and are supplied with fluid in blood circulation system continuously. The simplified conditions in typical 2D culture is incapable of mimicking a certain structure, biochemical and mechanical characteristics, as well as the interaction between cells as found in human bodies. In addition, 2D culture has the major drawback of being time-consuming and cost intensive. On the other hand, 3D cell cultures using scaffold mimicking the extracellular matrix (EMC) of proteins and other biological molecules promote cell adhesion and proliferation (Lutolf and Hubbell 2005). In the present chapter, the current progress and development of 2D and 3D perfused culture is reviewed and discussed.

3.1.1 Comparison between 2D and 3D Cell Cultures

Cell culture in 2D monolayer has received much attention in academic community. Theoretically, petri dishes or flasks are utilized as the platforms for cell growing and adhering. The cell culture's medium consists of simulating factors and nutrients. The nutrients are consumed whilst metabolic waste is generated by the successive cells that grow and adhere to the monolayer surface. Thus, the changes of fresh cell culture's medium are necessary. The mature proliferating cells spread and adhere to the surface, whereby the resultant cells could be harvested and the process of reseeding

[1] University of Nottingham Malaysia, Department of Mechanical, Materials and Manufacturing Engineering, Jalan Broga, 43500 Semenyih, Selangor, Malaysia.
[2] University of Nottingham Malaysia, Department of Chemical and Environmental Engineering, Jalan Broga, 43500 Semenyih, Selangor, Malaysia.
[3] Taylor's University Lakeside Campus, School of Engineering, No. 1 Jalan Taylor's, 47500 Subang Jaya, Selangor, Malaysia.
[*] Corresponding author: Kim-Yeow.Tshai@nottingham.edu.my

with reduced cell density take place. Over the past decades, 2D culture strategies and techniques have been widely explored. Nevertheless, typical 2D cell culture cannot reflect the real physiological complexity of human body's cells and inconsistencies were observed in the predicted tissue-specific response (Mitragotri and Lahann 2009).

In fact, biochemical cues from 3D environment are essential in forming relevant physiological tissue structures. Cells cultured in 3D matrix exhibit significant changes in morphology, functions and differentiation. Numerous literatures presented that fibroblasts cultured in 3D collagen achieved better *in vivo* phenotype and a high viability of hepatocytes was preserved. In this context, the interconnectivity of 3D structure improves the transportation of nutrients and metabolites in and out of the tissue model (Suuronen et al. 2005). In addition, a greater number of cells can be supported by 3D cell culture compared to 2D cell culture. Furthermore, a direct relationship between function and structure can be obtained via the well-defined architecture of 3D cell culture (Liu et al. 2007). As a result, a better understanding on the spatial resolution in human body is obtained.

3.1.2 The Ideal Scaffold for Cell Culture

The desirable 3D scaffold for cell culture should compose of several important criteria, including high biocompatibility, non-toxic, ease of removal or biodegradability, negligible to immune response and the implanted material must exhibit similar mechanical properties to the specific body tissues. The 3D scaffold architecture plays a significant role in depositing extracellular matrix (EMC) which replaces the scaffold structure over time. The high porosity in an interconnected structure assists in effective migration of nutrients to the cells (Webber et al. 2015). In the context of biocompatibility, 3D scaffold must be generated from non-inflammatory materials, releasing non-toxic breakdown residue with rate of degradation controllable by biological or enzymatic processes, which in turn allows the self-production of extracellular matrix (Dong et al. 2009). Furthermore, the bioactivity of 3D scaffold must be able to interact with host tissues, and stimulate cell growth, attachment and differentiation by combining the growth factors and other biological cues. One of the important criteria of desirable 3D scaffold is that it must exhibit similar mechanical properties to the host tissues to allow preservation of structural integrity and cell mechano-regulation *in vivo* (Chan and Leong 2008). Several of these important characteristics are discussed.

3.1.2.1 External Geometry

The design and development of scaffold capable of mimicking tissue structure at nanoscale remains one of the greatest challenges in tissue engineering. Among the important characteristics, external scaffold geometry having promising physical properties possesses greater capability to mimic natural ECM (Tran et al. 2004). For example, sufficient mechanical properties are crucial criteria to ensure structural support and cell attachment. Cell differentiation, tissue formation, homeostasis and regeneration processes could be completed accordingly if nanoscale structural architecture or external geometry of a scaffold can be developed and designed (Midwood et al. 2004). In this case, 3D nanofibrous scaffolds capable of maintaining a highly interconnected porous architecture provides an extremely large active surface area, favouring cell attachment, proliferation and growth for tissue and organ regeneration in their normal physiological shape.

3.1.2.2 Porosity

Scaffolds with high porosity and an open cell form, fully interconnected external geometry is preferred in tissue engineering, owing to their large active surface area for cell ingrowth, uniform distribution of cells and better facilitation of the structural neovascularization. In this aspect, the control of pore size, pore volume and pore size distribution are among the important features to determine the active surface area of a designed scaffold (Coombes et al. 2004). Besides, the scaffold porosity and pore interconnectivity are crucial for ensuring sufficient oxygen and nutrients transportation during the mass transfer process (Salgado et al. 2004).

3.1.2.3 Biocompatibility

Biocompatibility is defined as the ability of a biomaterial to support cellular activity without triggering any negative side effects in the patients (Nair and Laurencin 2007). Several crucial factors that control the biocompatibility of a scaffold are the chemistry, structure and morphology of the scaffold surface. Ammonia plasma treatment and oxidization of polystyrene surface are the two common surface modification methods used to improve the cell attachment and cell growth for polymeric scaffolds (Williams 2008).

3.1.2.4 Biodegradation

Biodegradation of a material involves cleavage of weaker chemical bonds hydrolytically or enzymatically (Katti et al. 2002). The biodegradation rate strongly depends on the chemical structure, surface hydrophilicity, percentage of crystallinity, molecular weight and glass transition temperature of the material (Middleton and Tipton 2000). For biodegradable scaffold, the degradation rate should be consistent with the cell growth rate. It is important to note that non-biodegradable scaffolds are biologically stable, and it can offer long term support and performance throughout its desired life span.

3.2 SCAFFOLD FABRICATION TECHNIQUES

3.2.1 Conventional Techniques

Conventional fabrication techniques are defined as processes that build scaffolds with bulk or porous (interconnected or non-interconnected) structure that lacks any long-range channeling microstructure (Moroni and Mirabella 2014). These techniques are often not sufficiently suitable to control scaffold structure to modulate mechanical properties. For this reason, conventional fabrication techniques, such as fibre bonding, solvent casting/particulate leaching, membrane laminations, soft lithography (Vozzi et al. 2003, Pimpin and Srituravanich 2012), melt molding (Borenstein et al. 2002, Von Der Mark et al. 2010), thermal induced phase separation (TIPS) (Liao et al. 2008, Liu and Ma 2009), and gas foaming (Murphy et al. 2002) have not been ideal in producing scaffolds with reproducible topology and mechanical properties. The main disadvantages of scaffold fabricated from these conventional techniques include lack of homogeneous structure, poor reproducibility, pore irregularity and insufficient pore interconnectivity. Several of the conventional methods are discussed in the following section.

3.2.1.1 Solvent-casting and Particulate-leaching

Solvent casting particulate leaching is a conventional way to produce scaffolds with high porosity, good degree of pore interconnectivity, controlled composition and pore size. The common chemical compounds used in this process include alcohol which serve as the solvent and sodium chloride (NaCl) powder that is used as the common porogen. The porous scaffold can be prepared by dissolving the polymer in alcohol and pouring the solution into a petri dish covered with sieved salt. After vacuum drying and thermal treatment, the porous scaffold needs to be immersed in water for 8 h to leach out the salts.

Studies done by Choudhury et al. (2015) showed the effect of different solvent in solvent casting particulate leaching fabrication method. Three different solvents, hexa-fluoroisopropanol (HFIP), dichloromethane (DCM) and chloroform (CF) were tested and compared. The PLA dissolved in dichloromethane produced more regular unit cell microstructure and hence showed higher mechanical strength. But the scaffold possesses lower porosity compared to those prepared with chloroform, owing to the rapid evaporation of dichloromethane around the salt particles (Choudhury et al. 2015).

Lu et al. (2011) fabricated biodegradable poly-(3-hydroxybutyrate-co-3-hydroxyvalerate) (PHBV) scaffolds into a meniscal shape by solvent casting particulate leaching technique. During the fabrication, the slurry of salt and chloroform dissolved PHBV were poured into a prefabricated mold in the shape of a rabbit meniscus (10 mm in length, 6 mm in width, and 2 mm in thickness). It was observed that 18 weeks after transplantation of proliferated meniscal cells seeded PHBV scaffolds into rabbit knee joints, the cell-seeded scaffolds maintained their approximate shape and showed ideal initial mechanical properties and biocompatibility (Lu et al. 2011).

3.2.1.2 Melt Molding

To produce porosity in the processed materials, polymer molding technologies have been combined with traditional scaffold fabrication technique especially with the particulate leaching. Melt molding with particulate leaching commonly involves the forming of porous polymer scaffold out of polymer powder and porogen. During the fabrication process, combination of polymer powder and porogen compound is subjected to positive applied pressure within the mold cavity and the mixtures are heated above the polymer glass transition temperature. The elevated temperature and applied pressure fuse and bind the mixtures together, allowing the molten mixture to conform to the shape of the mold cavity. Once the mold opened, the molded scaffold is separated from the mold and porogen can be leached out to yield porous scaffold architecture (Thomson et al. 1995).

Melt molding with porogen leaching is a relatively uncomplicated and economical route that offers advantages of producing 3D scaffolds without solvents, where the shapes and sizes can be defined by the geometry of the mold cavity. However, the process has its major drawbacks due to potential porogen residual and the usage of elevated temperature for melting may cause degradation of the biomaterials (Correlo et al. 2009).

3.2.1.3 Phase Separation

The fabrication of various polymeric or composite scaffolds with different pore size via phase separation method has been widely reported. This process takes advantage of the instability of the system after phase separation is induced thermally. The two main phases involved in this process are solid and liquid phases.

In solid-liquid phase separation, the first step is the preparation of polymer solution by dissolving polymer in a solvent. The polymer solution is then poured into a mold and cooled to a set quenching temperature until the solution becomes solidified. The separation of polymer-rich phase from polymer-lean phase occurs during quenching as the polymer solution becomes highly unstable. The frozen polymer solution is subsequently subjected to freeze-extraction or sublimation for the removal of polymer-lean phase (Ma and Zhang 1999, Ho et al. 2004). Ultimately, only polymer-rich phase remains as solidified matrix. The compressive modulus of PLA scaffolds prepared via phase separation method was reportedly 20 times higher than that of material prepared using salt-leaching technique (Ma et al. 2001). The profound modulus was attributed to the pore size of scaffolds prepared via solid-liquid phase separation ranging from 10-100 micron. To further increase the osteoconductivity of scaffolds, composite scaffolds of polymer and ceramic compound such as hydroxyapatite (HA) were also prepared by blending HA with polymer solution prior to phase separation. HA serving as reinforcing phase in polymeric matrix aided to enhance compressive modulus of the scaffolds which was also coupled with improved *in vitro* and *in vivo* efficacies promoting bone ingrowth (Wang et al. 2007).

Unlike solid-liquid phase separation, there are five steps involved in liquid-liquid phase separation. The first step is also the polymer dissolution and is followed by phase separation of polymer solution induced at its gel temperature. After gelation, solvent is extracted from the gel using water. The last two steps are freezing and freeze drying which aid to solidify the gel to generate porous matrix (Ma and Zhang 1999). The formation of gel is the key feature of liquid-liquid phase separation. This technique was applied by Ma and Zhang (1999) to fabricate nanofibrous scaffolds using biodegradable PLLA polymer. Through extensive investigation on the effects of few parameters before freeze drying on the pore structure of scaffolds, the authors demonstrated that nanofibrous scaffolds were only formed at low gelation temperature.

Regardless of phase separation method, the pore morphology and compressive modulus of the scaffolds are regulated by types and concentration of polymer, solvent, and quenching temperature. Scaffolds prepared via this method show good mechanical properties as the pore size falls in the range of 10-100 micron. However, it is not ideal for the penetration of cells. The pore size of scaffolds can be enlarged to more than 100 μm by using coarsening process in the later stage of phase separation. This resulted in the formation of bigger solvent crystallites (polymer-lean phase). After sublimation of these bigger solvent crystallites, bigger pores are then left in the scaffolds for better cell infiltration and vascularization (Wei and Ma 2004).

3.2.1.4 Freeze Drying

Freeze drying or lyophilization is a common processing method in pharmaceutical and food industries due to its ability to retain the stability of biological and heat sensitive products at low temperature and under vacuum pressure condition. With its versatility, it has readily been employed for the fabrication of various scaffolds in either two-dimension (2D) or three-dimension (3D) for tissue engineering applications.

Whang et al. (1995) successfully fabricated scaffolds with pore size of 20-200 μm by first creating emulsions in a mixture of polymer solution and water and freeze dry the frozen mixture (Whang et al. 1995). To prepare scaffolds via freeze drying, polymer solution is first frozen at a set range of temperature. After freezing, frozen solution is placed under vacuum and subjected to sublimation. During sublimation or first drying stage, ice crystals of solvent slowly vaporize leaving behind pores. The porous structure in scaffolds was retained owing to the low surface tension induced during drying. Finally, unfrozen solvent in scaffolds gets eliminated by desorption in the final stage of drying (Lu et al. 2013b). This method is not only suitable to process synthetic polymers, but also naturally delicate compounds like collagen and gelatin due to its aseptic processing nature and minimal destruction to samples.

Freezing temperature and polymer concentration prominently regulate the pore structure of scaffolds. Freezing of a polymer solution in liquid nitrogen caused the formation of small ice crystals. When the same polymer solution is subjected to freezing at –20°C, ice crystals tend to grow bigger due to slow ice nucleation leading to formation of bigger pores after freeze drying. A combination of lower polymer concentration and higher freezing temperature (–20°C) usually results in formation of scaffolds with pore ranging between 150 μm and 300 μm which was reported to favor cell infiltration (Madihally and Matthew 1999, Peter et al. 2010). Ikeda et al. (2014) generated chitosan scaffolds of porosity ranging from 75-85% with decreased concentration of chitosan. This study directly demonstrated the inverse correlation between the porosity of scaffolds and the concentration of chitosan. However, conventional freeze drying creates porous structures with random orientation mainly due to the isotropic ice formation during freezing (Ikeda et al. 2014). Ice crystals do not grow in a preferential direction as the polymer solution is cast onto a thermally insulating mold. This results in low mechanical performance of freeze dried scaffolds which limits their application in load bearing aspect. Therefore, post treatment like chemical crosslinking using glutaraldehyde was conducted for the reinforcement and biostability of freeze dried scaffolds (Ma et al. 2003, Adekogbe and Ghanem 2005).

3.2.1.5 Gas Foaming

Gas foaming technique is an organic solvent-free process. Low density, spongy, 3D polymer scaffolds can be produced with gas foaming agent which releases inert gas bubbles under an appropriate external stimuli, e.g. sodium bicarbonate releases N_2 and CO_2 in a moderately acidic environment (Dehghani and Annabi 2011). Other foaming techniques include the use of pressurized hydrogen (Joshi et al. 2015), N_2 and CO_2 (Di Maio et al. 2005), and water (Haugen et al. 2004) in molded polymers until saturation. The nucleation and growth of gas bubbles generate porous structure within the polymer, which yield 3D continuous phase of polymer scaffold upon removal of the dispersed gas phase. However, the process faced limitations as it tends to yield 3D scaffold architecture with unconnected pores and largely non-porous external skin (Quirk et al. 2004).

3.2.1.6 Electrospinning

In the fabrication of scaffold for tissue engineering applications, electrospinning stands out as one of the most facile and inexpensive method to produce non-woven nanofibrous scaffold structure (Dobrzański and Hudecki 2015). The most common electrospinning processes include the melt electrospinning and the solution electrospinning.

In the melt electrospinning process, polymeric materials are melted at an elevated temperature and the molten polymer liquid is subjected to the excitation of high voltage to produce non-woven fibers. The process is typically run at a much lower spinneret feed-rate compared to the solvent electrospinning and is performed in vacuum capillary to facilitate traveling of the charged melt fluid jet. Although melt electrospinning offers the advantages of minimizing chemical toxicity due to residual solvent accumulation in solvent electrospinning, the electrospun fiber strands are mainly in the submicron rather than in the nano-size range and the use of elevated temperature often precludes its popularity in tissue engineering or drug delivery applications (Larrondo and Manley 1981a, b, c, Dalton et al. 2006).

In solution electrospinning, polymers are first dissolved in a suitable solvent to produce liquid solution prior to the application of high voltage field to produce nanofiber. The dissolved polymer liquid is displaced at a controlled rate through a syringe securely held by a syringe pump, forming suspended droplet of polymer liquid at the spinneret tip commonly known as the Taylor's cone. The application of high voltage up to 40 kV charges the polymer droplet and imposes a strong electrostatic repulsive force in the charged polymer droplet. At a critical voltage when the electrostatic force is sufficiently high and exceeds the solution surface tension, a tiny jet emits from the Taylor's cone and accelerates from the spinneret towards a grounded metal collector placed at approximately 5-20 cm away. The jet stream is stretched and elongated while the solvent evaporates during the course of its travel, subsequently stretching the ideally solvent free polymer fiber experience further due to the Coulombic forces of the similarly charged fiber to yield continuous nanofibrous nonwoven membrane/scaffold to be collected at the collector. It is worth mentioning here that at a voltage much lower than the critical point, the polymer droplet eventually drips away from the spinneret tip as the voltage could not overcome the surface tension of the liquid to produce jet of liquid stream. On the other hand, if the excitation voltage is excessive and much higher than the critical point, a spray of jet stream commonly known as electrospraying occurs, causing intermittent discontinuous jet of liquid stream to be accelerated towards the metal collector, manifested by the formation of non-uniform nanofibers with scattered polymer beads. Figure 3.1 shows schematic of an electrospinning system with images of the electrospun nanofibers, as observed from scanning electron microscopy (Nuge et al. 2017).

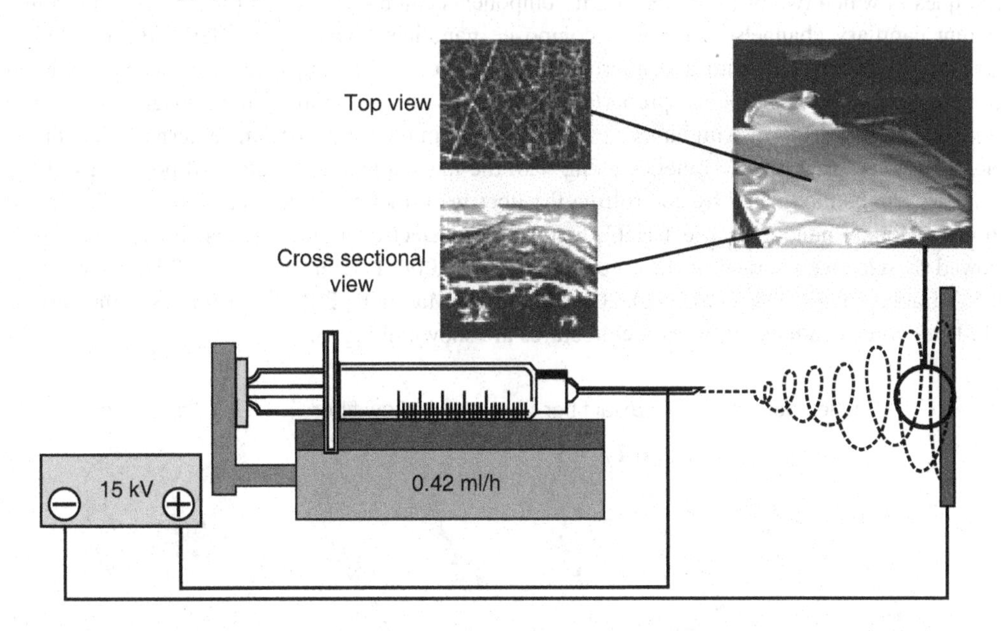

FIGURE 3.1 Schematic of an electrospinning system (Nuge et al. 2017).

The characteristic of fabricated nanofibrous scaffold is hugely dependent on the properties of dissolved polymer liquid (polymer concentration, viscosity, molecular weight, surface tension and charge), process parameters (applied voltage, solution feed rate, spinneret inner diameter, distance from collector) and ambient parameters (temperature, humidity and UV course). It has been reported in open literature that various engineered scaffold were electrospun from conventional polymer species for the purpose of tissue regeneration, e.g. electrospun poly(ε-caprolactone) PCL for bone engineering (Yoshimoto et al. 2003), electrospun collagen/poly(L-lactic-acid)-co-poly(3-caprolactone) (Coll/PLLCL) for skin engineering (Jin et al. 2011), electrospun PLLA for neural tissues engineering (Yang et al. 2005), electrospun nanohydroxyapatite/polycaprolactone (nHA/PCL) drug loaded membrane (Hassan and Sultana 2017) and electrospun silver-coated polycaprolactone/gelatine for antibacterial activity (Lim and Sultana 2016).

3.3 ADVANCED SCAFFOLD FABRICATION TECHNIQUES

3.3.1 Advanced Electrospinning

Over the past decade, electrospinning technique has gained research interest in tissue engineering and biomedical communities and there have been significant scientific advancements, which could cater to the fabrication of dedicated scaffold architectures (i.e. 3D nanofibers with hollow, core-sheath, dual/multicomponent and aligned nanofibers) from conventional and smart stimuli responsive polymers as well as their hybrids.

3.3.1.1 Core-sheath Electrospinning Technique

For the applications in tissue regeneration and controlled drug delivery, core-sheath electrospinning techniques in which two or more dissimilar components can be coaxially or triaxially fed through different capillary channels to produce composite nanofibers have been offered the possibility of forming multi-layered nano and micron diameter fibers with unique features such as tailored mechanical properties, biological properties and release of various drug and nutrient factors. For example, different drugs or stimulants can be embedded in the inner core and external layers of the nanofibers while drug release kinetics along with the mechanical and biological properties of the nanofibers can be modulated by controlling the fiber wall thickness and localization of the drugs/nutrients. Hollow nanofibers are feasible with coaxial electrospinning of two immiscible fluids followed by selective removal of the cores (Li and Xia 2004, Loscertales et al. 2004, Zhang et al. 2006, Varabhas et al. 2008, Wang et al. 2010, Khalf and Madihally 2017). Schematics of the various multi-layer core-sheath nanofibrous architectures are shown in Fig. 3.2.

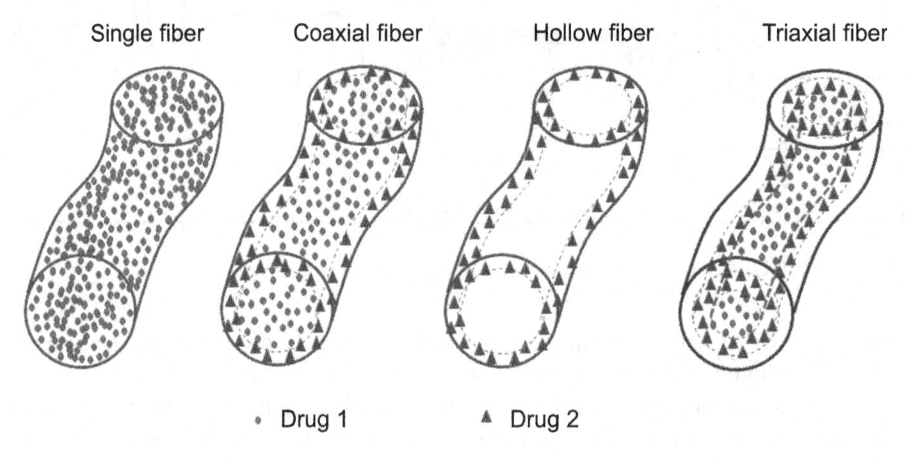

FIGURE 3.2 Schematic showing cross-section of single, coaxial and multiaxial fibers with various embedded stimulants/drugs.

In core-sheath electrospinning process, polymer fluid at the sheath layer received most of the charges from applied high voltage, which causes the formation of Taylor cone and, subsequently, jet initiation at a critical point when repulsive force is higher than the surface tension of the sheath fluid. The core fluid can be dragged along by the sheath jet stream owning to the viscous forces and friction between the sheath-core interfaces (Li and Xia 2004). Process parameters such as feed rate of polymer liquids at the inner core and outer layers, interfacial tension and viscoelasticity of the varying polymer fluids affect the uniformity of the core-sheath layers, morphological homogeneity and functionality along the nanofibrous scaffold (Jiang et al. 2006). It has been observed in the work of Li et al. (2004) that the formation of core-sheath architecture is largely relying on the immiscibility of the core and sheath polymers (Li et al. 2004). Figure 3.3 shows schematics of the setup for electrospinning nanofibers with core-sheath architectures.

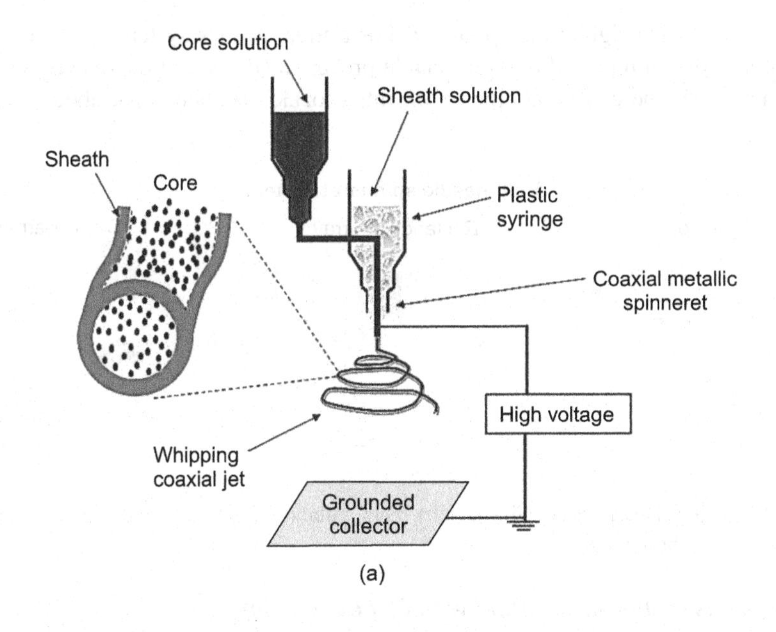

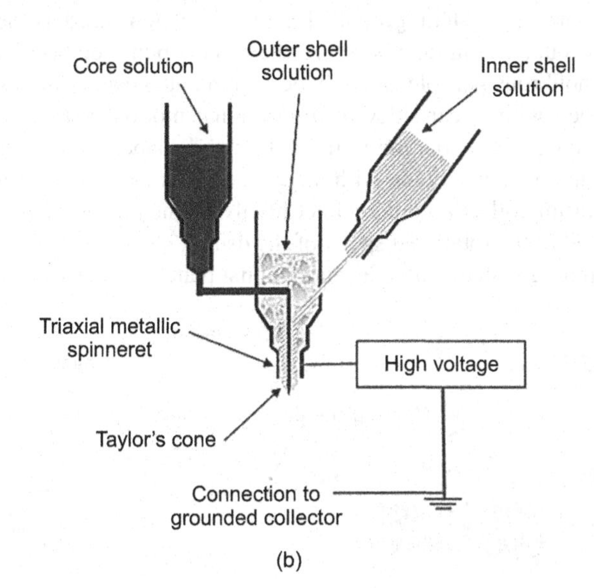

FIGURE 3.3 Schematics of electrospinning setup with coaxial system (a) and triaxial system (b).

3.3.1.2 Electrospinning Upscaling Technique

To increase cost effectiveness and fabrication rate of nanofibrous scaffold, upscaling of electrospinning system from single to multi-spinneret arranged in linear, rectangular and circular configurations have been proposed, similar to those shown in Fig. 3.4. The multi-spinneret approach can be effective in producing thicker membrane with larger coverage area and fabrication volume over a shorter electrospinning duration (Dosunmu et al. 2006, Persano et al. 2013, Tijing et al. 2017). It is worth noting here that there are limitations on the number and packing of spinneret in a multi-spinneret electrospinning system due to the influence of electrostatic charges. As the polymer jet emerges from the spinneret tip, it elongates and spread outs into an expanding cone shape, hence the multiple polymer jets may interfere with one another if there are too many spinnerets that are placed too close to their neighboring spinneret. For a nine nozzles system arranged in a square configuration, it has been reported that reasonable process stability and nanofibers' uniformity of the electrospun membrane can be attained with inter-nozzle distance set at about 1 cm (Theron et al. 2005).

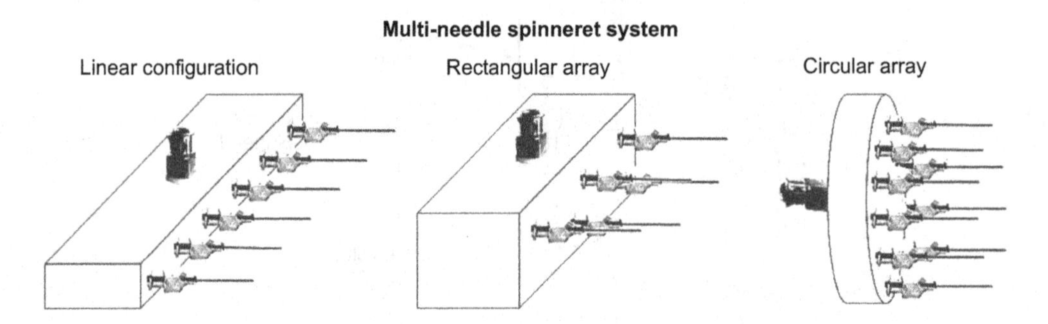

FIGURE 3.4 Schematic of multi-needle linear (left), rectangular (middle) and circular arrays (right) spinneret system for upscaling of throughput.

3.3.1.3 Aligned Nanofibrous Scaffold with Electrospinning

In conventional electrospinning system, grounded flat metal plate is used as the collector. The flat plate collector setup encounters limitations where the electrospun membrane assumes nonwoven architecture and the nanofibrous scaffold can only be accumulated over a small surface area, unless single spinneret equipped with linear sideway translational motion or multi-spinneret system is employed to widen the coverage. Evolution in the design and shape of collector has opened doors to the fabrication of nanofibers membrane with larger foot-print as well as preferential alignment. For example, rotating drum collector could collect highly aligned nanofibers where the degree of orientation can be related to the rotational speed of the drum (Zhu et al. 2010). Figure 3.5 presents schematic of electrospinning system with conventional flat plate vs. rotating drum collectors.

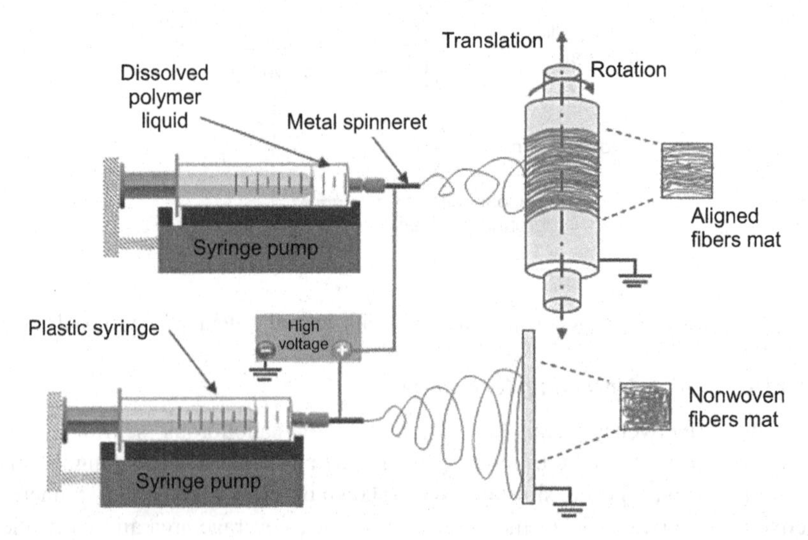

FIGURE 3.5 Electrospinning system with metal flat plate and rotating drum collectors.

With the emphasis on producing higher alignment in electrospun nanofibers, several researchers have devised a slightly modified collector system by replacing the rotating drum with rotating disc collector. Owing to higher field strength closer to the edge of rotating disc, charged fiber placement could be driven along the parallel edge with enhanced orientation over the rotating drum system (Zhong et al. 2006). Figure 3.6 shows schematic of electrospinning system with rotating disc collector.

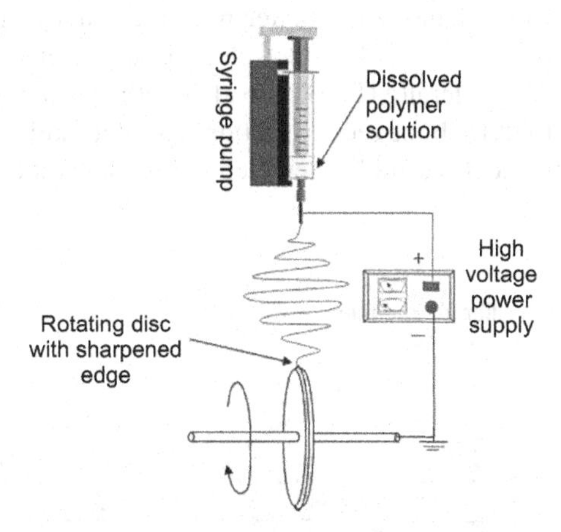

FIGURE 3.6 Schematic of electrospinning system equipped with metal rotating disc collector.

3.3.1.4 Simultaneous Electrospinning-electrospraying Technique

Several researchers investigated the possibility of functionalizing electrospun nanofibers with substances such as collagen coating on poly (ε-caprolactone) (PCL) nanofibers (Zhang et al. 2005), composite nanofibers containing metal particles (Shahi et al. 2011), ceramics (Li et al. 2006), carbon nanofibers and carbon nanotubes (Miao et al. 2016). In this context, multi-axial (coaxial/triaxial) electrospinning that offers the opportunity to simultaneously spin different polymer liquids deserve particular attention. Tijing et al. (2012) employed concurrent electrospinning of bi-polymer liquids containing polyurethane and Ag nanoparticle-decorated polyethylene oxide utilizing dual-spinneret electrospinning system produced antimicrobial hybrid nanofibers membrane with enhanced mechanical properties (Tijing et al. 2012). Birajdar and Lee (2016) utilized coaxial electrospinning and electrospraying system to produce electrospun core-sheath nanofibers function as encapsulation system with electrosprayed silica nanoparticles on the nanofibers surface which function as sonication sensitive cork release system (Birajdar and Lee 2016). Figure 3.7 shows schematic configuration of the combined electrospinning-electrospraying system.

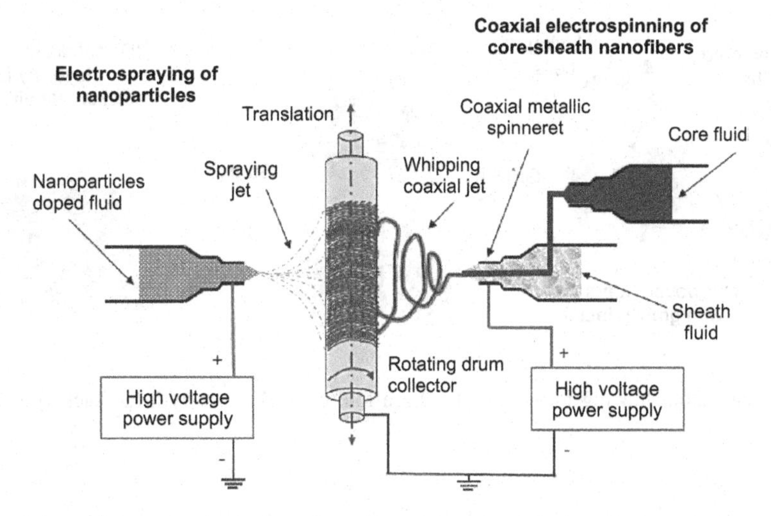

FIGURE 3.7 Schematic showing typical configuration of combined electrospinning-electrospraying system.

3.3.1.5 Electrospinning of 3D Scaffolds

Membranes or scaffolds produced through electrospinning typically resembled 2D geometry with thickness up to few mm. The lack of 3D architecture restricts cellular penetration and presents a challenging aspect to be used in tissue engineering for dressing large areas' critical defects. 3D electrospun scaffolds on the other hand could better mimic the native tissue hierarchy and offer higher mechanical properties. Lu et al. (2013a) rolled electrospun aligned and random-oriented poly (3-hydroxybutyrate-co-3-hydroxyvalerate) (PHBV) loaded with HA nanoparticles into rod like 3D scaffold which has positive effect for bone reconstruction in rabbit radius defects *in vivo*. Schematic of rolled electrospun 3D cylindrical scaffold is depicted in Fig. 3.8 (Lu et al. 2013a).

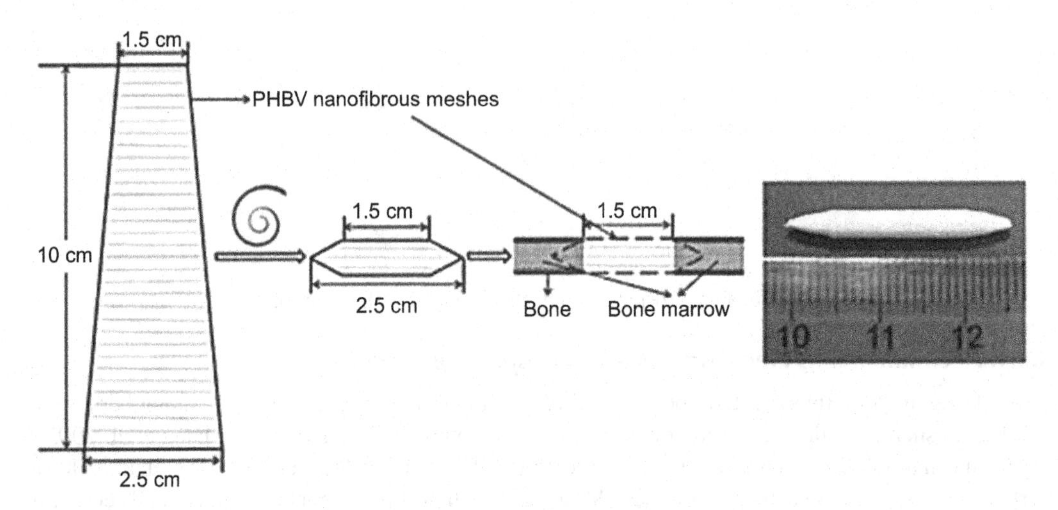

FIGURE 3.8 Schematic of rolled electrospun mat into 3D rod like shape scaffold. (Reprinted with permission from ACS Appl. Mater. Interfaces 2013, 5, 2, 319–330. Copyright 2013 American Chemical Society).

Wan et al. (2015) studied the layer-by-layer stacking of electrospun nanofibers to yield 3D nanofibrous scaffold which reported improved oxygen distribution, nutrition, and waste transportation. The 3D stacked scaffold offers advantages of having control over layer thickness, cell seeding density and drug/growth factor loading density in each layer as well as chemical gradients consisting of different polymers in different layers (Ge et al. 2014, Wan et al. 2015). Figure 3.9 shows schematic of the layer-by-layer stacking of electrospun membranes to yield 3D stacked scaffold.

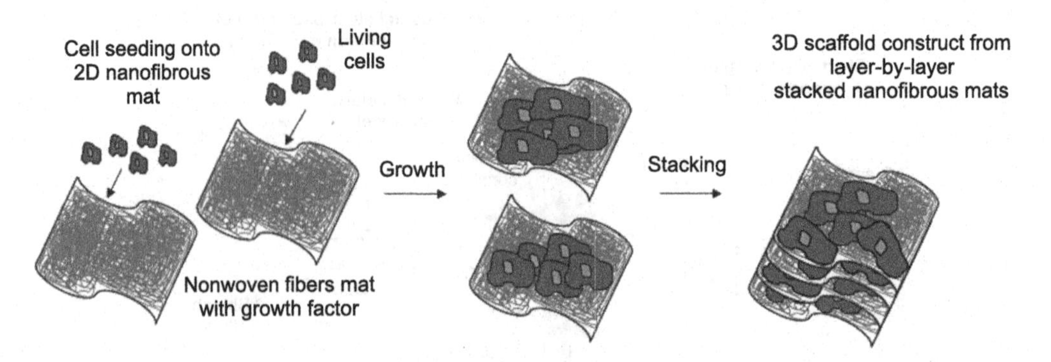

FIGURE 3.9 Schematic showing formation of 3D scaffold through layer-by-layer stacking of electrospun membranes.

3.3.2 Additive Manufacturing

Additive manufacturing (AM) is defined as a process where parts are fabricated layer-by-layer in an additive manner. The advancement of AM has enabled manufacturers to produce complex geometrical prototypes for rapid manufacturing. AM is a reliable process capable of providing low tooling costs, ease of fabrication, high accuracy and flexibility compared to other manufacturing processes such as CNC machining, molding or casting. The process begins by generation of solid 3D model using any computer aided engineering program. The computer aided drawing (CAD) file is converted to .STL ("Standard Triangle Language" or "Standard Tessellation Language") file format before it is used by the AM machine. Various AM techniques have been utilized to produce tissue engineering scaffold including: stereolithography, fused deposition modelling, selective laser sintering, 3D printing and 3D plotting. These manufacturing processes have shown great stability to manufacture replicas with almost 100% of likeness.

Common limitations of AM are long cycle times and high capital costs as compared with traditional processes such as injection molding. Various researches have been conducted to improve the application of AM in producing tissue engineering scaffold. Efforts include analyzing the effects of machine related parameters on the final properties of built parts, as well as the effects of efficiency factors such as material usage and build times. Table 3.1 summarizes the available technology and tolerance materials that have been applied in AM.

TABLE 3.1 A summary of technologies and tolerance materials for additive manufacturing

Technique	Method	Polymer	Ceramic	Metal	Live Cells
Photopolymerization	Stereolithography	✓			
Resin 3D printing	Digital light processing	✓			
Binder jetting	Polymer jetting	✓	✓	✓	
	Binder jetting	✓	✓	✓	
Power bed fusion	Selective laser sintering	✓	✓	✓	
	Selective laser melting	✓	✓	✓	
	Electron beam melting	✓	✓	✓	
Extrusion	Fused deposition modeling	✓			
3D bioprinting					✓

3.3.2.1 Fused Deposition Modeling

Fused deposition modeling (FDM) is the most commonly used AM technique since 2003. FDM was introduced by American company Stratasys in 1992 (Ahn et al. 2002). FDM was initially used to produce 3D parts directly from 3D computer aided design (CAD) files layer-by-layer for use in design verification, prototyping, development and manufacturing. The most considerable challenge for FDM is the selection of build parameters for optimization of performance in conjunction with cost minimization. The FDM machine input/control parameters are infill level (density), border for each layer, and layer height. The output will be measured based on the efficiency which includes total fabrication time, energy consumption, part weight and scrap weight (Zein et al. 2002, An et al. 2015).

Rabionet et al. (2018) successfully fabricated small diameter vascular graft and tubular scaffold from PCL by a novel 3D tubular printer equipped with rotatory platform, Fig. 3.10a, b. The filament was melted, introduced into an extruder nozzle and deposited onto a rotatory platform. The machine includes a nozzle with diameter of 0.4 mm that achieves a precision of 0.028 deg in *W*-axis. The highly porous scaffolds were designed to ensure blood circulation. Fibroblasts cultured on tubular scaffolds were maintained for 6 days. Fibroblasts proliferation exhibited a decreasing trend when printing temperature, speed and flow increased (Rabionet et al. 2018).

Two different extrusion heads specifically designed to handle grain/powder and wire/filament forms of input material were separately fitted on a modified FDM 3D printer by Cerretti et al. (2017), shown in Fig. 3.10c-e. Multi-layered PCL scaffolds were printed with extrusion temperature in the range of 80-90°C and with nozzle size 0.4 mm. Optical microscope measurement and statistical analysis with the aid of ANOVA demonstrated the grain extrusion head capable of producing more uniform printing geometry and higher accuracy of the deposited filament diameter. Human foreskin fibroblasts (HFF) were seeded and completely adhered on the PCL filaments after 24-h of seeding (Ceretti et al. 2017).

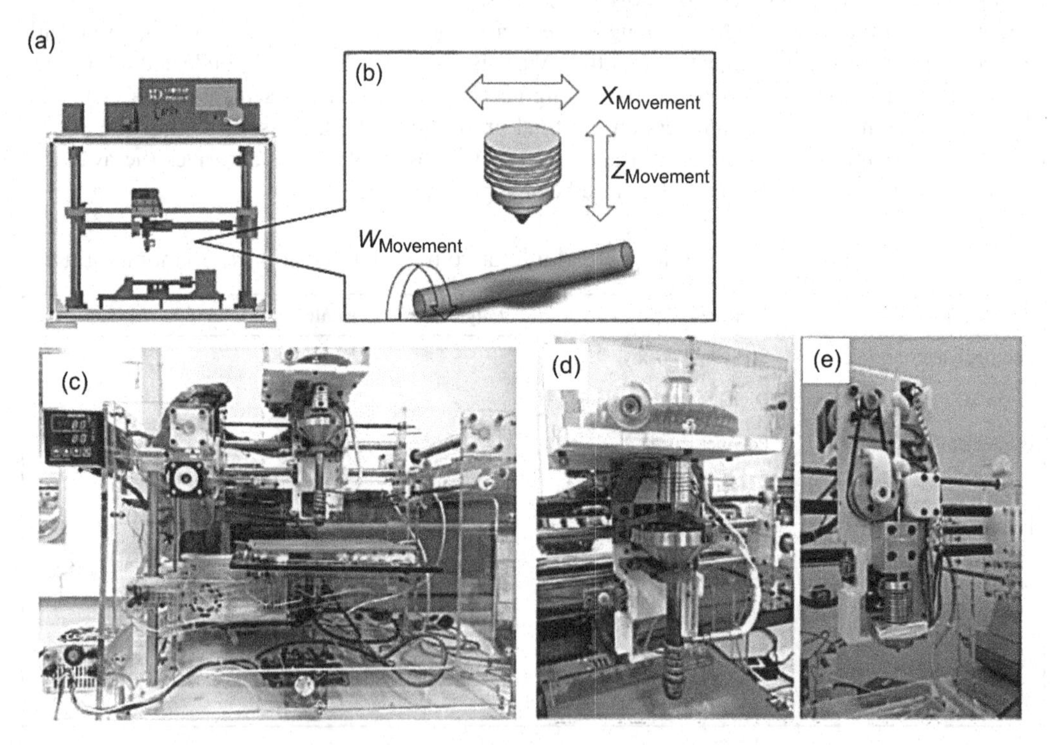

FIGURE 3.10 (a) and (b) FDM 3D tubular printer with custom-designed rotary platform; (c) FDM 3D printer fitted with grain extrusion head; (d) grain extrusion head and (e) filament extrusion head (a, b - Rabionet et al. 2018, c, d, e - Ceretti et al. 2017). (*Reproduced with permission from Elsevier Ltd.*).

3.3.2.2 Stereolithography

Stereolithography (SLA) is a fabrication method based on rapid prototyping, photopolymerization, computer driven and controlled by liquid resin irradiation (Melchels et al. 2010). SLA is a 3D printing process utilizing UV laser to cure photosensitive/photocurable liquid resin into hard polymer normally from the platform up-side down. Resin can be either poured into a reservoir or dispense automatically from a cartridge (Dhariwala et al. 2004). The process starts from a platform build lower into the resin leaving only a thin layer of liquid between the platform and the bottom of the resin tank. Galvanometer directs the UV laser to the transparent window at the bottom of the resin tank, drawing a cross-session of 3D model and selectively polymerizing the material through a layer-by-layer process, as show in Fig. 3.11 (Mouzakis 2018).

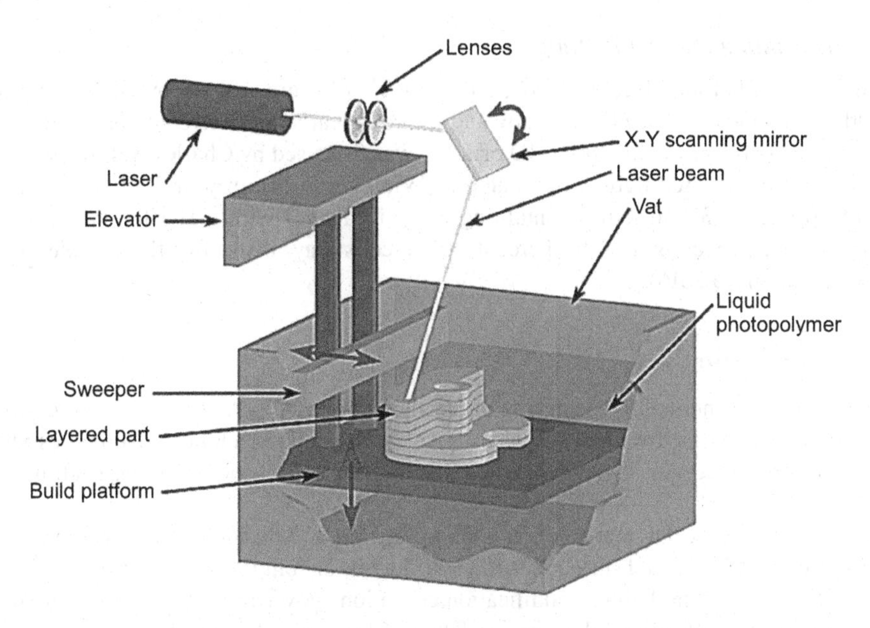

FIGURE 3.11 Schematic of the stereolithography apparatus (Mouzakis 2018). (Reprinted with permission from InTechOpen Ltd.).

3.3.2.3 Selective Laser Sintering

Selective laser sintering (SLS) applies powder bed fusion processes by using laser as an extreme heat source to sinter powder materials, Fig. 3.12. The powder will be melted and solidified layer-by-layer, eventually forming a 3D object. SLS printers are commonly equipped with two plates called pistons and a scanner system. In the beginning of the process, high power laser scans the first layer of powder that is laid onto the fabrication piston, selectively melting and sintering the powder material. Fabrication piston is moved lower after the first layer is solidified while the powder delivery bed is raised slightly to allow a roller to swipe another layer of powder on top of the previous solidified layer. This procedure is repeated to allow the laser to melt and solidify polymeric powder layer-by-layer, until the designed part has been finished bottom to top (Mazzoli 2013).

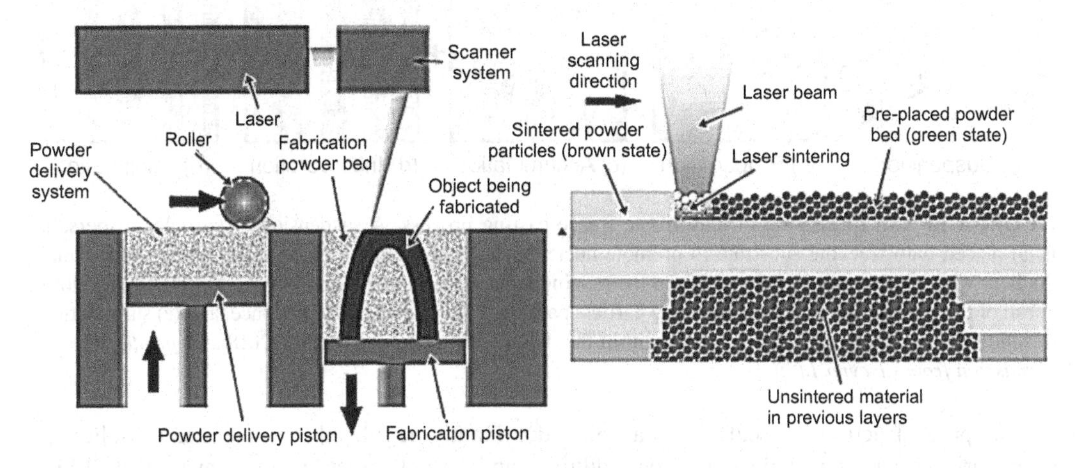

FIGURE 3.12 Schematic drawing of the selective laser sintering apparatus (Mouzakis 2018). (Reprinted with permission from InTechOpen Ltd.).

3.3.2.4 3D Printing and 3D Plotting

3D printing, also called solid freeform fabrication, is one of the greatest discoveries of 20th century in the field of biological science by Mironov and his co-researchers, University of Clemson, South Carolina in 2003 (Mironov et al. 2003). 3D printing was first used by Charles Hull in 1986 to build a 3D structure using layered light-cured material (Whitaker 2014). In year 2000, a research team successfully printed a 3D structure containing E. coli bacteria with a Hewlett-Packard DeskJet 550C that has been reconfigured. 90 percent of the cells stay alive after the completion of the printing process (Sirota 2016).

3.3.3 Freeze Casting

The limitation of conventional freeze drying in creating controlled porous structure in scaffolds has urged researchers to incline towards freeze casting as a fabrication technique. Numerous materials can be readily processed via few steps namely blending, casting, solidification, sublimation and sintering.

Particles suspension or polymer solution is first prepared and blended with or without additives such as dispersants, binders and surfactants to yield slurries or solution which is then cast in a mold for solidification or ice-templating. Solidification conditions govern resultant pores by controlling the orientation and rate of ice crystal formation. The mold is made of insulating sides and metal base responsible for promoting the nucleation and growth of ice in one direction, followed by propagation and accumulation. This is known as anisotropic freeze casting. After solidification is completed, vertically aligned ice crystals (solid grey) are formed and particles or solute (grey) are concentrated within interdendritic space as solidified fluid, Fig. 3.13 (Scotti and Dunand 2018). Subsequently, the solidified fluid undergoes sublimation yielding porous scaffolds. Only metal and ceramic materials are prompted to the final treatment of sintering by heating at elevated temperature below the melting point of the materials. Sintered metals, ceramics or hybrid metal-ceramics were reported to exhibit excellent fracture toughness and tensile strength for load bearing application (Launey et al. 2010). Freeze-cast and sintered HA scaffolds, known as "smart scaffolds", facilitated more effective cell penetration due to the capillary effect induced by the scaffolds (Bai et al. 2015).

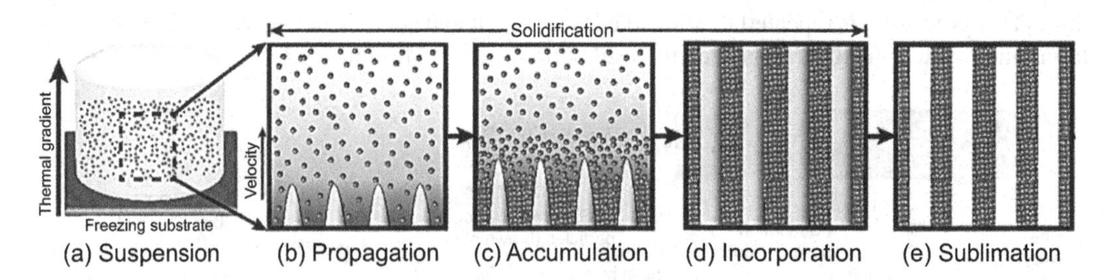

FIGURE 3.13 An illustration of anisotropic freeze casting process. A suspension of particles dispersant is (a) placed onto freezing substrate to promote nucleation, (b) propagation of dendrites along the thermal gradient while pushing particles away from the moving front, (c) accumulation of rejected particles develops ahead of the solid/liquid interface, inducing particle packing within inter-dendritic space, and (e) sublimation of solidified fluid leaving elongated pores within the dendrites (Scotti and Dunand 2018). (*Reproduced with permission from Elsevier Ltd.*).

The pore structure of scaffolds was also dominantly regulated by suspension (solution) characteristics which include fluid type, additives and particle fraction (Fukasawa et al. 2001, Munch et al. 2009, Naviroj et al. 2017). Aqueous suspension with viscosity slightly higher or close to water always yields lamellar pore structure, as the suspension develops perfectly vertical formation of ice crystals during solidification. The first freeze cast collagen sponges were reported with lamellar structure. They were not only biocompatible with preadipocytes, but also resulted in well-vascularized adipose-like tissue facilitating soft tissue regeneration (Von Heimburg et al. 2001, Kuberka et al. 2002).

The morphological transition from lamellar structure to dendritic structure was observed when the viscosity of suspension or solution is increased by adding more particles or additives or using viscous solvent (Scotti and Dunand 2018). Such anisotropic freeze casting causes ice crystals to branch out from the vertical pillar as shown in Fig. 3.14. Scaffolds with dendritic pore structure were reported to have higher wall interconnectivity and open porosity beneficial for cell infiltration (Zuo et al. 2010).

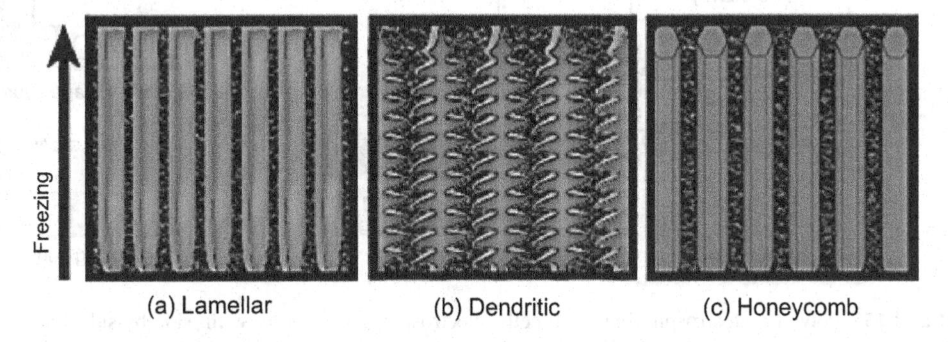

FIGURE 3.14 Freeze-cast materials showing commonly observed pore structures. (a) lamellar, (b) dendritic, and (c) elongated honeycomb (Scotti and Dunand 2018). (*Reproduced with permission from Elsevier Ltd.*).

3.4 EMERGING SCAFFOLD FABRICATION TECHNIQUES

3.4.1 Concurrent and Periodical Encapsulation of Living Cells through Electrospinning

Although 3D nanofibrous scaffolds are capable of closely mimicking the native tissue hierarchy and offering better mechanical support for in-growth of cells and new tissue, there are challenges such as the effectiveness of oxygen/nutrient distribution into the center and metabolic waste transportation from inner to outer layers of the scaffold, maintenance of a homeostasis environment and spatiotemporal chemical gradient throughout the living tissue. In addition, in conventional cell seeding method where random cells placement on external layer of the scaffold is employed, cell penetration became an issue as cells are unable to penetrate through the nanometer pore sizes of the nanofibrous 3D scaffold (Khetan and Burdick 2010, Seil and Webster 2011, Eng et al. 2013, Wan et al. 2015).

The approaches that aim to ensure proper cell distribution in 3D scaffold include parallel electrospinning of polymer solution and cell suspension with a side-by-side capillary configuration (Stankus et al. 2006), simultaneous deposition of cells and polymer using coaxial electrospinning (Andrew and Jayasinghe 2006), and periodical incorporation of living cells at fixed interim during fabrication of 3D scaffold (Seil and Webster 2011).

Stankus et al. (2006) employed concurrent electrospraying of vascular smooth muscle cells (SMCs) and electrospinning of biodegradable elastomeric poly(ester urethane) urea (PEUU) managed to achieve high level of cell viability through a side-by-side configuration of capillary system, Fig. 3.15. The electrosprayed SMCs spread and proliferated like the control unprocessed SMCs. The SMC was charged at 5 kV and fed at 15 mL/hr while the PEUU was charged at 10 kV and fed at 1.5 mL/hr. The flat aluminum collector plate is capable of linear horizontal translation of up to 8 cm along the x-y axes and was charged at −10 kV. The technique offers potential of application in tissue engineering of tubular and sheet structures, e.g. blood vessel, cardiovascular tube, soft tissue membrane for wound dressing, etc. (Stankus et al. 2006).

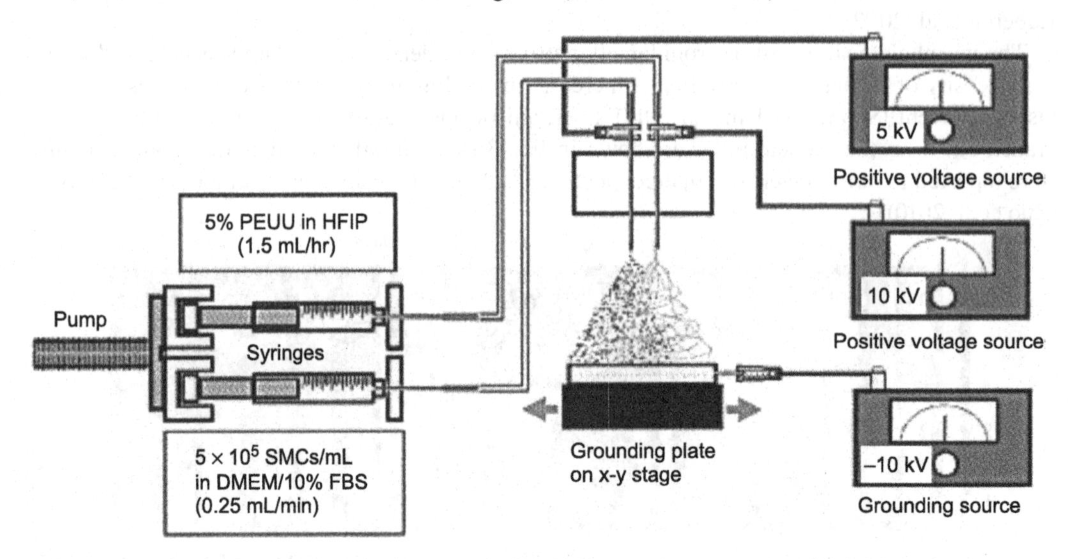

FIGURE 3.15 Polymer electrospinning and cells electrospraying system with side-by-side capillary configuration (Stankus et al. 2006). (*Reproduced with permission from Elsevier Ltd.*).

In the exploration of combining scaffold electrospinning and cells seeding into a single step, Andrew and Jayasinghe (2006) used a coaxial capillary configuration to feed a medical grade poly(dimethylsiloxane) (PDMS) through the outer needle while at the same time concentrated living biosuspension was supplied through the inner needle, as showed in Fig. 3.16. The PDMS has a low electrical conductivity of 10-15 $S{\cdot}m^{-1}$ and high viscosity of 12500 mPa·s. The post electrospun cell-bearing composite microthreads show minimum cellular damage and the collected cells have been viable during the fabrication process. The approach demonstrates feasibility of simultaneous placing of living cells into biocomposite scaffold with the aid of coaxial electrospinning (Andrew and Jayasinghe 2006).

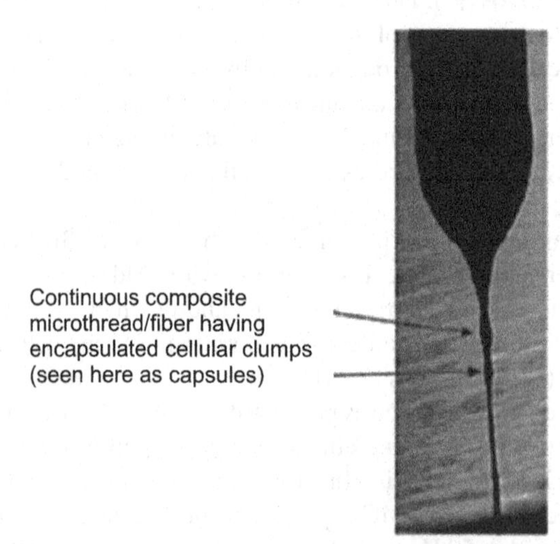

FIGURE 3.16 Composite microthread with living cells encapsulated into electrospun polymeric fibers. (Reprinted with permission from ACS Biomacromolecules 2006, 7, 12, 3364-3369. Copyright 2006 American Chemical Society).

In the work of Seil and Webster (2011), murine fibroblasts (embryo 3T3; CRL-1658; ATCC, Manassas, VA) were sprayed at regular intervals during the fabrication of the biodegradable and cytocompatible poly(DL-lactide-co-glycolide) (PLGA) 3D nanofibers scaffold. Evaluation of the cell viability with and without spraying showed that cells survive the spraying process and maintained viability, which demonstrate preliminary promise for such approaches. However, there are further challenges as the approach produces intermittent cells layer between ECM but natural tissue has a much more homogeneous distribution of cells throughout a tissue rather than layers of cells between layers of ECM. In addition, contamination may become apparent after extended periods of incubation (Seil and Webster 2011).

Although few of these cell electrospinning techniques were developed over the last decades, they show potential in tissue engineering applications and merits special attention for further investigation.

3.4.2 3D Bioprinting with Living Cells

The technology of bioprinting is an innovative cell printing tool to construct tissue and organ structures and enable drug delivery to patient. Similar to 3D printing, the bioprinting process uses computer aided engineering program that incorporates CAD and .STL files. The main difference of bioprinting is that the printing materials used by it are living cells and biological matrix (Munaz et al. 2016). Cells and necessary reagents are precisely allocated layer-by-layer on platform, building up anatomically correct biological structures. The biomaterials serve as a temporary or permanent support to the cells while building their own ECM. One step bioprinting can easily replace the two-step conventional 3D printing, which required construction of tissue engineering scaffold followed by cell seeding. By removing the additional step, bioprinting can effectively reduce time, cost and complexity during cells seeding (Ventola 2014).

Research in bioprinting is still considered new and keeps expanding in regenerative medicine. The challenges faced by bioprinting are finding the compatible biomaterials suitable for bioprinting and the control of fabrication parameters (Zhang et al. 2018). Currently, there is active on-going research on bioprinting technology and its potential as a future source for tissue implants (Li et al. 2016, Bishop et al. 2017, Aljohani et al. 2018, Derakhshanfar et al. 2018).

According to Murphy et al. (2002), the biomaterials that can function well with bioprinting need to show good deposition and biocompatibility *in vitro* by performing cross-linking mechanisms (Murphy et al. 2002). Deposition of living cells into hydrogel to bioprint organ have been quite successful recently (Wu et al. 2016, Lee and Dai 2017, Gao et al. 2018, Giuseppe et al. 2018, Li et al. 2018). This technology of organ printing is mainly deployed for chemical-free phase changing hydrogels. After the biomaterials are deposited on platform, liquid phase is solidified by UV light, temperature, pH, or ion concentrations. During the printing process, living cells and growth factors can be mixed with biomaterials and form biomimicry native tissue constitution. Table 3.2 shows different complexes and important organs that have been synthesized.

TABLE 3.2 Examples of organs that have been successfully synthesized through 3D bioprinting

Organs	Ref.
Kidney	(Rezende et al. 2013)
Liver	(Kizawa et al. 2017)
Blood vessels	(Li et al. 2018)

Research team with the long-term goal of producing human organs-on-demand created commercial bioprinter called CELLINK Bio X, which is capable of utilizing 13 types of bio-ink and 3 different types of extrusion methods. This commercial bioprinter showed high potential in the formation of artificial organs, and may even be used as bioprinting substitutes for bone (Ozbolat et al. 2014).

3.4.2.1 *Stereolithography Bioprinting*

Compared with other AM methods, SLA bioprinting offers capability to produce complex 3D biological construct with high speed—the highest resolution printing attributed to the accuracy of laser and cell viability (Yi et al. 2017, Derakhshanfar et al. 2018). The process typically uses ultraviolet, infrared, or visible wavelengths to cure photo-sensitive/photo-curable liquid resin into 3D architecture built-up through a layer-by-layer process (Park et al. 2017). Figure 3.17 shows schematic of SLA bioprinting for tissue engineering applications.

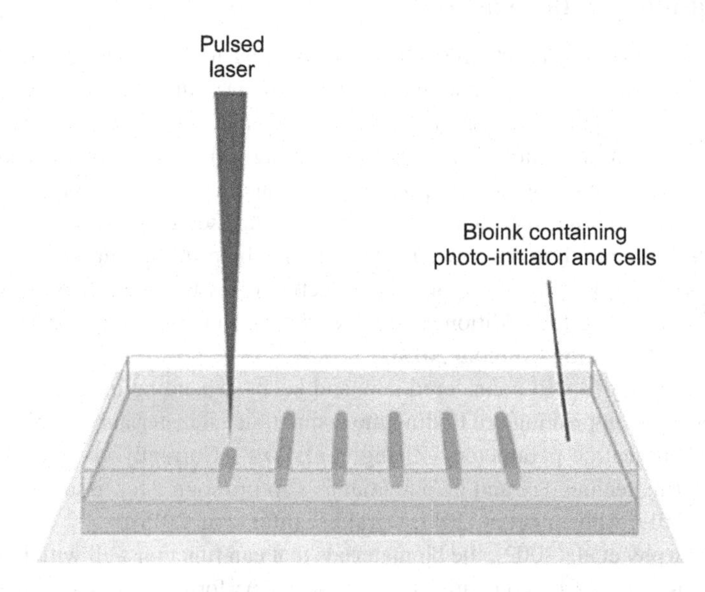

FIGURE 3.17 Schematic of stereolithography bioprinting for tissue engineering. Adapted from Jang et al. 2016 reprinted with permission from ACS Biomater. Sci. Eng. 2016, 2, 10, 1722-1731.

Despite its advantages, several researchers have found that utilization of UV as the polymerization catalyst could be harmful to DNA cells and even cause skin cancer. To address this issue, visible light stereolithography (VL-STL) bioprinting systems have garnered much attention. An ideal VL-STL bioprinting system shall establish uniform distribution of living cells within a dedicated scaffold through utilization of non-cytotoxic, water soluble, visible light sensitive initiator capable of generating sustainable source of radicals within the initiator, monomer and living cells suspension system during the VL-STL process. The bioprinted scaffold shall be hydrophilic and non-cytotoxic to support cell viability (Lin et al. 2013). Figure 3.18 shows schematic of typical VL-STL bioprinting setup with cell fabrication capability.

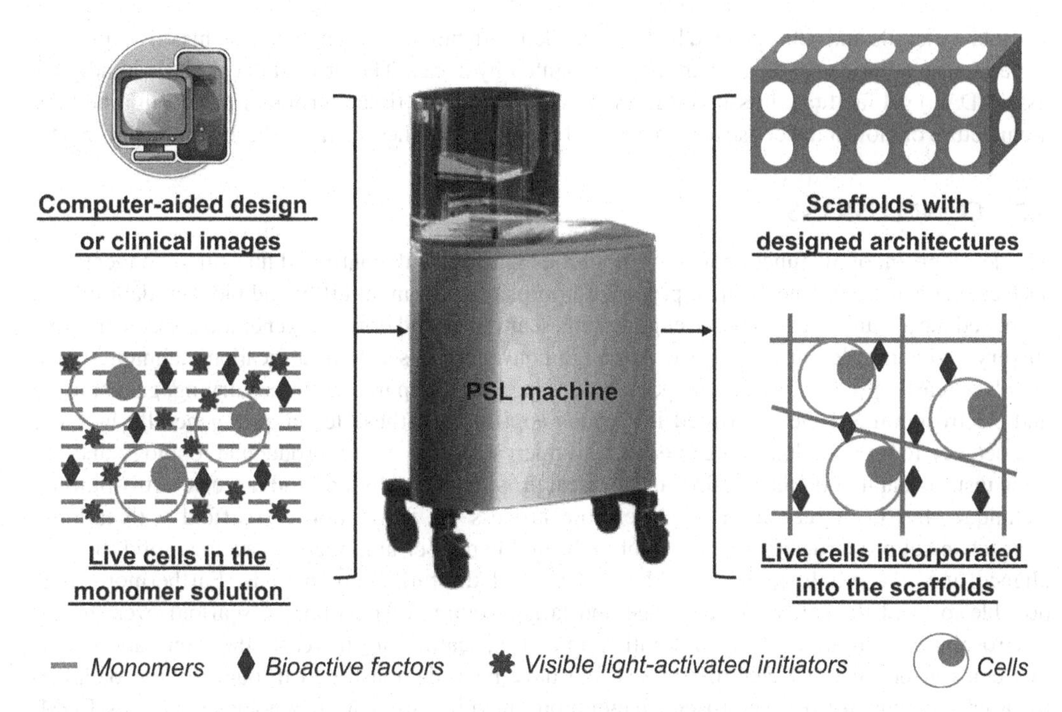

FIGURE 3.18 Schematic of visible light stereolithography (VL-STL) system with live cell-fabrication capability. (Lin et al. 2013). (*Reproduced with permission from Elsevier Ltd.*).

3.4.2.2 Extrusion-based Bioprinting

Typically, an extrusion-based 3D bioprinting machine is installed with a gantry robot and a *Z*-axis printing stage. The gantry robot extrudes any printable hydrogels at *X* and *Y* axes. A deposition tool is connected to a syringe with interchangeable tips of various diameters. The speed and position of syringe tip can be precisely controlled by computer aided manufacturing (CAM) system.

Highly printable hydrogel like alginate, hyaluronic acid, collagen or gelatin is first dissolved in biocompatible solvent and filtered with sterile membrane before mixing with cell pellet collected by centrifugation (Chung et al. 2013, Park et al. 2014, Kundu et al. 2015). Trace amount of growth factors can be mixed gently with the hydrogel suspension and cells before dispensing (Cohen et al. 2006). Subsequently, hydrogel with laden cells is loaded into the deposition tool and extruded as filaments which are deposited on the printing stage in layer-by-layer manner. It is quite common to find extrusion-based 3D printing machines installed with multi-printing syringe for construction of tissue of high complexity (Kundu et al. 2015). For reinforcement, extruded filaments are directly deposited in crosslinking reagent. Alternatively, scaffold constructs can be subjected to photo-crosslinking by UV light source (Bian et al. 2013). These attempts usually take 10-15 min resulting in longer structural and mechanical integrity of pre-seeded scaffolds than those without treatment.

Certain cell sources such as liver cells, chondrocytes, osteoblasts, fibroblasts and etc. were successfully loaded on aforementioned hydrogels for respective tissue regeneration. Billiet et al. (2014) successfully bioprinted an artificial liver tissue with 100% interconnected pore network by using hepatocarcinoma cells (HepG2) loaded in gelatin methacrylamide hdyrogel. In particular, they investigated how different photo-initiators affected the cell viability and concluded that VA-086 photo-initiator was the best photo-crosslinker which ensured high cell viability of more than 97% (Billiet et al. 2014). Advantages of interaction between host tissue and extrusion based bio-printed scaffolds were highlighted in another work by Pati et al. (2015). Despite using same scaffold materials and cells, printed hydrogel loaded with human mesenchymal stem cells expressed more intense adipogenic genes than the non-printed hydrogel. The printed decellularized adipose tissue (DAT) gel facilitated tissue infiltration, remodeling and tissue formation which signified the architecture of bioprinted constructs in terms of promoting higher *in vitro* efficacy (Pati et al. 2015).

3.5 CONCLUSIONS

The development of 3D functionalized nanofibrous scaffold with controlled fibers interconnectivity, surface characteristics, mechanical properties, porosity, biocompatibility and biodegradability has propelled the advancement in tissue engineering, scar repair and organs regeneration, reconstructive surgery and controlled drug delivery. Although conventional scaffold fabrication techniques such as solvent casting/particulate leaching, melt molding, phase separation, freeze drying, gas foaming and electrospinning have been used in various applications, these techniques generally lack the competency to produce long-range channeling microstructure with reproducible topology and are often insufficient to modulate desired mechanical properties. Evolution in advanced electrospinning techniques has been seen as a more versatile process for fabrication of scaffold with tailored biomechanical properties from a range of dedicated nanofiber architectures such as solid, hollow, aligned and/or coaxial/triaxial core-sheath layers of dissimilar compounds. Furthermore, it is possible to yield thicker, wider coverage and larger volume of nanofibrous scaffold over shorter electrospinning duration through multi-spinneret upscaling to increase the fabrication rate. Strategies to produce 3D electrospun scaffold have been demonstrated through rolling or layer-by-layer stacking of post electrospun nanofibrous membranes. AM technologies such as FDM, SLA, SLS and 3D printing are gaining momentum in changing the manner in which structural 3D objects are produced. The integration of these AM technologies with advanced scaffold fabrication techniques such as concurrent electrospinning/electrospraying and 3D bioprinting of living cells have revolutionized the cell seeding mechanism from the conventional random cell placement on exterior surfaces of post fabricated 3D scaffold to concurrent deposition and in-line periodical encapsulation of living cells throughout the scaffold fabrication process, hence overcoming the issues associated with cell penetration through nanosized pore opening of nanofibrous scaffold. However, the technology is still at its early development stage and there are existing challenges in the development of these emerging techniques, e.g. from the viewpoint of cell viability, homogeneity in cell distribution, contamination over extended period of incubation, limited availability of compatible biomaterials and reproducibility in simultaneous one step fabrication of biological cells along with biomaterials compared to the two-step synthesis of biomaterial scaffold followed by cell seeding. Further development and widespread adoption of the advanced techniques unavoidably require close collaboration amongst material scientists, biotechnologists and medical practitioners to realize the full potential of enhanced scaffold fabrication techniques, such as on-demand synthesis of patient specific organ and bone amongst others.

REFERENCES

Adekogbe, I. and A. Ghanem. 2005. Fabrication and characterization of DTBP-crosslinked chitosan scaffolds for skin tissue engineering. Biomaterials 26: 7241-7250. doi: 10.1016/j.biomaterials.2005.05.043.

Ahn, S.-H., M. Montero, D. Odell, S. Roundy and P.K. Wright. 2002. Anisotropic material properties of fused deposition modeling ABS. Rapid Prototyping Journal 8(4): 248-257. doi: 10.1108/13552540210441166.

Aljohani, W., M.W. Ullah, X. Zhang and G. Yang. 2018. Bioprinting and its applications in tissue engineering and regenerative medicine. International Journal of Biological Macromolecules 107: 261-275.

An, J., J.E.M. Teoh, R. Suntornnond and C.K. Chua. 2015. Design and 3D printing of scaffolds and tissues. Engineering 1(2): 261-268. doi: 10.15302/J-ENG-2015061.

Andrew, T.-N. and S.N. Jayasinghe. 2006. Cell electrospinning: a unique biotechnique for encapsulating living organisms for generating active biological microthreads/scaffolds. Biomacromolecules 7(12): 3364-3369. doi: 10.1021/bm060649h.

Bai, H., D. Wang, B. Delattre, W. Gao, J. de Coninck, S. Li and A.P. Tomsia. 2015. Biomimetic gradient scaffold from ice-templating for self-seeding of cells with capillary effect. Acta Biomaterialia 20: 113-119. doi: 10.1016/j.actbio.2015.04.007.

Bian, L., C. Hou, E. Tous, R. Rai, R.L. Mauck and J.A. Burdick. 2013. The influence of hyaluronic acid hydrogel crosslinking density and macromolecular diffusivity on human MSC chondrogenesis and hypertrophy. Biomaterials 34: 413-421. doi: 10.1016/j.biomaterials.2012.09.052.

Billiet, T., E. Gevaert, T. de Schryver, M. Cornelissen and P. Dubruel. 2014. The 3D printing of gelatin methacrylamide cell-laden tissue-engineered constructs with high cell viability. Biomaterials 35: 49-62. doi: 10.1016/j.biomaterials.2013.09.078.

Birajdar, M.S. and J. Lee. 2016. Sonication-triggered zero-order release by uncorking core-shell nanofibers. Chemical Engineering Journal 288: 1-8. doi: https://doi.org/10.1016/j.cej.2015.11.095.

Bishop, E.S., S. Mostafa, M. Pakvasa, H.H. Luu, M.J. Lee, J.M. Wolf, G.A. Ameer, T.-C. He and R.R. Reid. 2017. 3-D bioprinting technologies in tissue engineering and regenerative medicine: current and future trends. Genes & Diseases 4(4): 185-195. doi: https://doi.org/10.1016/j.gendis.2017.10.002.

Borenstein, J.T., H. Terai, K.R. King, E.J. Weinberg, M.R. Kaazempur-Mofrad and J.P. Vacanti. 2002. Microfabrication technology for vascularized tissue engineering. Biomedical Microdevices 4(3): 167-175. doi: 10.1023/A:1016040212127.

Ceretti, E., P. Ginestra, P. Neto, A. Fiorentino and J. Silva. 2017. Multi-layered scaffolds production via fused deposition modeling (FDM) using an open source 3D printer: process parameters optimization for dimensional accuracy and design reproducibility. Procedia CIRP 65: 13-18. doi: 10.1016/j.procir.2017.04.042.

Chan, B.P. and K.W. Leong. 2008. Scaffolding in tissue engineering: general approaches and tissue-specific considerations. European Spine Journal 17: 467-479. doi: 10.1007/s00586-008-0745-3.

Choudhury, M., S. Mohanty and S. Nayak. 2015. Effect of different solvents in solvent casting of porous pla scaffolds – in biomedical and tissue engineering applications. Journal of Tissue Science & Engineering 6(142): 1-7. doi: 10.4172/2157-7552.1000142.

Chung, J.H.Y., S. Naficy, Z. Yue, R. Kapsa, A. Quigley, S.E. Moulton and G.G. Wallace. 2013. Bio-ink properties and printability for extrusion printing living cells. Biomaterials Science 1(7): 763-773. doi: 10.1039/c3bm00012e.

Cohen, D.L., E. Malone, H. Lipson and L.J. Bonassar. 2006. Direct freeform fabrication of seeded hydrogels in arbitrary geometries. Tissue Engineering 12: 1325-1335. doi: 10.1089/ten.2006.12.1325.

Coombes, A.G.A., S.C. Rizzi, M. Williamson, J.E. Barralet, S. Downes and W.A. Wallace. 2004. Precipitation casting of polycaprolactone for applications in tissue engineering and drug delivery. Biomaterials 25: 315-325. doi: 10.1016/S0142-9612(03)00535-0.

Correlo, V.M., L.F. Boesel, E. Pinho, A.R. Costa-Pinto, M.L. Alves da Silva, M. Bhattacharya, J.F. Mano, N.M. Neves and R.L. Reis. 2009. Melt-based compression-molded scaffolds from chitosan-polyester blends and composites: morphology and mechanical properties. Journal of Biomedical Materials Research – Part A 91(2): 489-504. doi: 10.1002/jbm.a.32221.

Dalton, P.D., J.L. Calvet, A. Mourran, D. Klee and M. Möller. 2006. Melt electrospinning of poly-(ethylene glycol-block-ε-caprolactone). Biotechnology Journal 1(9): 998-1006. doi: 10.1002/biot.200600064.

Dehghani, F. and N. Annabi. 2011. Engineering porous scaffolds using gas-based techniques. Current Opinion in Biotechnology 22: 661-666. doi: 10.1016/j.copbio.2011.04.005.

Derakhshanfar, S., R. Mbeleck, K. Xu, X. Zhang, W. Zhong and M. Xing. 2018. 3D bioprinting for biomedical devices and tissue engineering: a review of recent trends and advances. Bioactive Materials 3(2): 144-156. doi: https://doi.org/10.1016/j.bioactmat.2017.11.008.

Dhariwala, B., E. Hunt and T. Boland. 2004. Rapid prototyping of tissue-engineering constructs, using photopolymerizable hydrogels and stereolithography. Tissue Engineering 10: 1316-1322. doi: 10.1089/ten.2004.10.1316.

Di Maio, E., G. Mensitieri, S. Iannace, L. Nicolais, W. Li and R.W. Flumerfelt. 2005. Structure optimization of polycaprolactone foams by using mixtures of CO2 and N2 as blowing agents. Polymer Engineering and Science 45(3): 432-441. doi: 10.1002/pen.20289.

Dobrzański, L.A. and A. Hudecki. 2015. Polymer nanofibers materials, fabrication technologies and research methods. pp. 83-126. *In*: L.A. Dobrzański [ed.]. Polymer Nanofibers Produced by Electrospinning Applied in Regenerative Medicine. Gliwice, Poland: Narodowe Centrum Nauki.

Dong, Z., Y. Li and Q. Zou. 2009. Degradation and biocompatibility of porous nano-hydroxyapatite/polyurethane composite scaffold for bone tissue engineering. Applied Surface Science 255(12): 6087-6091. doi: https://doi.org/10.1016/j.apsusc.2009.01.083.

Dosunmu, O.O., G.G. Chase, W. Kataphinan and D.H. Reneker. 2006. Electrospinning of polymer nanofibers from multiple jets on a porous tubular surface. Nanotechnology 17: 1123-1127. doi: 10.1088/0957-4484/17/4/046.

Eng, G., B.W. Lee, H. Parsa, C.D. Chin, J. Schneider, G. Linkov, S.K. Sia and G. Vunjak-Novakovic. 2013. Assembly of complex cell microenvironments using geometrically docked hydrogel shapes. Proceedings of the National Academy of Sciences 110(12): 4551-4556. doi: 10.1073/pnas.1300569110.

Fukasawa, T., M. Ando, T. Ohji and S. Kanzaki. 2001. Synthesis of porous ceramics with complex pore structure by freeze-dry processing. Journal of the American Ceramic Society 84: 230-232. doi: 10.1111/j.1151-2916.2001.tb00638.x.

Gao, Q., H.M. Zhao, F.F. Yang, J.-Z. Fu and Y. He. 2018. Practical laboratory methods for 3D bioprinting. pp. 7-32. *In*: D.J. Thomas, Z.M. Jessop and I.S. Whitaker [eds.]. 3D Bioprinting for Reconstructive Surgery Techniques and Applications. Duxford, United Kingdom: Woodhead Publishing, Elsevier. https://doi.org/10.1016/B978-0-08-101103-4.00003-X.

Ge, L., Q. Li, Y. Huang, S. Yang, J. Ouyang, S. Bu, W. Zhong, Z. Liu and M.M.Q. Xing. 2014. Polydopamine-coated paper-stack nanofibrous membranes enhancing adipose stem cells adhesion and osteogenic differentiation. Journal of Materials Chemistry B 2: 6917-6923. doi: 10.1039/c4tb00570h.

Giuseppe, M. Di, N. Law, B. Webb, R.A. Macrae, L.J. Liew, T.B. Sercombe, R.J. Dilley and B.J. Doyle. 2018. Mechanical behaviour of alginate-gelatin hydrogels for 3D bioprinting. Journal of the Mechanical Behavior of Biomedical Materials 79: 150-157. doi: https://doi.org/10.1016/j.jmbbm.2017.12.018.

Hassan, M.I. and N. Sultana. 2017. Characterization, drug loading and antibacterial activity of nanohydroxyapatite/polycaprolactone (nHA/PCL) electrospun membrane. 3 Biotech 7(4): 249. doi: 10.1007/s13205-017-0889-0.

Haugen, H., V. Ried, M. Brunner, J. Will and E. Wintermantel. 2004. Water as foaming agent for open cell polyurethane structures. Journal of Materials Science: Materials in Medicine 15: 343-346. doi: 10.1023/B:JMSM.0000021099.33619.ac.

Ho, M.H., P.Y. Kuo, H.J. Hsieh, T.Y. Hsien, L.T. Hou, J.Y. Lai and D.M. Wang. 2004. Preparation of porous scaffolds by using freeze-extraction and freeze-gelation methods. Biomaterials 25: 129-138. doi: 10.1016/S0142-9612(03)00483-6.

Ikeda, T., K. Ikeda, K. Yamamoto, H. Ishizaki, Y. Yoshizawa, K. Yanagiguchi, S. Yamada and Y. Hayashi. 2014. Fabrication and characteristics of chitosan sponge as a tissue engineering scaffold. BioMed Research International 786892: 8. doi: 10.1155/2014/786892.

Jang, J., H.-G. Yi and D.-W. Cho. 2016. 3D printed tissue models: present and future. ACS Biomaterials Science & Engineering 2(10): 1722-1731. https://doi.org/10.1021/acsbiomaterials.6b00129.

Jiang, H., Y. Hu, P. Zhao, Y. Li and K. Zhu. 2006. Modulation of protein release from biodegradable core-shell structured fibers prepared by coaxial electrospinning. Journal of Biomedical Materials Research - Part B. Applied Biomaterials 79A: 50-57. doi: 10.1002/jbm.b.30510.

Jin, G., Prabhakaran, M.P. and S. Ramakrishna. 2011. Stem cell differentiation to epidermal lineages on electrospun nanofibrous substrates for skin tissue engineering. Acta Biomaterialia 7(8): 3113-3122. doi: 10.1016/j.actbio.2011.04.017.

Joshi, M.K., H.R. Pant, A.P. Tiwari, H.J. Kim, C.H. Park and C.S. Kim. 2015. Multi-layered macroporous three-dimensional nanofibrous scaffold via a novel gas foaming technique. Chemical Engineering Journal 275: 79-88. doi: https://doi.org/10.1016/j.cej.2015.03.121.

Katti, D.S., S. Lakshmi, R. Langer and C.T. Laurencin. 2002. Toxicity, biodegradation and elimination of polyanhydrides. Advanced Drug Delivery Reviews. 54: 933-961. doi: 10.1016/S0169-409X(02)00052-2.

Khalf, A. and S.V. Madihally. 2017. Recent advances in multiaxial electrospinning for drug delivery. European Journal of Pharmaceutics and Biopharmaceutics 112: 1-17. doi: 10.1016/j.ejpb.2016.11.010.

Khetan, S. and J.A. Burdick. 2010. Patterning network structure to spatially control cellular remodeling and stem cell fate within 3-dimensional hydrogels. Biomaterials 31(32): 8228-8234. doi: 10.1016/j.biomaterials.2010.07.035.

Kizawa, H., E. Nagao, M. Shimamura, G. Zhang and H. Torii. 2017. Scaffold-free 3D bio-printed human liver tissue stably maintains metabolic functions useful for drug discovery. Biochemistry and Biophysics Reports 10: 186-191. doi: 10.1016/j.bbrep.2017.04.004.

Kuberka, M., D. Von Heimburg, H. Schoof, I. Heschel and G. Rau. 2002. Magnification of the pore size in biodegradable collagen sponges. International Journal of Artificial Organs 25: 67-73. doi: 10.1177/039139880202500111.

Kundu, J., J.H. Shim, J. Jang, S.W. Kim and D.W. Cho. 2015. An additive manufacturing-based PCL-alginate-chondrocyte bioprinted scaffold for cartilage tissue engineering. Journal of Tissue Engineering and Regenerative Medicine 9(11): 1286-1297. doi: 10.1002/term.1682.

Larrondo, L. and R.S.J. Manley. 1981a. Electrostatic fiber spinning from polymer melts. I. Experimental observations on fiber formation and properties. Journal of Polymer Science: Polymer Physics Edition 19(6): 909-920. doi: 10.1002/pol.1981.180190601.

Larrondo, L. and R.S.J. Manley. 1981b. Electrostatic fiber spinning from polymer melts. II. Examination of the flow field in an electrically driven jet. Journal of Polymer Science: Polymer Physics 19: 921-932. doi: 10.1002/pol.1981.180190602.

Larrondo, L. and R.S.J. Manley. 1981c. Electrostatic fiber spinning from polymer melts. III. Electrostatic deformation of a pendant drop of polymer melt. Journal of Polymer Science: Polymer Physics Edition 19(6): 933-940. doi: 10.1002/pol.1981.180190603.

Launey, M.E., E. Munch, D.H. Alsem, E. Saiz, A.P. Tomsia and R.O. Ritchie. 2010. A novel biomimetic approach to the design of high-performance ceramic-metal composites. Journal of The Royal Society Interface 7: 741-753. doi: 10.1098/rsif.2009.0331.

Lee, V.K. and G. Dai. 2017. Printing of three-dimensional tissue analogs for regenerative medicine. Annals of Biomedical Engineering 45: 115-131. doi: 10.1007/s10439-016-1613-7.

Li, D. and Y. Xia. 2004. Direct fabrication of composite and ceramic hollow nanofibers by electrospinning. Nano Letters 4(5): 933-938. doi: 10.1021/nl049590f.

Li, D., A. Babel, S.A. Jenekhe and Y. Xia. 2004. Nanofibers of conjugated polymers prepared by electrospinning with a two-capillary spinneret. Advanced Materials 16(22): 2062-2066. doi: 10.1002/adma.200400606.

Li, Dan, J.T. McCann, Y. Xia and M. Marquez. 2006. Electrospinning: a simple and versatile technique for producing ceramic nanofibers and nanotubes. Journal of the American Ceramic Society 89(6): 1861-1869. doi: 10.1111/j.1551-2916.2006.00989.x.

Li, H., Y.J. Tan and L. Li. 2018. A strategy for strong interface bonding by 3D bioprinting of oppositely charged κ-carrageenan and gelatin hydrogels. Carbohydrate Polymers 198: 261-269. doi: https://doi.org/10.1016/j.carbpol.2018.06.081.

Li, J., M. Chen, X. Fan and H. Zhou. 2016. Recent advances in bioprinting techniques: approaches, applications and future prospects. Journal of Translational Medicine 14(1): 271. doi: 10.1186/s12967-016-1028-0.

Li, X., L. Liu, X. Zhang and T. Xu. 2018. Research and development of 3D printed vasculature constructs. Biofabrication 10(3): 32002. http://stacks.iop.org/1758-5090/10/i=3/a=032002.

Liao, S., C.K. Chan and S. Ramakrishna. 2008. Stem cells and biomimetic materials strategies for tissue engineering. Materials Science and Engineering: C 28(8): 1189-1202. doi: https://doi.org/10.1016/j.msec.2008.08.015.

Lim, M.M. and N. Sultana. 2016. *In vitro* cytotoxicity and antibacterial activity of silver-coated electrospun polycaprolactone/gelatine nanofibrous scaffolds. 3 Biotech 6(2): 211. doi: 10.1007/s13205-016-0531-6.

Lin, H., D. Zhang, P.G. Alexander, G. Yang, J. Tan, A.W.M. Cheng and R.S. Tuan. 2013. Application of visible light-based projection stereolithography for live cell-scaffold fabrication with designed architecture. Biomaterials. Elsevier Ltd 34(2): 331-339. doi: 10.1016/j.biomaterials.2012.09.048.

Liu, C., Z. Xia and J.T. Czernuszka. 2007. Design and development of three-dimensional scaffolds for tissue engineering. Chemical Engineering Research and Design 85: 1051-1064. doi: 10.1205/cherd06196.

Liu, X. and P.X. Ma. 2009. Phase separation, pore structure, and properties of nanofibrous gelatin scaffolds. Biomaterials 30(25): 4094-4103. doi: 10.1016/j.biomaterials.2009.04.024.

Loscertales, I.G., A. Barrero, M. Márquez, R. Spretz, R. Velarde-Ortiz and G. Larsen. 2004. Electrically forced coaxial nanojets for one-step hollow nanofiber design. Journal of the American Chemical Society 126(17): 5376-5377. doi: 10.1021/ja049443j.

Lu, H., D. Cai, G. Wu, K. Wang and D. Shi. 2011. Whole meniscus regeneration using polymer scaffolds loaded with fibrochondrocytes. Chinese Journal of Traumatology 14(4): 195-204. doi: https://doi.org/10.3760/cma.j.issn.1008-1275.2011.04.001.

Lu, L.-X., X.-F. Zhang, Y.-Y. Wang, L. Ortiz, X. Mao, Z.-L. Jiang, Z.-D. Xiao and N.-P. Huang. 2013a. Effects of hydroxyapatite-containing composite nanofibers on osteogenesis of mesenchymal stem cells *in vitro* and bone regeneration *in vivo*. ACS Applied Materials and Interfaces 5(2): 319-330. https://doi.org/10.1021/am302146w.

Lu, T., Y. Li and T. Chen. 2013b. Techniques for fabrication and construction of three-dimensional scaffolds for tissue engineering. International Journal of Nanomedicine 8: 337-350. https://doi.org/10.2147/IJN.S38635.

Lutolf, M.P. and J.A. Hubbell. 2005. Synthetic biomaterials as instructive extracellular microenvironments for morphogenesis in tissue engineering. Nature Biotechnology 23: 47-55. doi: 10.1038/nbt1055.

Ma, L., C. Gao, Z. Mao, J. Shen, X. Hu and C. Han. 2003. Thermal dehydration treatment and glutaraldehyde cross-linking to increase the biostability of collagen-chitosan porous scaffolds used as dermal equivalent. Journal of Biomaterials Science, Polymer Edition 14: 861-874. doi: 10.1163/156856203768366576.

Ma, P.X., R. Zhang, G. Xiao and R. Franceschi. 2001. Engineering new bone tissue *in vitro* on highly porous poly(alpha-hydroxyl acids)/hydroxyapatite composite scaffolds. Journal of Biomedical Materials Research 54(2): 284-293.

Ma, R. and P. Zhang. 1999. Poly(alpha-hydroxyl acids)/hydroxyapatite porous composites for bone-tissue engineering. Journal of Biomedical Materials Research 44: 446-455.

Madihally, S.V. and H.W.T. Matthew. 1999. Porous chitosan scaffolds for tissue engineering. Biomaterials 20: 1133-1142. doi: 10.1016/S0142-9612(99)00011-3.

Mazzoli, A. 2013. Selective laser sintering in biomedical engineering. Medical & Biological Engineering & Computing 51(3): 245-256. doi: 10.1007/s11517-012-1001-x.

Melchels, F.P.W., K. Bertoldi, R. Gabbrielli, A.H. Velders, J. Feijen and D.W. Grijpma. 2010. Mathematically defined tissue engineering scaffold architectures prepared by stereolithography. Biomaterials 31(27): 6909-6916. doi: 10.1016/j.biomaterials.2010.05.068.

Miao, F., C. Shao, X. Li, K. Wang, N. Lu and Y. Liu. 2016. Electrospun carbon nanofibers/carbon nanotubes/polyaniline ternary composites with enhanced electrochemical performance for flexible solid-state supercapacitors. ACS Sustainable Chemistry and Engineering 4: 1689-1696. doi: 10.1021/acssuschemeng.5b01631.

Middleton, J.C. and A.J. Tipton. 2000. Synthetic biodegradable polymers as orthopedic devices. Biomaterials 21: 2335-2346. doi: 10.1016/S0142-9612(00)00101-0.

Midwood, K.S., L.V. Williams and J.E. Schwarzbauer. 2004. Tissue repair and the dynamics of the extracellular matrix. International Journal of Biochemistry and Cell Biology 36: 1031-1037. doi: 10.1016/j.biocel.2003.12.003.

Mironov, V., T. Boland, T. Trusk, G. Forgacs and R.R. Markwald. 2003. Organ printing: computer-aided jet-based 3D tissue engineering. Trends in Biotechnology 21: 157-161. doi: 10.1016/S0167-7799(03)00033-7.

Mitragotri, S. and J. Lahann. 2009. Physical approaches to biomaterial design. Nature Materials 8: 15-23. doi: 10.1038/nmat2344.

Moroni, F. and T. Mirabella. 2014. Decellularized matrices for cardiovascular tissue engineering. American Journal of Stem Cells 3(1): 1-20. doi: 10.1517/14712598.2010.534079.

Mouzakis, D.E. 2018. Advanced technologies in manufacturing 3D-layered structures for defense and aerospace. pp. 89-113. *In*: C. Osheku [ed.]. IntechOpen. doi: 10.5772/intechopen.74331.

Munaz, A., R.K. Vadivelu, J. St. John, M. Barton, H. Kamble and N.-T. Nguyen. 2016. Three-dimensional printing of biological matters. Journal of Science: Advanced Materials and Devices 1(1): 1-17. doi: https://doi.org/10.1016/j.jsamd.2016.04.001.

Munch, E., E. Saiz, A.P. Tomsia and S. Deville. 2009. Architectural control of freeze-cast ceramics through additives and templating. Journal of the American Ceramic Society 92: 1534-1539. doi: 10.1111/j.1551-2916.2009.03087.x.

Murphy, W.L., R.G. Dennis, J.L. Kileny and D.J. Mooney. 2002. Salt Fusion: an approach to improve pore interconnectivity within tissue engineering scaffolds. Tissue Engineering 8(1): 43-52. doi: 10.1089/107632702753503045.

Nair, L.S. and C.T. Laurencin. 2007. Biodegradable polymers as biomaterials. Progress in Polymer Science (Oxford) 32: 762-798. doi: 10.1016/j.progpolymsci.2007.05.017.

Naviroj, M., P.W. Voorhees and K.T. Faber. 2017. Suspension- and solution-based freeze casting for porous ceramics. Journal of Materials Research 32: 3372-3382. doi: 10.1557/jmr.2017.133.

Nelson, C.M. and M.J. Bissell. 2006. Of extracellular matrix, scaffolds and signaling: tissue architecture regulates development, homeostasis, and Cancer. Annual Review of Cell and Developmental Biology 22: 287-309. doi: 10.1146/annurev.cellbio.22.010305.104315.

Nuge, T., K.Y. Tshai, S.S. Lim, N. Nordin and M.E. Hoque. 2017. Preparation and characterization of CU-, FE-, AG-, ZN- and NI- doped gelatin nanofibers for possible applications in antibacterial nanomedicine. Journal of Engineering Science and Technology 12(Special Issue 1): 68-81.

Ozbolat, I.T., H. Chen and Y. Yu. 2014. Development of 'Multi-arm Bioprinter' for hybrid biofabrication of tissue engineering constructs. Robotics and Computer-Integrated Manufacturing 30(3): 295-304. doi: https://doi.org/10.1016/j.rcim.2013.10.005.

Park, J.H., J. Jang, J.-S. Lee and D.-W. Cho. 2017. Three-dimensional printing of tissue/organ analogues containing living cells. Annals of Biomedical Engineering 45(1): 180-194. doi: 10.1007/s10439-016-1611-9.

Park, J.Y., J.C. Choi, J.H. Shim, J.S. Lee, H. Park, S.W. Kim, J. Doh and D.W. Cho. 2014. A comparative study on collagen type i and hyaluronic acid dependent cell behavior for osteochondral tissue bioprinting. Biofabrication 6(3): 035004. doi: 10.1088/1758-5082/6/3/035004.

Pati, F., D.-H. Ha, J. Jang, H.H. Han, J.-W. Rhie and D.-W. Cho. 2015. Biomimetic 3D tissue printing for soft tissue regeneration. Biomaterials 62: 164-175. doi: https://doi.org/10.1016/j.biomaterials.2015.05.043.

Persano, L., A. Camposeo, C. Tekmen and D. Pisignano. 2013. Industrial upscaling of electrospinning and applications of polymer nanofibers: a review. Macromolecular Materials and Engineering 298: 504-520. doi: 10.1002/mame.201200290.

Peter, M., N.S. Binulal, S.V. Nair, N. Selvamurugan, H. Tamura and R. Jayakumar. 2010. Novel biodegradable chitosan–gelatin/nano-bioactive glass ceramic composite scaffolds for alveolar bone tissue engineering. Chemical Engineering Journal 158(2): 353-361. doi: https://doi.org/10.1016/j.cej.2010.02.003.

Pimpin, A. and W. Srituravanich. 2012. Reviews on micro- and nanolithography techniques and their applications. Engineering Journal 16: 37-55. doi: 10.4186/ej.2012.16.1.37.

Quirk, R.A., R.M. France, K.M. Shakesheff and S.M. Howdle. 2004. Supercritical fluid technologies and tissue engineering scaffolds. Current Opinion in Solid State and Materials Science 8(3): 313-321. doi: https://doi.org/10.1016/j.cossms.2003.12.004.

Rabionet, M., A.J. Guerra, T. Puig and J. Ciurana. 2018. 3D-printed tubular scaffolds for vascular tissue engineering. Procedia CIRP 68: 352-357. doi: https://doi.org/10.1016/j.procir.2017.12.094.

Rezende, R.A., F.D.A.S. Pereira, V. Kasyanov, D.T. Kemmoku, I. Maia, J.V.L. da Silva and V. Mironov. 2013. Scalable biofabrication of tissue spheroids for organ printing. Procedia CIRP 5: 276-281. doi: 10.1016/j.procir.2013.01.054.

Salgado, A.J., O.P. Coutinho and R.L. Reis. 2004. Bone tissue engineering: State of the art and future trends. Macromolecular Bioscience 4: 743-765. doi: 10.1002/mabi.200400026.

Scotti, K.L. and D.C. Dunand. 2018. Freeze casting – a review of processing, microstructure and properties via the open data repository, FreezeCasting.net. Progress in Materials Science 94: 243-305. doi: https://doi.org/10.1016/j.pmatsci.2018.01.001.

Seil, J.T. and T.J. Webster. 2011. Spray deposition of live cells throughout the electrospinning process produces nanofibrous three-dimensional tissue scaffolds. International Journal of Nanomedicine 6: 1095-1099. doi: 10.2147/IJN.S18803.

Shahi, M., A. Moghimi, B. Naderizadeh and B. Maddah 2011. Electrospun PVA–PANI and PVA–PANI–AgNO3 composite nanofibers. Scientia Iranica 18(6): 1327-1331. doi: https://doi.org/10.1016/j.scient.2011.08.013.

Sirota, C. 2016. 3D organ printing. The Science Journal of the Lander College of Arts and Sciences 10(1): 66-72.

Stankus, J.J., J. Guan, K. Fujimoto and W.R. Wagner. 2006. Microintegrating smooth muscle cells into a biodegradable, elastomeric fiber matrix. Biomaterials 27(5): 735-744. doi: 10.1016/j.biomaterials.2005.06.020.

Suuronen, E.J., H. Sheardown, K.D. Newman, C.R. McLaughlin and M. Griffith. 2005. Building *in vitro* models of organs. International Review of Cytology 244: 137-173. doi: 10.1016/S0074-7696(05)44004-8.

Theron, S.A., A.L. Yarin, E. Zussman and E. Kroll. 2005. Multiple jets in electrospinning: experiment and modeling. Polymer 46(9): 2889-2899. doi: https://doi.org/10.1016/j.polymer.2005.01.054.

Thomson, R.C., M.C. Wake, M.J. Yaszemski and A.G. Mikos. 1995. Biodegradable polymer scaffolds to regenerate organs. pp. 245-274. *In*: N. Peppas and R. Langer [eds.]. Biopolymers II. Advances in Polymer Science. Berlin, Heidelberg: Springer. doi: 10.1007/3540587888_18.

Tijing, L.D., M.T.G. Ruelo, A. Amarjargal, H.R. Pant, C.-H. Park and C.S. Kim. 2012. One-step fabrication of antibacterial (silver nanoparticles/poly (ethylene oxide)) – Polyurethane bicomponent hybrid nanofibrous mat by dual-spinneret electrospinning. Materials Chemistry and Physics 134(2): 557-561. doi: https://doi.org/10.1016/j.matchemphys.2012.03.037.

Tijing, L.D., Y.C. Woo, M. Yao, J. Ren and H.K. Shon. 2017. Electrospinning for membrane fabrication: strategies and applications. pp. 418-444. *In*: E. Drioli, L. Giorno and E. Fontananova [eds.]. Comprehensive Membrane Science and Engineering, 2nd Ed. Oxford: Elsevier.

Tran, K.T., L. Griffith and A. Wells. 2004. Extracellular matrix signaling through growth factor receptors during wound healing. Wound Repair and Regeneration: Official Publication of the Wound Healing Society [and] the European Tissue Repair Society 12: 262-268. doi: 10.1111/j.1067-1927.2004.012302.x.

Varabhas, J.S., G.G. Chase and D.H. Reneker 2008. Electrospun nanofibers from a porous hollow tube. Polymer 49(19): 4226-4229. doi: https://doi.org/10.1016/j.polymer.2008.07.043.

Ventola, C.L. 2014. Medical Applications for 3D Printing: Current and Projected Uses. Pharmacy and Therapeutics. MediMedia USA, Inc., 39(10): 704-711.

Von Der Mark, K., J. Park, S. Bauer and P. Schmuki. 2010. Nanoscale engineering of biomimetic surfaces: cues from the extracellular matrix. Cell and Tissue Research 339: 131-153. doi: 10.1007/s00441-009-0896-5.

Von Heimburg, D., S. Zachariah, H. Kühling, I. Heschel, H. Schoof, B. Hafemann and N. Pallua. 2001. Human preadipocytes seeded on freeze-dried collagen scaffolds investigated *in vitro* and *in vivo*. Biomaterials 22: 429-438. doi: 10.1016/S0142-9612(00)00186-1.

Vozzi, G., C. Flaim, A. Ahluwalia and S. Bhatia. 2003. Fabrication of PLGA scaffolds using soft lithography and microsyringe deposition. Biomaterials 24(14): 2533-2540. doi: 10.1016/S0142-9612(03)00052-8.

Wan, W., S. Zhang, L. Ge, Q. Li, X. Fang, Q. Yuan, W. Zhong, J. Ouyang and M. Xing. 2015. Layer-by-layer paper-stacking nanofibrous membranes to deliver adipose-derived stem cells for bone regeneration. International Journal of Nanomedicine 10: 1273-1290. doi: 10.2147/IJN.S77118.

Wang, C., K.-W. Yan, Y.-D. Lin and P.C.H. Hsieh. 2010. Biodegradable core/shell fibers by coaxial electrospinning; processing fiber characterization and its application in sustained drug release. Macromolecules 43(15): 6389-6397.

Wang, H., Y. Li, Y. Zuo, J. Li, S. Ma and L. Cheng. 2007. Biocompatibility and osteogenesis of biomimetic nano-hydroxyapatite/polyamide composite scaffolds for bone tissue engineering. Biomaterials 28(22): 3338-3348. doi: https://doi.org/10.1016/j.biomaterials.2007.04.014.

Webber, M.J., O.F. Khan, S.A. Sydlik, B.C. Tang and R. Langer. 2015. A perspective on the clinical translation of scaffolds for tissue engineering. Annals of Biomedical Engineering 43: 641-656. doi: 10.1007/s10439-014-1104-7.

Wei, G. and P.X. Ma. 2004. Structure and properties of nano-hydroxyapatite/polymer composite scaffolds for bone tissue engineering. Biomaterials 25(19): 4749-4757. doi: https://doi.org/10.1016/j.biomaterials.2003.12.005.

Whang, K., C.H. Thomas, K.E. Healy and G. Nuber. 1995. A novel method to fabricate bioabsorbable scaffolds. Polymer 36: 837-842. https://doi.org/10.1016/0032-3861(95)93115-3.

Whitaker, M. 2014. The history of 3D printing in healthcare. The Bulletin of the Royal College of Surgeons of England 96(7): 228-229. doi: 10.1308/147363514X13990346756481.

Williams, D.F. 2008. On the mechanisms of biocompatibility. Biomaterials 29: 2941-2953. doi: 10.1016/j.biomaterials.2008.04.023.

Wu, Z., X. Su, Y. Xu, B. Kong, W. Sun and S. Mi. 2016. Bioprinting three-dimensional cell-laden tissue constructs with controllable degradation. Scientific Reports 6: doi: 10.1038/srep24474.

Yang, F., R. Murugan, S. Wang and S. Ramakrishna. 2005. Electrospinning of nano/micro scale poly(l-lactic acid) aligned fibers and their potential in neural tissue engineering. Biomaterials 26(15): 2603-2610. doi: 10.1016/j.biomaterials.2004.06.051.

Yi, H.-G., H. Lee and D.-W. Cho. 2017. 3D Printing of Organs-On-Chips. Bioengineering 4(1): 10. doi: 10.3390/bioengineering4010010.

Yoshimoto, H., Y.M. Shin, H. Terai and J.P. Vacanti. 2003. A biodegradable nanofiber scaffold by electrospinning and its potential for bone tissue engineering. Biomaterials 24(12): 2077-2082. doi: 10.1016/S0142-9612(02)00635-X.

Zein, I., D.W. Hutmacher, K.C. Tan and S.H. Teoh. 2002. Fused deposition modeling of novel scaffold architectures for tissue engineering applications. Biomaterials 23(4): 1169-1185. doi: 10.1016/S0142-9612(01)00232-0.

Zhang, B., Y. Luo, L. Ma, L. Gao, Y. Li, Q. Xue, H. Yang and Z. Cui. 2018. 3D bioprinting: an emerging technology full of opportunities and challenges. Bio-Design and Manufacturing. Springer Singapore 1(1): 2-13. doi: 10.1007/s42242-018-0004-3.

Zhang, Y.Z., J. Venugopal, Z.M. Huang, C.T. Lim and S. Ramakrishna. 2005. Characterization of the surface biocompatibility of the electrospun PCL-Collagen nanofibers using fibroblasts. Biomacromolecules 6: 2583-2589. doi: 10.1021/bm050314k.

Zhang, Y.Z., X. Wang, Y. Feng, J. Li, C.T. Lim and S. Ramakrishna. 2006. Coaxial electrospinning of (fluorescein isothiocyanate-conjugated bovine serum albumin)-encapsulated poly(ε-caprolactone) nanofibers for sustained release. Biomacromolecules 7: 1049-1057. doi: 10.1021/bm050743i.

Zhong, S., W.E. Teo, X. Zhu, R.W. Beuerman, S. Ramakrishna and L.Y.L. Yung. 2006. An aligned nanofibrous collagen scaffold by electrospinning and its effects on *in vitro* fibroblast culture. Journal of Biomedical Materials Research - Part A 79A: 456-463. doi: 10.1002/jbm.a.30870.

Zhu, Y., Y. Cao, J. Pan and Y. Liu. 2010. Macro-alignment of electrospun fibers for vascular tissue engineering. Journal of Biomedical Materials Research - Part B Applied Biomaterials 92B: 508-516. doi: 10.1002/jbm.b.31544.

Zuo, K.H., Y.P. Zeng and D. Jiang. 2010. Effect of polyvinyl alcohol additive on the pore structure and morphology of the freeze-cast hydroxyapatite ceramics. Materials Science and Engineering C 30: 283-287. doi: 10.1016/j.msec.2009.11.003.

4

Next Generation Tissue Engineering Strategies by Combination of Organoid Formation and 3D Bioprinting

Shikha Chawla, Juhi Chakraborty and Sourabh Ghosh*

4.1 INTRODUCTION

Organogenesis, during embryonic stage, is governed by a constellation of complex processes, involving cell-cell and cell-matrix protein interactions, cell migration, regulation of large number of signaling molecules and signaling pathways. Progenitor cells differentiate to specific phenotypes, and produce organ specific ECM. At the same time, according to the embryonic developmental plan and anatomical architecture, concerted cellular self-assembly leads to formation of the "organ germ". These rudimentary organ germs then undergo organ-specific morphogenesis to meet the requirements of biological as well as mechanical functionality (Sasai 2013, Sasai et al. 2012). Since past few decades, tissue engineers have tried to recapitulate these complex developmental biology signaling cascades and morphogenesis by combining progenitor cells or primary cells, various polymeric scaffolds, and bioactive molecules or growth factors by using tissue engineering techniques *in vitro*. But in the past decades, very few tissue engineered constructs could achieve desired level of success in human clinical trials.

There are several reasons for this disappointing performance. Tissue engineered constructs invariably fail to replicate cellular alignment, biological and mechanical functionality to a large extent. There is compromised reproducibility owing to donor-to-donor variation and/or differentiation of cells towards unstable phenotypic populations. Various natural and synthetic polymers are used for scaffold fabrication, but we have poor understanding of how cellular signaling pathways get modulated. There are still major gaps remaining in understanding of immune response to such engineered constructs and how resultant immune responses can modulate pathways of organ regeneration. By culturing the cells over scaffolds *in vitro* for only few weeks, it may not be justified to expect similar dynamic mechanical properties of an adult tissue.

Department of Textile Technology, Indian Institute of Technology Delhi, India-110016.
* Corresponding author: sghosh08@textile.iitd.ac.in

So far, the main focus was to simulate the architecture, mechanical properties and biochemical composition of adult human organs. Now tissue engineering strategies are undergoing a paradigm shift to recapitulate the stages of tissue development. Rather than entirely focusing on engineering an adult tissue/organ, now the focus is shifting towards replicating the engineering of a developmental or regeneration process. In order to achieve that, tissue engineers are exploring various other strategies, such as 3D bioprinting, or organoid formation. 3D bioprinting can enable us to recapitulate strategic cellular arrangement, in order to recapitulate tissue and organ level complexities, spatial and temporal control of intrinsic positioning of the cells in a construct. This is particularly interesting because even if we start with homogeneous cell population in a bioprinted construct, based on the positioning, cells may attain varied properties, in a rationally designed spatial manner. Fascinating evidences are emerging on how recapitulation of cellular pattern formation can augment organ regeneration (Takagi et al. 2016). Parallely, other groups of researchers are exploring organoid development, to develop the organ germs of retina, eye cup, lung, thymus, pituitary gland, pancreas etc. (Eiraku et al. 2008). In this chapter, we will discuss about the future directions of tissue regeneration by combining concepts of developmental biology, organoid formation and 3D bioprinting (Fig. 4.1). Importantly, these insights will lead to improved tissue regeneration strategy, only when we can understand exactly how these techniques can modulate expression of various transcription factors and multiple signaling pathways.

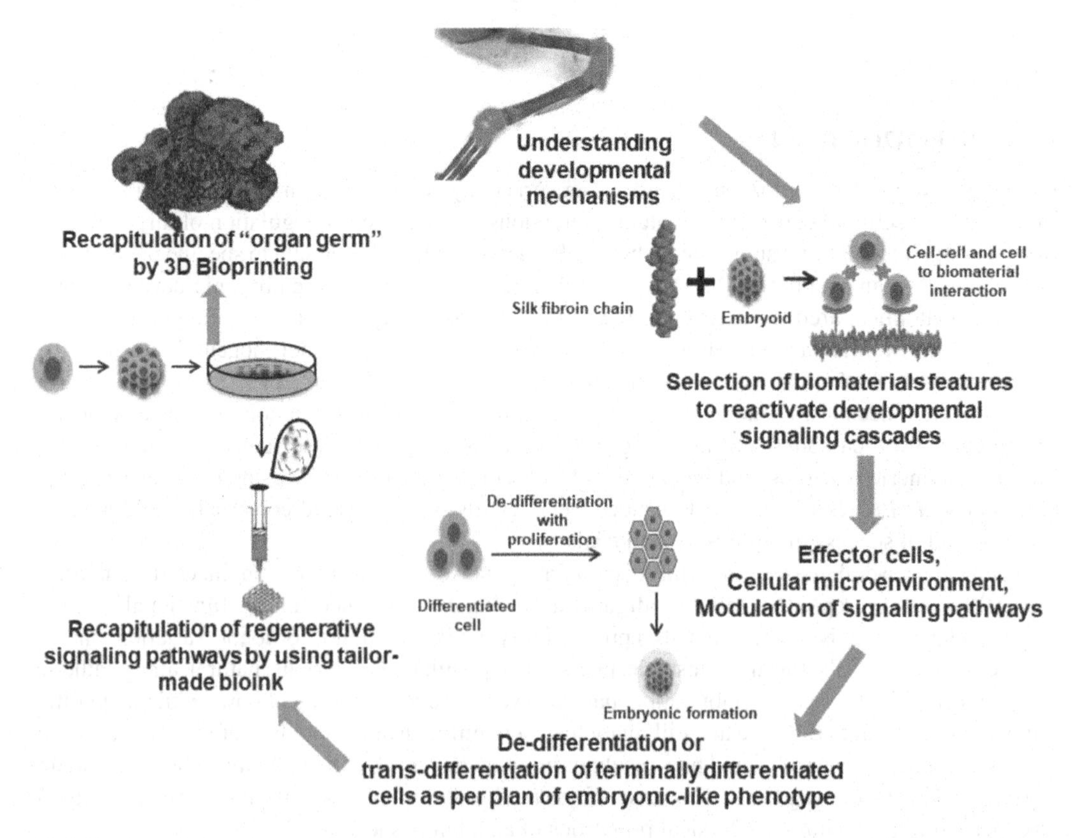

FIGURE 4.1 Combining concepts of developmental biology, organoid formation and bioink regulated signaling events of 3D bioprinting to develop "organ germ", which can be used to develop fully functional organ for transplantation, or as *in vitro* tissue model.

4.2 STRATEGIES TOWARDS ORGAN REGENERATION

4.2.1 Scaffold and Bioreactor-based Tissue Engineering

In past two decades, tissue engineering strategies were based on cell-scaffold-bioreactor system, aimed to replicate the *in vivo* tissue microenvironment. Scaffold acts as a structural template for the attachment of cells and development of new tissue, and provides various cues (chemical, biophysical, architectural) to the cells, thus regulating their gene expression and protein production. The important aspects of scaffold design include cytocompatibility, material chemistry, porosity and pore geometry, degradation rate, and mechanical properties. Additionally, several modifications could be performed to enhance the bioactivity of the scaffolds, for instance, covalent/non-covalent attachment of cell-adhesion ligands to enhance cell-matrix interaction, incorporation of additional structural motifs to increase degradation and tissue remodeling (Rice et al. 2013) . It is important to mention here that benefits of cells-scaffold interaction can only be fully understood by correlating the functional and physiological changes to the cellular signaling events, in terms of stiffness and architecture of scaffold and scaffold chemistry. For instance, human mesenchymal stem cells (hMSCs) could react to even subtle changes in the matrix stiffness of nano-fibrillar electrospun silk matrices. hMSCs encounter different extent of migratory and adhesive signals based on the stiffness of the silk scaffolds and could expedite the early events of cellular condensation and chondrogenic differentiation (Ghosh et al. 2009). Further, we investigated how scaffold dimensionality and architecture (3D micro-porous collagen scaffold and cell-encapsulated collagen hydrogel) control the cell-scaffold interaction and modulate the cellular response (Chawla et al. 2015). That study reported that the fibroblasts-laden collagen hydrogel showed comparatively higher proliferation, metabolic activity, and matrix remodeling than 3D micro-porous collagen scaffold. The differences observed could be attributed to increased adhesive interactions amongst cells embedded in 3D collagen gel as compared to 3D porous collagen matrix, where transmembrane cellular receptors are limited to 2D interface of cells and the surface of pores of scaffold. Another study from our laboratory demonstrated the significance of the optimization of scaffold chemistry in terms of retaining the extracellular matrix proteins produced endogenously by the cultured cells (Chameettachal et al. 2015). We demonstrated that the newly synthesized glycosaminoglycan (GAG) molecules leached out rapidly into the medium in case of silk-only constructs, while the chitosan-reinforced silk scaffolds helped to retain statistically significant higher GAG content within the constructs. The tremendous potential of scaffolds in terms of providing the template, ease of chemical and physical modification and the advent of possibility to mimic the cellular signaling events makes these 3D *in vitro* models beneficial to develop strategies for successful regeneration of tissues.

Bioreactors play a crucial role in maintaining precise culture conditions like pH, osmolality, temperature, and oxygen concentration required for cell proliferation and differentiation. Furthermore, there is a possibility to facilitate even mass and nutrient transfer for the cultured cells. Bioreactors have been utilized since long to provide additional control over physical signals corresponding to tissue of interest (physiological mechanical loading, compression, shear and interstitial fluid flow). Development of anatomically relevant intervertebral tissue construct is an interesting example, where bioreactor induced fluid flow modulated the constructs towards attaining an *in vivo* like tissue gradient. The interior region of the constructs showed stiffer compression along with the overexpression of collagen II, GAG, while the outer region showed stiffer tension and overexpressed collagen I (Bhattacharjee et al. 2014). Proper combination of cell-scaffold-bioreactor system has been also used to develop some disease models like the 3D tumor models, where this system helps in stimulating tumor characteristics *in vitro* for an extended time (Guller et al. 2016). Such successful models could be utilized for the research, development and diagnosis of new strategies for cancer treatment.

4.2.2 Organoid Formation Strategies

Embryoid bodies are prepared using embryonic stem cells and induced pluripotent cells, using selective cocktails of cytokines. In several cases, preparation of embryoid bodies is a crucial first step in lineage-specific differentiation protocols for stem cells (Brickman and Serup 2017). In this process, relatively large numbers of cells (>1000 to 1 million cells per well) are aggregated, using microwells and hanging drops, and induced to differentiate towards specific cell type. Gradually with increasing culture time, within the embryoid body, differentiated cells seem to assume different morphologies, to develop disorganized cell mass. Reaggregation of embryoid bodies, followed by cellular self-assembly, leads to development of organoids. Recently, self-assembled embryonic stem cells based organoid formation has been acknowledged as one of the key *in vitro* methods to generate mini organs like mini-gut, brain-like organoids (Warmflash et al. 2014) and liver organoids (Takebe et al. 2013) etc. Morphogen signaling gradient plays a critical role in positioning and patterning of the whole embryos and tissues during normal development, leading to final body parts compartmentalization. Thus, quest for modalities to include combination of cytokines to reproduce such embryonic development based on spatio-temporal morphogen signaling is the need of current developmental biology inspired tissue engineering. Apart from cellular self-assembly tissue, organ formation may involve other complex phenomenon, for instance, varied affinity amongst distinct elements (expression of hemophilic or heterophilic adhesion molecules), hypoxia or mass transport issues, cellular rearrangements and active cell migration. Several organoid formation strategies have been utilized in the past to develop mini-organs. Few examples include the development of two layered optic cup-like organoid using 3D aggregates of self-assembled mouse pluripotent embryonic stem cells (Eiraku et al. 2011). Nelson et al. developed a 3D *in vitro* model of mouse mammary epithelial tubules; they utilized micropatterning technology to guide the 3D structure of **the developed organoid. They successfully demonstrated that the geometry of multicellular tubules,** in addition to the availability of the spatially controlled levels of autocrine inhibitory morphogen concentration, can regulate the branch positioning during mammary epithelial cells and primary organoids morphogenesis (Nelson et al. 2006). Another interesting example of morphogen induced control of organoid formation was demonstrated by Montesano et al. (Montesano et al. 1991). Using 3D culture of Madin–Darby canine kidney cells, Montesano et al. demonstrated voluntary tubular formation from the epithelial cyst in the presence of hepatocyte growth factor. Thus, conclusively, developmental biology inspired morphogen signaling in 3D organoids tissue and organ is a promising **approach for developing tissue-engineered equivalents (Fig. 4.2). It is pertinent to mention here** that since this strategy is scaffold/biomaterial-free, there is no concern how polymer/biomaterial will change cell behavior and thus may closely mimic the *in vivo* like self-driven cellular/tissue morphogenesis.

The most interesting application of these organoids is possibly the design and development of *in vitro* disease model systems, degenerative conditions, developmental disease and tumors (Clevers 2016). Furthermore, in future such *in vitro* organoid models may offer additional prospect to derive tissue/mini-organs from patient's own cells to provide substitute to organ replacement (Lancaster and Knoblich 2014).

Further, it would be intriguing to explore the prospect of combination of organoid formation strategies and biomaterials response. For the first time, a study from our laboratory provided necessary insights into this approach (Gupta et al. 2018). The study focused on development of spheroid based organoid model of human hair follicle, in which first we developed cellular spheroid of human dermal papilla cells and then coated the developed spheroid with silk-gelatin hydrogel, in which two other cell population (human hair follicle stem cells and keratinocytes) were embedded. Silk-gelatin hydrogel provided necessary microenvironmental features as well as inhibited the

BMP signaling and showed upregulated expression of β-catenin; these characteristics have been comprehensively observed *in vivo* for improved hair follicle regeneration, leading to increased cell proliferation and enhanced synthesis of matrix proteins.

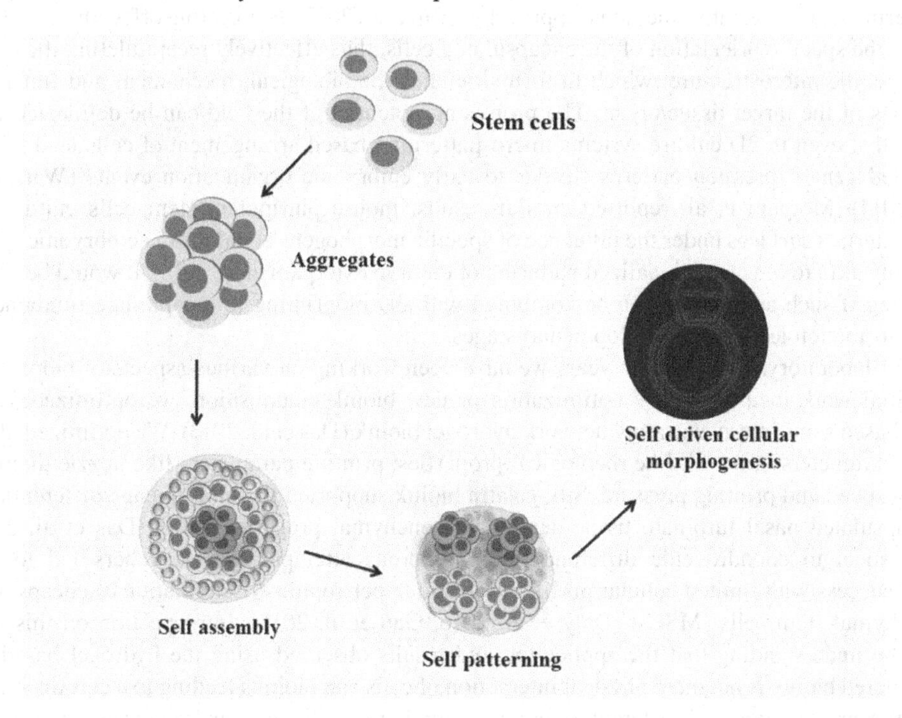

FIGURE 4.2 Developmental biology inspired morphogen signaling leading to organoid development and gradual progression towards organ germ formation, by recapitulation of establishment of body axes during embryogenesis (gastrulation). We need to develop clear understanding about the spontaneous breaking of symmetry/uniformity to predict and govern self-organization of organoids within the bioprinted constructs.

Despite these fascinating prospects offered by the field of organoid development, this field of research suffers from many limitations. Firstly, to replicate the dynamic and complex course of *in vivo* like spatio-temporal delivery of cytokine/morphogens in these 3D organoids is still an unachievable goal. Secondly, mostly organoids are made up of only few thousand cells. These organoids offer limited control over dimensional and structural features of the individual tissue/organ; even if larger organoids are developed, there are limitations with respect to supply of nutrients and oxygen (Gjorevski et al. 2014). Thus, there is a need for combinatorial bioengineering methodologies that can narrow down the discontinuity between *in vitro* organoid based tissue culture models and *in vivo* morphogenesis. Such combinatorial approaches have the potential to deduce the complex mechanistic insight about developmental organogenesis thus opening new avenues for drug discovery and rapid drug testing. In future probably, tissue engineers will be able to generate new strategies for tissue regeneration in the clinic using such culture technologies.

4.2.3 Bioprinting

3D Bioprinting has evolved dramatically during the last decade as an approach that is at the crossroad of bioengineering and regenerative medicine. It offers humongous potential to design patient-specific and tissue/organ specific tissue engineered equivalents. The advent of rapid manufacturing technologies associated with the engineering aspect of bioprinting offers precise control over architecture and resolution, which makes this field way ahead of the current tissue engineering and regenerative medicine approaches (Cui et al. 2017). Bioprinting offers the possibility to guide the specific orientation of the encapsulated cells, thus effectively recapitulating the tissue/organ specific microstructure, which in turn simulates the biological, mechanical and functional properties of the target tissue/organ. The promising potential of the field can be delineated from the fact that even in 2D culture systems micro-patterning based arrangement of cells lead to self-assembled gene expression patterns similar to early embryonic organization events (Warmflash et al. 2014). Morgani et al. reported similar results: mouse pluripotent stem cells cultured on micropatterned surfaces under the influence of specific morphogens could mimic embryonic spatial patterning akin to *in vivo* regionalized pattering of embryo (Morgani et al. 2018). It would be highly interesting if such approaches can be combined with 3D bioprinting to recapitulate fundamental pattern formation features of developmental stages.

In our laboratory, since last ten years we have been working on various aspects of bioprinting. Our initial work focused on the optimization of new bioink composition, we optimized a silk-gelatin based semi-interpenetrating network hydrogel bioink (Das et al. 2013). We optimized all the critical parameters of bioink like rheological properties, printing parameters like nozzle diameter, printing speed and printing pressure. Silk-gelatin bioink supported the multilineage differentiation of encapsulated nasal turbinate tissue-derived mesenchymal progenitor cells (Das et al. 2015). With respect to chondrogenic differentiation, numerous attempts by researchers led to only limited success with limited cellular proliferation and hypertrophic differentiation of encapsulated mesenchymal stem cells (MSCs) (Daly et al. 2016, Gao et al. 2015). Introspection on this issue led to the understanding that the application and results observed using the hydrogel based cell encapsulated bioink is not mere physical interaction of cells and bioinks leading to a certain amount of cellular proliferation and differentiation, but a number of cellular signaling events are initiated upon such interactions. Thus, the aim of tissue engineers should be to utilize the advantages of such a proposition where bioink holds the capacity to modulate cellular signaling events in the cell-encapsulated 3D bioprinted constructs.

Since past 30 years, researchers are trying to develop articular cartilage; however, development of functional, clinically conformant cartilage tissue equivalent for load bearing applications still could not be achieved using traditional tissue engineering approaches. The main limitation in this context is the inability of tissue engineers to distinguish the developed cartilage tissue equivalent in terms of articular or transient cartilage differentiation of progenitor cells. Instead of developing stable articular cartilage, they invariably prepared transient (i.e. the cartilage that undergoes hypertrophic differentiation) cartilage which cannot withstand load. We on the other hand tried to develop phenotypically stable articular cartilage utilizing the ability of our silk-gelatin bioink to modulate Wnt signaling. We perceived that a major emphasis should be on to understand how scaffold/bioink can modulate cellular signaling pathways. Such understanding combined with 3D bioprinting can be extended and utilized further to replicate developmental level cellular signaling events to generate physiologically and anatomically relevant 3D bioprinted tissue equivalents. In that context, we could develop cartilage tissue equivalents using MSC encapsulated silk-gelatin bioink that demonstrated the expression of Autotaxin, lubricin, which are definite markers of articular cartilage formation during embryonic cartilage development (Chawla et al. 2017). Through this study, we also identified that the silk-gelatin bioink negatively regulated Indian Hedgehog signaling pathway and Wnt signaling pathway leading to control of hypertrophic differentiation of MSCs.

As a next step in this direction, we tried to combine the developmental biology cues and 3D bioprinting to develop a tissue engineered bone tissue equivalents (Chawla et al. 2018). Using our optimized silk-gelatin bioink and mimicking the *in vivo* bone development inspired endochondral ossification protocol (3 weeks of culture of the bioprinted construct in chondrogenic media, followed by 2 weeks of culture in osteogenic media), we could develop 3D bioprinted bone tissue equivalents that showed close similarity with temporal gene expression specific to early and terminal osteogenic differentiation events. Based on the success of our study, we very strongly feel that a combinatorial approach combining the principles of developmental biology and 3D bioprinting can generate an approach that has tremendous potential to deliver necessary instructive cues to activate the organogenesis pathways in patient-specific manner.

The spatio-temporal levels of Wnt signaling are one of the key signaling events during embryogenesis and adult tissue homeostasis. *In vivo*, 19 Wnt family genes are associated with regulating cell proliferation, gene expression, differentiation, migration, adhesion, maintenance of cell polarity and apoptosis etc. (de Lau et al. 2014). Wnt Signaling can be categorized into two pathways: canonical and non-canonical pathways (Veeman et al. 2003). β-catenin, Cytoplasmic Dishevelled protein (Dsh) and Frizzled (Fz) transmembrane receptor initiates the canonical Wnt pathway to initiate transcription of target genes, for example, tight control of genes involved in vertebrate development (Wodarz and Nusse 1998). Wnt signaling also plays a critical role in maintaining the stem cell microenvironment by conserving their self-renewal capability, as well as also helps to define the lineage commitment in other cell types.

Consequently, stimulation of these signals seems the most significant approach in regenerative medicine and regulation of *in vitro* stem cell fate. Thus, immobilization of Wnt molecules and temporal activation of related pathways have profound influence on MSC differentiation and migration in 3D *in vitro* models (Lowndes et al. 2016). Further, it is intriguing to explore the *in vivo* like signaling crosstalks amongst Wnt and the transforming growth factor β(TGF-β) signaling, bone morphogenetic proteins (BMPs) signaling, and Hedgehog and Notch signaling in an *in vitro* system to attenuate the potential of tissue engineering based regenerative approaches. Conclusively, as a perspective, cell polarization, patterning, modification of matrix organization, chemistry and mechanical signals with the aim to precisely simulate the embryonic developmental program can thus advance the structural reliability of 3D bioprinted constructs and physiological applicability of encapsulated cells via activation of multiple morphogen signaling events.

The final goal of 3D bioprinting is the design and development of functional anatomically relevant human tissues/organs to replace the diseased, damaged tissues/organs, to develop person-specific, defect-site specific constructs. Nevertheless, there are many unsolved challenges to resolve. For example, development of fully vascularized 3D bioprinted structures with anatomically relevant thickness and rigidity to ensure integration with the native tissue post-implantation is still a challenge. No studies till date could recapitulate complex anatomical architecture of kidney, liver, cardiac tissues. In such scenario, combination of organoid development strategies combined with 3D bioprinting and developmental biology inspired *in vitro* approaches might be an answer to prepare "organ germ" for organ replacement.

4.3 DEVELOPMENTAL ENGINEERING

The main challenge of tissue engineers is to fabricate clinically relevant patient-specific constructs. Recently, conventional tissue engineering is experiencing a paradigm shift toward "developmental engineering" (Lenas et al. 2009). Now, tissue regeneration strategies are directed towards understanding how precisely *in vitro* tissue engineering strategies can mimic *in vivo* developmental processes (Rivron et al. 2009). Some of the initial steps in this direction include the study by Scotti et al. which utilized the capacity of human MSC to generate bone tissue simulating *in vivo* like endochondral program to develop a model to study mechanisms of bone development using transwell culture (Scotti et al. 2010). The idea behind their approach was to replicate the process of bone development (endochondral ossification, a process via which most of the long bones of the limbs, vertebrae and rib-cage develop *in vivo*). The key event in the process of endochondral ossification is the development of cartilaginous template that later develops into bone as opposed to intramembranous ossification that is involved with craniofacial bone development that occurs without any intermediate cartilaginous template. Till now, the bone tissue engineering strategies mostly used direct differentiation protocols. Such bone models generated via a "developmental engineering" paradigm could generate advanced grafts for bone regeneration following architecture of woven bone or cancellous bone, if combined with bioprinting strategies (Scotti et al. 2010). However, this study was based on spheroids and transwell based approaches and thus could not replicate the anatomical architecture and thickness of native bone tissue. Thus, to further extend this understanding, recently, in our laboratory we utilized such a strategy where similar endochondral ossification inspired protocol was utilized and combined with 3D bioprinting to develop bone tissue equivalents. Simulation of the developmental-biology-inspired endochondral ossification route in the MSC laden silk-gelatin 3D bioprinted constructs triggered the Wnt/β-catenin, IHH and parathyroid hormone (PTH), signaling *in vitro* osteogenic differentiation, leading to improved osteogenic differentiation potential of MSCs and augmented mineral deposition (Chawla et al. 2018).

Another significant event during embryogenesis is gastrulation that involves patterning of pluripotent epiblast into the three germ layers that later develops into the embryo. This event involves a signaling pathway involving the BMP, Wnt and Nodal pathways (Chhabra et al. 2018). Thus, replication of such events seems quintessential to the approach of developmental biology inspired tissue engineering strategies (Fig. 4.2). This section would be incomplete without the mention of another interesting phenomenon of developmental biology, that is, directed tissue assembly, a process where closely placed tissue spheroids undergo fusion to replicate this fundamental biophysical and biological principle of directed tissue assembly (Mironov et al. 2009). Tissue engineers tried to take cues from this process and incorporated those features with 3D bioprinting, that led to the emergence of the new field of organ printing that holds promise to design and fabricate engineered tissue/mini organs for repair, regeneration and replacement of injured or damaged organs. One of our studies represents a true example of such an approach, where we first developed 3D spheroids of MSCs and chondrocytes for successful replication of mesenchymal condensation involving cell-cell adhesion formation through neural cell adhesion molecule (NCAM). These spheroids probably provided *in vivo* like microenvironment for development of stable cartilage tissue equivalent. Then these spheroids were combined with silk-gelatin hydrogel to develop 3D bioprinted cartilage tissue equivalents (Chameettachal et al. 2016) (Fig 4.3).

The idea behind such an approach was to replicate the phenomenon of mesenchymal condensation followed by chondrogenic differentiation. Our MSC spheroids-encapsulated silk-gelatin 3D bioprinted constructs demonstrated upregulated expression of Hypoxia Inducible Factor 1 Subunit Alpha as compared to single cell suspension, which is a marker of hypoxia thus leading to overexpression of chondrogenic markers like Cartilage oligomeric matrix protein precursor and Aggrecan. Therefore, our study represents the basic organ printing strategy inspired from developmental biology based directed tissue fusion. An interesting application of this strategy could be the development of organ-on-chip platforms for new drug efficacy and toxicity testing. We will discuss few examples in the next section to shed some light on such applications.

Conclusively, we can state that understanding and replication of the embryonic level tissue and organ development is certainly the most promising and upcoming approach in tissue engineering. Such an approach holds the potential to provide successful alternatives to solid-scaffold based strategies.

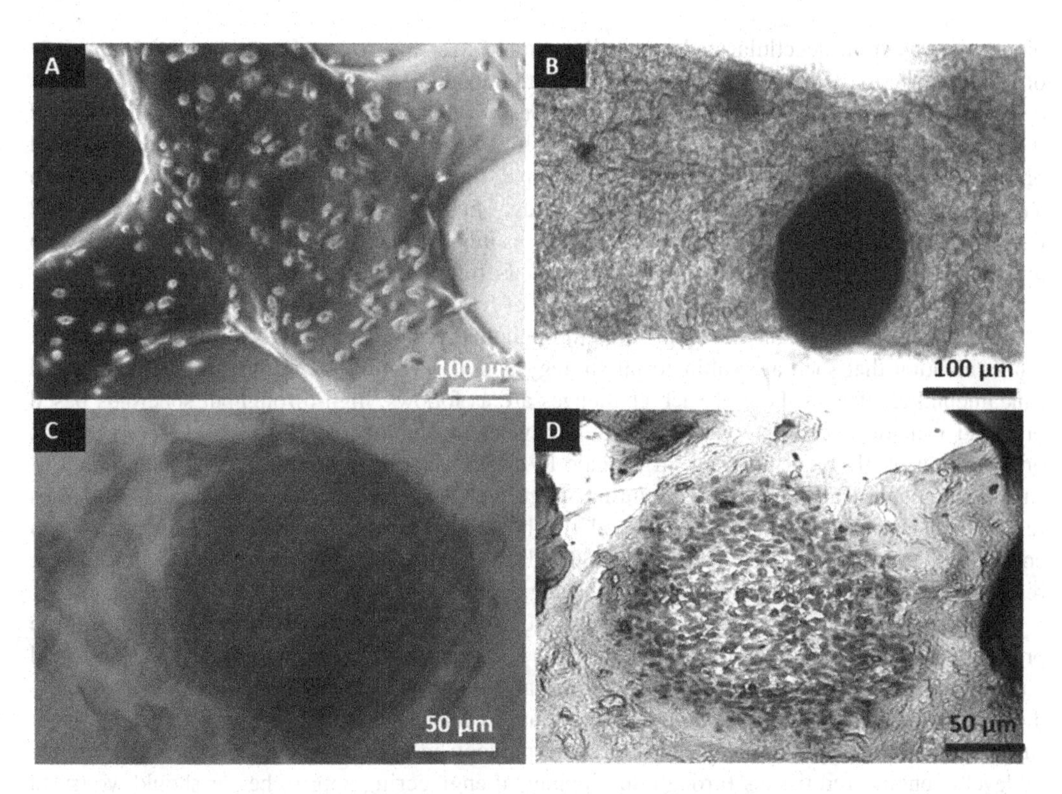

FIGURE 4.3 Human mesenchymal stem cells laden 3D bioprinted silk-gelatin construct with (A) single dispersed cells, or (B) spheroids. With the progression in culture time culture of cells migrate from the spheroids and undergo *in vivo* like cellular self assembly in 3D bioprinted construct (C) - as evident from the phase contrast image and (D) Hematoxylin and eosin staining of the spheroids in the bioprinted construct.

4.4 PERSPECTIVES

It is clear that the combination of the understanding of developmental biology and 3D bioprinting should be the final road map to meet tissue engineers' demand to simulate and reproduce the key structure-function relationship of the native human tissues/organs. However, it is still debatable that how far tissue engineers should try to replicate embryonic morphogenesis and organogenesis. The logistics and monetary limitations further restrict the development of living tissues and organs, creating a rational thought that organ printing will possibly not allow the replication of all the characteristics of functional living human organs. However, development of vascularized and functional tissue engineered mini-organs proficient of reestablishing the key functions of damaged or lost organ, if successful, would be a true representation of technological aim and epitomize a pronounced achievement.

As stated earlier, the most promising aspect of organ printing would be to develop organ-on-a-chip platforms that could be utilized to develop 3D *in vitro* mini-organ models. Such models would prove to be the answer to the need of researchers to extend the understanding of developmental processes and disease progression and would provide successful platform for rapid drug and toxicity testing. A recent insightful study meticulously outlined this strategy of organ-on-a-chip (Skardal et al. 2017). This study describes the development of liver-on-chip and heart-on-chip platforms where first the spheroids that represent liver organoids and cardiac organoids were established using primary human stellate cells, Kupffer cells, hepatocytes, and induced pluripotent stem (iPS) cells, respectively. This was followed by 3D bioprinting of these organoids into the microreactors using bioink derived from decellularized extracellular matrix for liver-on-a-chip and bioink comprised of gelatin fibrin for heart-on-a chip platform. Another example of liver-on-a-chip development was described by Bhise et al. where liver-on-a-chip was developed using hepatic spheroids bioprinted with gelatin methacryloyl bioink (Bhise et al. 2016). The authors also successfully demonstrated proof of concept toxicity testing of the developed liver-on-chip platform using acetaminophen, with a response similar to animal testing of the same drug. Thus, other than for organ transplantation and tissue regeneration, there is a vast potential use — as *in vitro* disease models, and as alternatives to animal models for preclinical testing or screening of drug molecules.

Having discussed the promising prospects of combinatorial approach of developmental reengineering, organoids and 3D bioprinting throughout this chapter, it will be profoundly incorrect not to mention that such a combinatorial strategy is still surrounded by many technological and fundamental challenges. Few of such challenges are as follows: high-throughput spheroids based organoid formation technologies by microfluidic devices, availability of industry level 3D bioprinters, optimization of 3D bioprinting of large scale living tissues and mini-organs, vascularization of such thick anatomically relevant tissue/mini-organs etc. Nevertheless, the certainty that numerous researchers have accepted and understood the benefits and thus are trying to focus on such combinatorial strategies over traditional scaffold based strategies offers confidence for further research, reconnoitring this new field in tissue engineering.

It is also essential to mention that, so far in tissue engineering, we do not discriminate cell culture protocols as per age of the recipient, whether it is paediatric or geriatric tissue regeneration. Aged patients have less chance of regeneration, as availability of the stem cell pool would be drastically different from young patients. It would be important to consider whether we should try to engineer fully adult/matured adult tissue (architecture, morphology, mechanical properties) or should we try to develop engineered tissues through developmental engineering approaches or should we try to rather develop embryonic ("organ germ") or tissues present in small children (early form of developed tissue), to ignite regeneration/ tissue integration to host tissues. Therefore, 3D bioprinted engineered constructs should ideally be designed by considering the genetic or epigenetic constitution of the patient, and their age-dependent immune status.

REFERENCES

Bhattacharjee, M., S. Chameettachal, S. Pahwa, A.R. Ray and S. Ghosh. 2014. Strategies for replicating anatomical cartilaginous tissue gradient in engineered intervertebral disc. ACS Applied Material and Interfaces 6: 183-193. doi: 10.1021/am403835t.

Bhise, N.S., V. Manoharan, S. Massa, A. Tamayol, M. Ghaderi, M. Miscuglio, Q. Lang, Y.S. Zhang, S.R. Shin, G. Calzone, N. Annabi, T.D. Shupe, C.E. Bishop, A. Atala, M.R. Dokmeci and A. Khademhosseini. 2016. A liver-on-a-chip platform with bioprinted hepatic spheroids. Biofabrication. doi: 10.1088/1758-5090/8/1/014101.

Brickman, J.M. and P. Serup. 2017. Properties of embryoid bodies. Wiley Interdisciplinary Reviews: Developmental Biology doi: 10.1002/wdev.259.

Chameettachal, S., S. Murab, R. Vaid, S. Midha and S. Ghosh. 2015. Effect of visco-elastic silk-chitosan microcomposite scaffolds on matrix deposition and biomechanical functionality for cartilage tissue engineering. Journal of Tissue Engineering and Regenerative Medicine doi: 10.1002/term.2024.

Chameettachal, S., S. Midha and S. Ghosh. 2016. Regulation of chondrogenesis and hypertrophy in silk fibroin-gelatin-based 3D bioprinted constructs. ACS Biomaterial Science and Engineering 2: 1450-1463. doi: 10.1021/acsbiomaterials.6b00152.

Chawla, S., S. Chameettachal and S. Ghosh. 2015. Probing the role of scaffold dimensionality and media composition on matrix production and phenotype of fibroblasts. Material Science and Engineering C 49: 588-596. doi: 10.1016/j.msec.2015.01.059.

Chawla, S., A. Kumar, P. Admane, A. Bandyopadhyay and S. Ghosh. 2017. Elucidating role of silk-gelatin bioink to recapitulate articular cartilage differentiation in 3D bioprinted constructs. Bioprinting 7: 1-13. doi: 10.1016/j.bprint.2017.05.001.

Chawla, S., A. Sharma, A. Bandyopadhyay and S. Ghosh. 2018. Developmental biology-inspired strategies to engineer 3D bioprinted bone construct. ACS Biomater. Sci. Eng. doi: 10.1021/acsbiomaterials.8b00757.

Chhabra, S., L. Liu, R. Goh and A. Warmflash. 2018. The timing of signaling events in the BMP, WNT, and Nodal cascade determines self-organized fate patterning in human gastruloids. BioRxiv. doi: 10.1101/440164.

Clevers, H. 2016. Modeling Development and Disease with Organoids. Cell. doi:10.1016/j.cell.2016.05.082.

Cui, H., M. Nowicki, J.P. Fisher and L.G. Zhang. 2017. 3D Bioprinting for Organ Regeneration. Advanced Healthcare Materials doi: 10.1002/adhm.201601118.

Daly, A.C., S.E. Critchley, E.M. Rencsok and D.J. Kelly. 2016. A comparison of different bioinks for 3D bioprinting of fibrocartilage and hyaline cartilage. Biofabrication. doi: 10.1088/1758-5090/8/4/045002.

Das, S., F. Pati, S. Chameettachal, S. Pahwa, A.R. Ray, S. Dhara and S. Ghosh. 2013. Enhanced redifferentiation of chondrocytes on microperiodic silk/gelatin scaffolds: toward tailor-made tissue engineering. Biomacromolecules 14: 311-321. doi: 10.1021/bm301193t.

Das, S., F. Pati, Y.J. Choi, G. Rijal, J.H. Shim, S.W. Kim, A.R. Ray, D.W. Cho and S. Ghosh. 2015. Bioprintable, cell-laden silk fibroin-gelatin hydrogel supporting multilineage differentiation of stem cells for fabrication of three-dimensional tissue constructs. Acta Biomaterialia 11: 233-246. doi: 10.1016/j.actbio.2014.09.023.

de Lau, W., W.C. Peng, P. Gros and H. Clevers. 2014. The R-spondin/Lgr5/Rnf43 module: regulator of wnt signal strength. Genes & Development doi: 10.1101/gad.235473.113.

Eiraku, M., K. Watanabe, M. Matsuo-Takasaki, M. Kawada, S. Yonemura, M. Matsumura, T. Wataya, A. Nishiyama, K. Muguruma and Y. Sasai. 2008. Self-Organized formation of polarized cortical tissues from ESCs and its active manipulation by extrinsic signals. Cell Stem Cell. doi: 10.1016/j.stem.2008.09.002.

Eiraku, M., N. Takata, H. Ishibashi, M. Kawada, E. Sakakura, S. Okuda, K. Sekiguchi, T. Adachi and Y. Sasai 2011. Self-organizing optic-cup morphogenesis in three-dimensional culture. Nature. doi: 10.1038/nature09941.

Gao, G., A.F. Schilling, K. Hubbell, T. Yonezawa, D. Truong, Y. Hong, G. Dai and X. Cui. 2015. Improved properties of bone and cartilage tissue from 3D inkjet-bioprinted human mesenchymal stem cells by simultaneous deposition and photocrosslinking in PEG-GelMA. Biotechnology Letters 37: 2349-2355. doi: 10.1007/s10529-015-1921-2.

Ghosh, S., M. Laha, S. Mondal, S. Sengupta and D.L. Kaplan. 2009. *In vitro* model of mesenchymal condensation during chondrogenic development. Biomaterials 30: 6530-6540. doi:10.1016/j.biomaterials.2009.08.019.

Gjorevski, N., A. Ranga and M.P. Lutolf. 2014. Bioengineering approaches to guide stem cell-based organogenesis. Development. doi: 10.1242/dev.101048.

Guller, A.E., P.N. Grebenyuk, A.B. Shekhter, A.V. Zvyagin and S.M. Deyev. 2016. Bioreactor-based tumor tissue engineering. Acta Naturae 8: 44-58.

Gupta, A.C., S. Chawla, A. Hegde, D. Singh, B. Bandyopadhyay, C.C. Lakshmanan, G. Kalsi and S. Ghosh. 2018. Establishment of an *in vitro* organoid model of dermal papilla of human hair follicle. Journal of Cellular Physiology doi: 10.1002/jcp.26853.

Lancaster, M.A. and J.A. Knoblich. 2014. Organogenesisin a dish: Modeling development and disease using organoid technologies. Science 345. doi: 10.1126/science.1247125.

Lenas, P., M.J. Moos, F.P. Luyten, P. Lenas and M.J. Moos. 2009. Developmental Engineering: a new paradigm for the design and manufacturing of cell based products. Part I: From three-dimensional cell growth to biomimetics of *in vivo* development. Tissue Engineering Part B 15: 381-394. doi: 10.1089/ten.TEB.2008.0575.

Lowndes, M., M. Rotherham, J.C. Price, A.J. El Haj and S.J. Habib. 2016. Immobilized WNT proteins act as a stem cell niche for tissue engineering. Stem Cell Reports. doi: 10.1016/j.stemcr.2016.06.004.

Mironov, V., R.P. Visconti, V. Kasyanov, G. Forgacs, C.J. Drake and R.R. Markwald. 2009. Organ printing: tissue spheroids as building blocks. Biomaterials. doi: 10.1016/j.biomaterials.2008.12.084.

Montesano, R., K. Matsumoto, T. Nakamura and L. Orci. 1991. Identification of a fibroblast-derived epithelial morphogen as hepatocyte growth factor. Cell. doi: 10.1016/0092-8674(91)90363-4.

Morgani, S.M., J.J. Metzger, J. Nichols, E.D. Siggia and A.K. Hadjantonakis. 2018. Micropattern differentiation of mouse pluripotent stem cells recapitulates embryo regionalized cell fate patterning. Elife. doi: 10.7554/eLife.32839.

Nelson, C.M., M.M. VanDuijn, J.L. Inman, D.A. Fletcher and M.J. Bissell. 2006. Tissue geometry determines sites of mammary branching morphogenesis in organotypic cultures. Science 314: 298-300. doi: 10.1007/s40265-017-0856-4.

Rice, J.J., M.M. Martino, L. De Laporte, F. Tortelli, P.S. Briquez and J.A. Hubbell. 2013. Engineering the regenerative microenvironment with biomaterials. Advanced Healthcare Materials 2: 57-71. doi: 10.1002/adhm.201200197.

Rivron, N.C., J. Rouwkema, R. Truckenmüller, M. Karperien, J. De Boer and C.A. Van Blitterswijk. 2009. Tissue assembly and organization: Developmental mechanisms in microfabricated tissues. Biomaterials 30: 4851-4858. doi: 10.1016/j.biomaterials.2009.06.037.

Sasai, Y., M. Eiraku and H. Suga. 2012. *In vitro* organogenesis in three dimensions: self-organising stem cells. Development. doi: 10.1242/dev.079590.

Sasai, Y. 2013. Cytosystems dynamics in self-organization of tissue architecture. Nature. doi: 10.1038/nature11859.

Scotti, C., B. Tonnarelli, A. Papadimitropoulos, A. Scherberich, S. Schaeren, A. Schauerte, J. Lopez-Rios, R. Zeller, A. Barbero and I. Martin. 2010. Recapitulation of endochondral bone formation using human adult mesenchymal stem cells as a paradigm for developmental engineering. Proceedings of the National Academy of Sciences of the United States of America 107: 7251-7256. doi: 10.1073/pnas.1000302107.

Skardal, A., S.V. Murphy, M. Devarasetty, I. Mead, H.W. Kang, Y.J. Seol, Y.S. Zhang, S.R. Shin, L. Zhao, J. Aleman, A.R. Hall, T.D. Shupe, A. Kleensang, M.R. Dokmeci, S. Jin Lee, J.D. Jackson, J.J. Yoo, T. Hartung, A. Khademhosseini, S. Soker, C.E. Bishop and A. Atala. 2017. Multi-tissue interactions in an integrated three-tissue organ-on-a-chip platform. Scientific Reports doi: 10.1038/s41598-017-08879-x.

Takagi, R., J. Ishimaru, A. Sugawara, K.E. Toyoshima, K. Ishida, M. Ogawa, K. Sakakibara, K. Asakawa, A. Kashiwakura, M. Oshima, R. Minamide, A. Sato, T. Yoshitake, A. Takeda, H. Egusa and T. Tsuji. 2016. Bioengineering a 3D integumentary organ system from iPS cells using an *in vivo* transplantation model. Science Advances doi: 10.1126/sciadv.1500887.

Takebe, T., K. Sekine, M. Enomura, H. Koike, M. Kimura, T. Ogaeri, R.R. Zhang, Y. Ueno, Y.W. Zheng, N. Koike, S. Aoyama, Y. Adachi and H. Taniguchi. 2013. Vascularized and functional human liver from an iPSC-derived organ bud transplant. Nature. doi: 10.1038/nature12271.

Veeman, M.T., J.D. Axelrod and R.T. Moon. 2003. A second canon: functions and mechanisms of β-catenin-independent Wnt signaling. Developmental Cell doi: 10.1016/S1534-5807(03)00266-1.

Warmflash, A., B. Sorre, F. Etoc, E.D. Siggia and A.H. Brivanlou. 2014. A method to recapitulate early embryonic spatial patterning in human embryonic stem cells. Nature Methods doi: 10.1038/nMeth.3016.

Wodarz, A. and R. Nusse 1998. Mechanisms of wnt signaling in development. Annual Review of Cell and Developmental Biology doi: 10.1146/annurev.cellbio.14.1.59.

5

A Strategy for Regeneration of Three-Dimensional (3D) Microtissues in Microcapsules: Aerosol Atomization Technique

Chin Fhong Soon[1,*], Wai Yean Leong[2], Kian Sek Tee[2],
Mohd Khairul Ahmad[1] and Nafarizal Nayan[1]

5.1 INTRODUCTION

Culturing monolayers of cells in plastic dishes is routinely performed in life sciences and cell biological studies. Currently, scientific committee has begun to realize the many limitations of monolayer or two-dimensional (2D) culture model (Antoni et al. 2015, Souza et al. 2010). 2D cell model is missing accurate representation of physiological origins in terms of the proliferation, differentiation, gene and protein expression, functionality and morphology of cells (Edmondson et al. 2014). Contrarily, the three-dimensional (3D) cell culture creates extracellular matrix where cells are permitted to grow or interact with its surroundings. 3D cell culture regenerates biological relevant tissue model that restores specific cellular activities, signaling molecules and morphological structures similar to those *in vivo* (Kunz-Schughart et al. 2004). The cell interactions, responses and organization occurring within a 3D context demonstrated more native like and the severe limitations of 2D culture (Edmondson et al. 2014, Soon et al. 2016). 3D cell culture is part of the effort in regenerative medicine or biotechnology to recreate living and functional tissues *in vitro*, in which they are needed for replacement of damaged tissues (Kang et al. 2014), cancer research, application in tissue engineering (Stevens et al. 2004), pharmacological testing and stem cell research (Sugiura et al. 2005). Microencapsulation is an intensive research area to create cell and tissue model for rehabilitation of functional tissues (Zhao et al. 2017) and therapeutics purpose (da Rocha et al. 2014, Shin et al. 2013).

A microcapsule is a hollow particle with solid shell with a diameter ranging from a few to thousands of micrometers (Gasperini et al. 2014, Sun 1997). Hydrogel based microcapsules are semipermeable that could enable the passage of proteins, nutrients, drugs, and allow the diffusion of oxygen, nutrients, therapeutic products and wastes, while blocking the entry of antibodies and immunocytes (Paredes Juarez et al. 2014). In tissue transplantation, microcapsule functions as an

---

[1] Biosensor and Bioengineering Laboratory, Microelectronic and Nanotechnology-Shamsuddin Research Centre (MiNT-SRC), 86400, Batu Pahat, Johor, Malaysia.
[2] Faculty of Electrical and Electronic Engineering, Universiti Tun Hussein Onn Malaysia, Batu Pahat, Johor, Malaysia.
[*] Corresponding author: soon@uthm.edu.my

immune-protection. The islet cells placed inside the tiny capsules created a physical barrier to protect the islets from the immune system as reported previously (Vaithilingam and Tuch 2011). Therefore, cell encapsulation in biocompatible and semipermeable biopolymeric membranes is an effective method to overcome rejection of the implanted organ (Rabanel et al. 2009).

Various types of biopolymers such as chitosan, alginate, collagen, gelatin and agarose have been used for encapsulation of cells and particles (Gasperini et al. 2014, Hunt and Grover 2010). They differ in the solidification processes and hence, influence the design of the microencapsulation system. Alginate is an anionic polysaccharide derived from seaweed and it can be easily polymerized in divalent ionic solution at room temperature (de Vos et al. 2003, Lee and Mooney 2012, Sugiura et al. 2007). Due to its biodegradability and biocompatibility properties, alginate has been successfully applied in encapsulating cells and tissues to be transplanted into human body (Ghidoni et al. 2008). Moreover, alginate is currently recognised as a clinically ready biomaterial by the United States Food and Drug Administration (US FDA) (de Vos et al. 2014, Paredes Juarez et al. 2014).

Several methods had been developed for the microencapsulation of living organism and chemicals such as simple dripping (Chan et al. 2009, Swioklo et al. 2016), micromolding (Khademhosseini et al. 2006, Koh et al. 2002), extrusion (Martinez et al. 2012), microfluidic device (Hu et al. 2012, Huang et al. 2007), electrostatic droplet generation (Lewinska et al. 2008, Li et al. 2013, Zhang and He 2009), coaxial air-flow (Cui et al. 2001, Herrero et al. 2006, Martin-Banderas et al. 2010), vibration (Whelehan and Marison 2011) and jet cutting techniques (Gao et al. 2015, Herran and Huang 2012, Schwinger et al. 2002). Dripping method is simple without the involvement of chemical or mechanical treatment but the diameter of the capsules produced is huge ranging from 600-1000 μm (Gautier et al. 2011). Smaller alginate capsules in a few hundreds micrometers are desirable because this range of microcapsules can equilibrate rapidly across the ultrathin membrane with larger surface to volume relationship, and hence provide better transport of gases and nutrients for the encapsulated cells (Chicheportiche and Reach 1988). Other alternatives such as microfluidic and electrostatic dropping methods are able to produce microcapsules in smaller size ranging from 200 to 600 μm (Martin-Banderas et al. 2010). However, these techniques are considered unfavourable to the survival of cells due to the processing procedures that require the use of oil phase, organic solvent, high voltage or ultra-violet treatment (Martinez et al. 2012, Sugiura et al. 2005, Wan 2012). Hence, the simpler the production process (without harsh and post-processing treatment), the lesser the threat to the cells whilst ensuring cells to proliferate in the encapsulations.

Amongst the previous methods discussed (Lewinska et al. 2008, Herran and Huang 2012, Hu et al. 2012, Khademhosseini et al. 2006, Koh et al. 2002, Li et al. 2013, Martinez et al. 2012, Whelehan and Marison 2011, Prüsse et al. 2008, Schwinger et al. 2002, Zhang and He 2009), aerosol atomization technique is a simple and efficient method to generate high throughput microcapsules with well-controlled size and shape without the use of hostile chemicals (Herrero et al. 2006, Sugiura et al. 2007, Tendulkar et al. 2012). In this work, an electronic aerosol atomization system was applied to generate the desired size of calcium alginate microcapsules for the microencapsulation of cells. Although the aerosol atomization method has been developed previously (Ahn et al. 2012, Sohail et al. 2011), but the microcapsules' size ranged from 10 to 40 μm which is too small and not suitable for cells encapsulation. Current applications (air jets, fuel injection and spray coating) based on aerosol atomization technique require high air flow rate (50-600 l/min) and large volume (millilitre) of solution to create small beads size (approximately 1-3.5 μm) (Ahn and Kim 2015). Thus, an adjustable electronic aerosol atomization system was designed to produce different airflow rates (0.2-0.5 l/min), in which it can be used to disperse small volume of cells-alginate suspension in microlitres and to generate narrow range of microcapsules ranging from 80 to 360 μm. Instead of using compressed air from a gas cylinder (Herrero et al. 2006, Sugiura et al. 2007) which is costly, the proposed electronic aerosol atomization system presented a different approach in the generation of airflow by using a direct current (dc) air pump. This technique was successfully demonstrated to regenerate microtissues of keratinocyte cell lines (HaCaT). The encapsulated HaCaT cells grown into microtissues presented potential applications for tissue padding and pharmacology study.

5.2 PROCEDURES TO MICROENCAPSULATE CELLS AND CHARACTERIZATION

5.2.1 Cell Culture and Preparation

Human keratinocyte cell lines (HaCaT) were purchased from cell line services (CLS, Eppelheim, Germany) and maintained in Dulbecco's Modified Eagle Medium (DMEM, Gibco®, Life Technologies, USA), supplemented with 10% fetal bovine serum (FBS, Invitrogen, California, USA), penicillin (100 units/ml, Sigma Aldrich, Dorset, UK) and streptomycin (100 mg/ml, Sigma Aldrich, Dorset, UK), Amphotericin-B (2.5 mg/l, Sigma Aldrich, Dorset, UK), at 37°C in a 5% carbon dioxide (CO_2) humidified air-jacketed incubator. When the cells had grown to 80% confluency, the media was discarded from the cell culture flask and the flask was washed three times with Hank's Balanced Salt Solution (HBSS, Invitrogen, California, USA). Subsequently, the HBSS was removed and the cells were treated with 1 ml of trypsin (0.5 mg/ml, Sigma Aldrich, Dorset, UK) at 37°C for 5 minutes. After the cells were detached, the cells were centrifuged into cell pellet, and old media was discarded. The cell pellet was resuspended in supplemented culture media and was readied for microencapsulation.

5.2.2 Preparation of Cell-alginate and Calcium Chloride Solutions

HaCaT cells at a density of 3×10^7 cells/ml were added to 100 μl of 1.5% wt/v alginate solution (Fig. 5.1). The cell-alginate suspension was filled in a 0.5 ml syringe (Becton, Dickinson and Company, Franklin Lakes, NJ, USA) having a 29-gauge insulin needle (Fig. 5.2) and the syringe was fitted to a syringe pump (NE-4002X, New Era, Farmingdale, NY, USA) to extrude the cells-alginate suspension. Subsequently, the needle of the insulin syringe was inserted to the air tube extending from the air pump system. The overall system setup is as shown in Fig. 5.3. The electronic aerosol atomization system was set to extrude microdroplets of cells-alginate at an air flow rate of 0.3 l/min and an extrusion rate of 20 μl/min. Subsequently, 4 ml of filtered calcium chloride ($CaCl_2$) solution at 1% w/v for crosslinking the microdropets of cells-alginate was then prepared in a sterilised petri dish with a diameter of 6 cm. The petri dish was placed at a drop distance D = 6 cm under the insulin needle as shown in Fig. 5.3. The dispersion of the microbeads from the needle nozzle is as illustrated in Fig. 5.4a. Figure 5.4b shows the actual configuration of the aerosol nozzle.

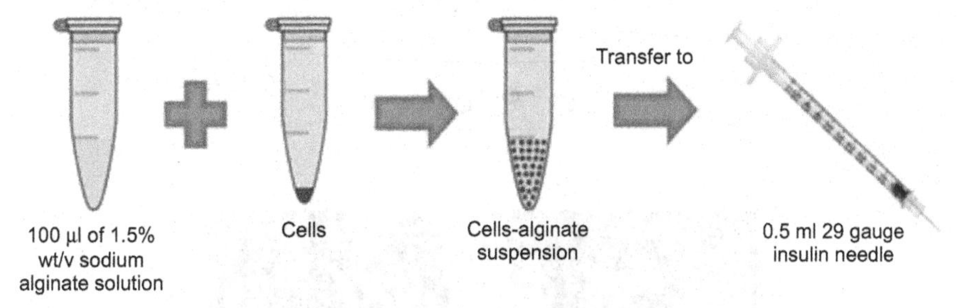

FIGURE 5.1 Preparation of 1.5% wt/v cell-alginate suspension.

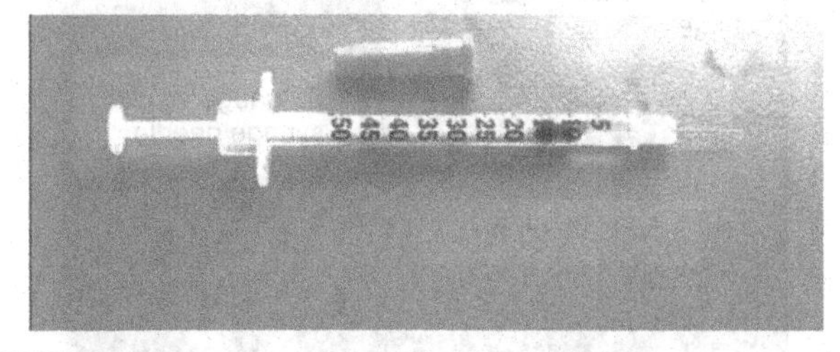

FIGURE 5.2 Insulin needle with 100 μl cell-alginate suspension.

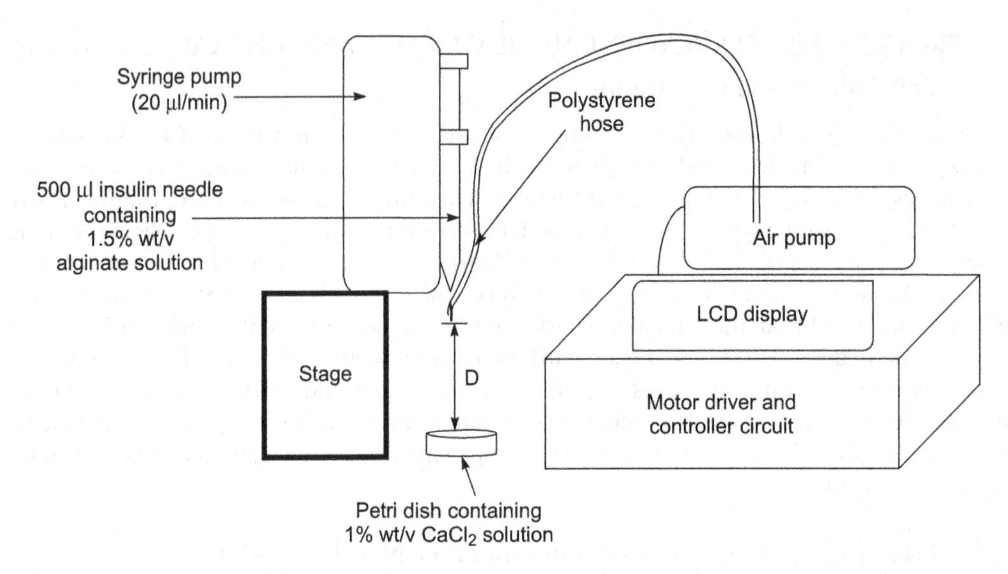

FIGURE 5.3 A schematic illustration of the experimental setup of an electronic aerosol atomization system for generating calcium alginate microcapsules.

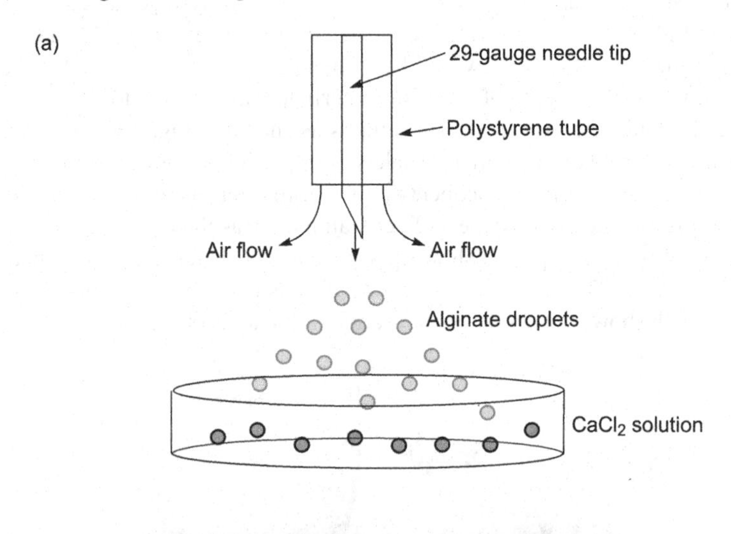

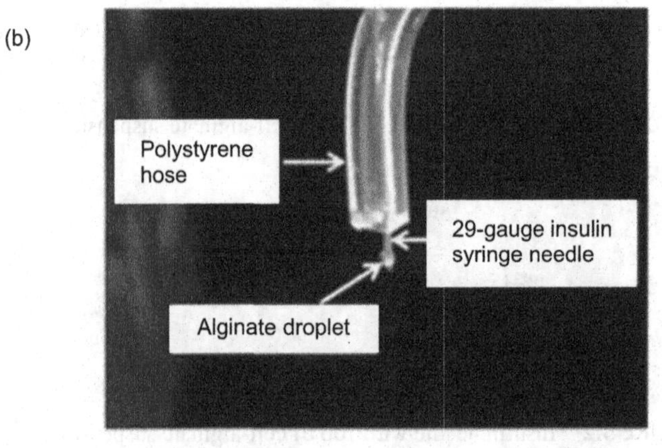

FIGURE 5.4 An illustration of the aerosol nozzle. (a) The schematic diagram of the insulin syringe needle nozzle and (b) the image of the insulin syringe needle.

5.2.3 Validation of Microcapsules Drop Distance

During initial atomizing test, it is essential to determine the drop distance between the nozzle of the needle and the surface of the calcium chloride fluid. The dispersed droplets need to overcome the surface tension of the fluid to form spherical capsules. Hence, the penetration force of the droplets is associated with the drop distance. Figure 5.5a shows the experiment setup to investigate the effect of drop distance to the shape of the microcapsules. In this experiment, the extrusion rate and airflow rate were fixed at 20 μl/min and 0.3 l/min, respectively. The drop distance of the microcapsules was measured between the insulin needle tip and the surface of calcium chloride solution in a petri dish (Fig. 5.5a). The drop distances were set at 3, 6 and 9 cm height to determine the effects of different drop distances to the shape of the microcapsules. The coverage of the aerosoling (Fig. 5.5b) at different drop distances was calculated based on Equation 1:

$$C = 2D \tan\left(\frac{\theta}{2}\right) \qquad (1)$$

where C is the coverage in the unit of millimetre (mm), D is the drop distance in the unit of millimetre (mm) and θ is the angle in the unit of degree (°). The microcapsules were observed using a BX60M optical microscope (Olympus, Tokyo, JAPAN) linked to a U-PMTVC CCD camera (Olympus, Tokyo, JAPAN). The images of the microcapsules were captured using the MaterialPlus image analysis software (PACE Technologies, Version 4.2, USA).

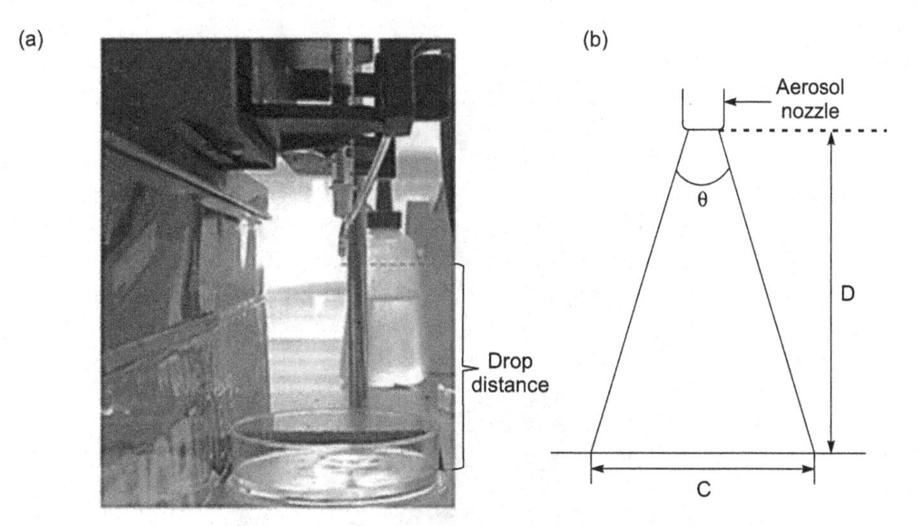

FIGURE 5.5 The microcapsules drop distance validation setup and (b) the schematic diagram of the dispersed coverage (C), angle (θ) and drop distance (D).

5.2.4 Microencapsulation of 3D Cells using Aerosol Atomization

During the experiment, the aerosol system dispersed the microdroplets of cell-alginate from the aperture of the needle, and the microdroplets dropped into the petri dish followed by polymerization in calcium alginate solution for approximately 10 min. The standard polymerization time was determined by the mean absorbance at 330 nm using Multiskan™ GO Microplate Spectrophotometer (Thermo Fisher Scientific, MA, USA). Subsequently, the solution in the petri dish was carefully discarded, leaving only the polymerized microcapsules of HaCaT cells. The microcapsules containing cells were rinsed three times with HBSS solution followed by incubation in 2 ml of DMEM at 37°C in a 5% CO_2 humidified incubator. The media was replenished every two days to provide enough nutrients for the growth of the encapsulated cells. All experiments were performed in a SC2-4A1 biological safety cabinet (ESCO, Singapore). The growth of HaCaT cells encapsulated in the microcapsules was monitored every two days up to 16 days of culture. Photomicrographs of the cells were captured using an inverted phase contrast microscope (TS100, Nikon, Tokyo, Japan) coupled with a Go-5 CCD digital camera (QImaging, Surrey, Canada).

5.2.5 Fourier Transform Infrared (FTIR) Spectroscopy

The biochemical compounds of the calcium alginate microcapsules and calcium alginate laden cells capsule were analyzed using a Fourier transform infrared (FTIR) spectrometer (Perkin Elmer Spectrum 100, Shelton, CT, USA) in transmission mode. FTIR spectroscopy is a useful tool to characterise the interaction as well as infrared-spectra shift due to the interactions between biomaterials and cells (Notingher et al. 2003). In this experiment, the samples were prepared and placed onto a small window of small attenuated total reflection (ATR) crystal followed by engaging a pressure arm over the sample (Fig. 5.6). A force, F ($70 \text{ N} < F < 100 \text{ N}$ force transmission), was applied to the sample by pushing it onto the crystal surface. The spectra of the samples were collected for 32 scans at a resolution of 4.0 cm^{-1}. In the PerkinElmer Spectrum Express software, the FTIR spectra of the measurement were plotted over a spectral (wave number) range of 4000-600 cm^{-1}. The spectrum was taken at three different measurements sites of the sample and the average of the measurements was determined. Similar measurements were repeated three times for each sample.

FIGURE 5.6 The samples placed on a window of ATR crystal in the FTIR machine.

5.2.6 Alginate Lyase Activity

The alginate shell of the microcapsules of HaCaT were removed by using alginate lyase (Sigma Aldrich, St. Louis, MO, USA) at 0.2 mg/ml prepared in a media after 16 days of culture. Alginate lyase catalyse the biodegradation of complex structured alginate by cleaving the glycosidic bond via a β-elimination reaction (Zhu and Yin 2015). Within 1 to 2 min of immersion in the alginate lyase media, the 3D microtissues released from the alginate shell were collected and washed three times in HBSS. The microtissues should not be kept too long in the alginate lyase solution. The purified microtissues were ready for physical examination using a field emission-scanning electron microscopy (FE-SEM).

5.2.7 FE-SEM Imaging of the Calcium Alginate Microcapsules and 3D Microtissues

The surface structures of the calcium alginate microcapsules and HaCaT microtissues were examined using a JSM-7600F field emission-scanning electron microscope (FE-SEM) (JOEL, Tokyo, Japan) with an upper secondary electron imaging (SEI) detector. Before the imaging, the calcium alginate microcapsules were collected from the gelation bath (petri dish) and left for air dried on a microscope slides for 15 minutes. The microtissues of HaCaT were fixed in 4% formaldehyde (Sigma Aldrich, St. Louis, MO, USA) for 24 hrs at 5°C and allowed to be air dried. Subsequently, the glass slide containing microcapsules and 3D microtissues were coated with conductive gold coatings in a JFC-1600 Auto Fine Coater (JEOL, Tokyo, Japan) powered at 20 mA for 30 s. Then, the gold coated glass slides containing the microcapsules and microtissues were mounted to the mounting stub using a double-sided carbon tape before loading into the FE-SEM for imaging as shown in Fig. 5.7. The mounting stub was then inserted into the specimen holder and pushed into the specimen chamber using a stainless-steel rod. After the specimen was loaded, the vent valve was shut, and the chamber was pumped into vacuum. During the FE-SEM scanning, the microcapsules and the microtissues of HaCaT were exposed to an accelerated voltage beam of 5 kV. The images obtained were magnified at 150 ×, 300 × and 10,000 ×, respectively.

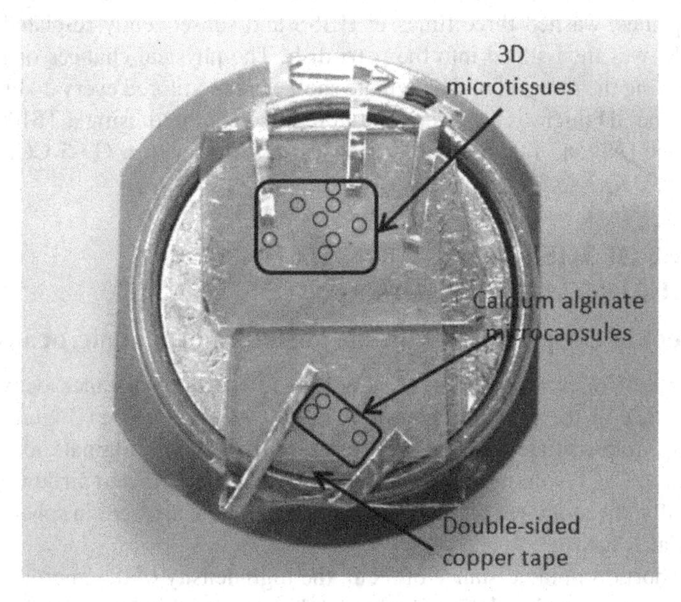

FIGURE 5.7 The mounting stub with microcapsules and 3D microtissues. The circles indicate the location of the microcapsules and microtissues.

5.2.8 DAPI Staining

DAPI (4', 6-diamidino-2-phenylindole dihydrochloride) staining was performed to investigate the cells or nuclei distribution in the microtissues after reaching dormant phase. For staining agent preparation, DAPI (0.1 µg/ml) (Sigma Aldrich, St. Louis, MO, USA) was diluted in HBSS. Then, the microcapsules of HaCaT were washed in HBSS and then incubated in DAPI solution for 20 min in the dark. Subsequently, the DAPI stain was removed and the stained microtissues were washed with HBSS solution. The stained microtissues images were viewed and captured using a BX53 fluorescence microscope (Olympus, Tokyo, Japan) mounted with a DP73 CCD camera (Olympus, Tokyo, Japan).

5.2.9 Live and Dead Cell Staining

Live/dead® viability kit for mamalian cells (Invitrogen, Paisley, UK) was used to stain the live and dead cells of the microtissues formed in the calcium alginate microcapsules. The live/dead® cell viability kit can differentiate live cells from the dead cells by double staining the HaCaT cells in the microtissues with green-fluorescent Calcein-AM (Invitrogen, Paisley, UK), which indicates intracellular esterase activity and red-fluorescent Ethidium homodimer-1 (EthD-1) (Invitrogen, Paisley, UK), which in turn indicates the loss of plasma membrane integrity. After 16 days of culture, the HaCaT microtissues formed in the calcium alginate microcapsules were incubated in mixed solutions of 2 μM of Calcein-AM and 4 μM of EthD-1 for 20 min in the dark. Subsequently, the stain solutions were removed and the microtissues were washed three times in HBSS solution. The photomicrographs of stained microtissues were captured using a BX53 fluorescence microscope (Olympus, Tokyo, Japan) mounted with a DP73 digital camera.

5.2.10 Re-seeding Microtissues

The way to examine the physical changes or viability of cells without staining is by studying the cells with the simple re-seeding of expriment. The microtissues of HaCaT were removed from the alginate microcapsules, washed three times in HBSS and subsequently re-plated in a petri dish. Two ml of DMEM was then added into the petri dish. The physical changes or migratory of the microtissues following the removal of the alginate shell were monitored every 24 hr up to 72 hr. The transformation of the 3D microtissues were monitored and captured using a TS100 inverted phase contrast microscope (Nikon, Tokyo, Japan) that was installed with a Go-5 CCD digital camera (QImaging, Surrey, UK).

5.3 OUTCOME OF THE ATOMIZATION TECHNIQUE IN PRODUCING MICROTISSUES

5.3.1 The Effect of Drop Distance on the Structure of the Microcapsules

Different structures of microcapsules were subjected to the drop distance between the aerosol nozzle and the surface of the calcium chloride ($CaCl_2$) bath (Fig. 5.8a-c). Figure 5.8d-f show the effects of different drop distances on the structures of the calcium alginate microcapsules after being polymerised in $CaCl_2$ solution at an extrusion rate of 20 μl/min and airflow rate of 0.3 l/min, respectively. Initially, the atomized alginate microdroplets in the air were in spherical shape before dipping into the $CaCl_2$ bath.

However, at a short dropping distance of 3 cm, the high density of the microdroplets streaming at a high flow rate split the ionic solution and forced the alginate droplets to the bottom of the petri dish (Fig. 5.8a). The split ionic solution appeared as a void circular pool at the centre of the ionic solution (Fig. 5.8a). The dispersed alginate microdroplets were then accumulated and merged at the bottom of the petri dish and turned to a thin sheet of calcium alginate by the surrounding ionic solutions as shown in Fig. 5.8d.

At a drop distance of 6 cm (Fig. 5.8b), the air flow exerted less dynamic pressure to the $CaCl_2$ solution (Elger et al. 2014) but sufficiently to break the surface tension of the ionic solution, allowing the alginate microcapsules to be successfully polymerized into calcium alginate microcapsules in the ionic bath (Fig. 5.8e). The results suggested that 6 cm was the suitable drop distance to generate spherical calcium alginate microcapsules at a flow rate of 0.3 l/min.

When the drop distance increased to 9 cm (Fig. 5.8c), the microcapsules failed to form into spheres but distorted microcapsules (Fig. 5.8f). This was due to the decreased air velocity which was not sufficient to overcome the surface tension of the ionic bath and thus, decreased the success to form perfect microcapsules (Aly et al. 2013, Elger et al. 2014, Watson 2010, Zhao and Yang 2012). In addition, some of the alginate microdroplets were found dispersed out of the collection zone (petri dish) due to the wider coverage at a drop distance of 9 cm as illustrated in Fig. 5.9.

Based on Equation 1, the dispersed angle (θ) was equal to 45.2° when the drop distance (D) and the coverage (C) were at 3 and 2.5 cm, respectively. The coverage of 2.5 cm was within the diameter of the petri dish (Fig. 5.9). Subsequently, when D = 9 cm and = 45.2°, the coverage (C) was calculated as equal to 7.5 cm, which was larger than the diameter of the petri dish. Hence, some of the alginate microdroplets were dispersed out of the petri dish. As a result, the alginate microdroplets were not suitable to disperse at a drop distance of 9 cm. Therefore, the drop distance between the aerosol nozzle and the $CaCl_2$ hardening bath was suggested to be 6 cm at a flow rate of 0.3 l/min.

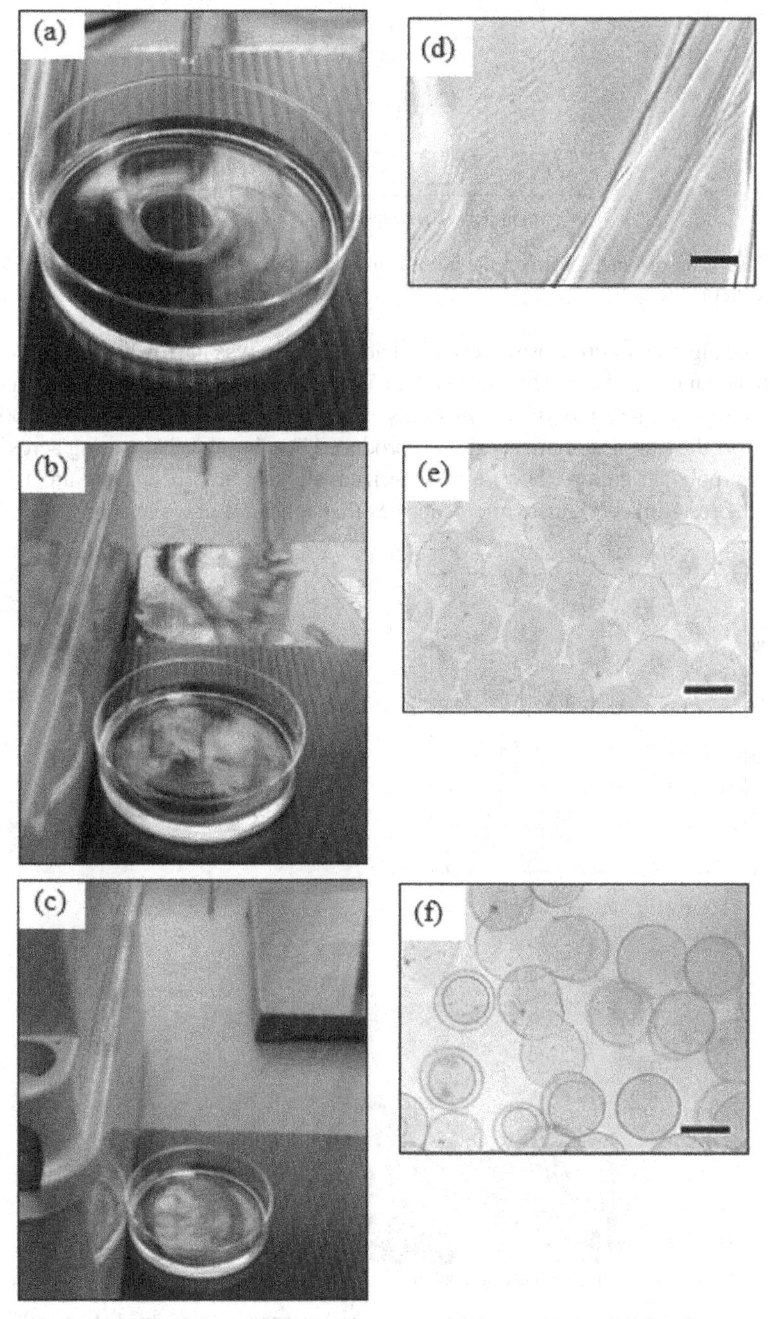

FIGURE 5.8 Photographs of changing the distance between needle tip and $CaCl_2$ bath surface at (a) 3, (b) 6 and (c) 9 cm and the photomicrographs of polymerised calcium alginate formed at drop distance of (d) 3, (e) 6 and (f) 9 cm (scale bar: 200 μm).

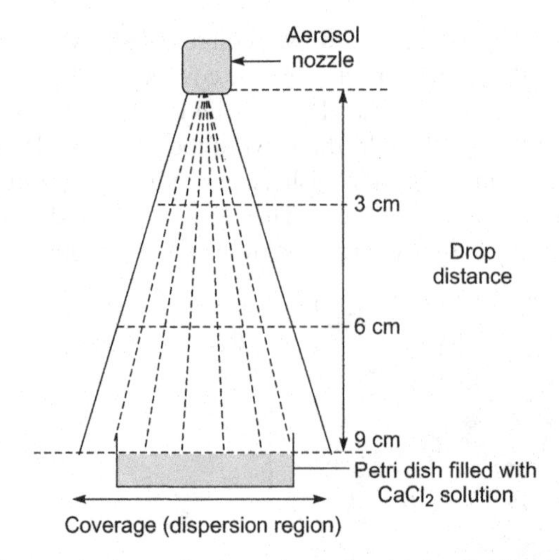

FIGURE 5.9 The drop distance between the aerosol nozzle and the $CaCl_2$ solution surface determine the microdroplets' coverage region.

The size of the alginate microcapsules extruded at 20 µl/min was narrowly distributed at 234.51 ± 18.52 µm as shown in Fig. 5.10a-b. This range falls within the recommended size of microcapsules for epithelial cells to grow to the approximate thickness of the epidermis which is between 200 to 300 µm. Figure 5.10c shows the samples of microcapsules produced in spherical shape using aerosol technique before and after polymerization. The un-polymerized alginate microcapsules are transparent in comparison with cross-linked alginate microcapsules that appeared in white (Fig. 5.10c).

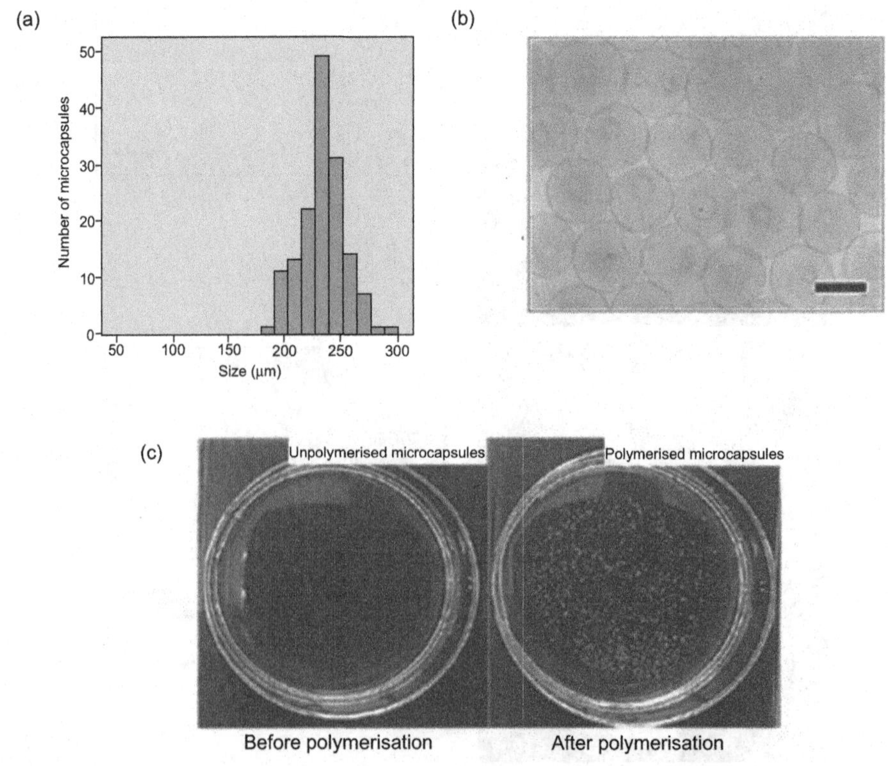

FIGURE 5.10 (a) The polymerization absorbance of the microcapsules upon irradiation at 330 nm (b) The illustration of before and after polymerization calcium alginate microcapsules. Adapted from Leong et al. 2017, by permission of the *Bioengineering* under the Creative Commons Attribution License.

Figure 5.11 shows the gradual development of a 3D HaCaT microtissue in the calcium alginate microcapsule over a period of 16 days. On the first day of culture (Day 0), individual HaCaT cells were randomly distributed in the calcium alginate microcapsules. From Day 2 onwards, the quantity of cells encapsulated in the calcium alginate microcapsules increased and continued to form aggregates in the encapsulations (Fig. 5.11). It is an indication of cell proliferation in the capsule. The cells continued to proliferate, and extracellular matrix proteins were secreted by the cells in the encapsulation. Approximately after 14 days of culture, the clusters of cells in the microcapsules grew into microtissues and the masses of cells completely filled the microcapsules. Towards day 16 of culture, full grown microtissues of HaCaT can be clearly observed in dark aggregations.

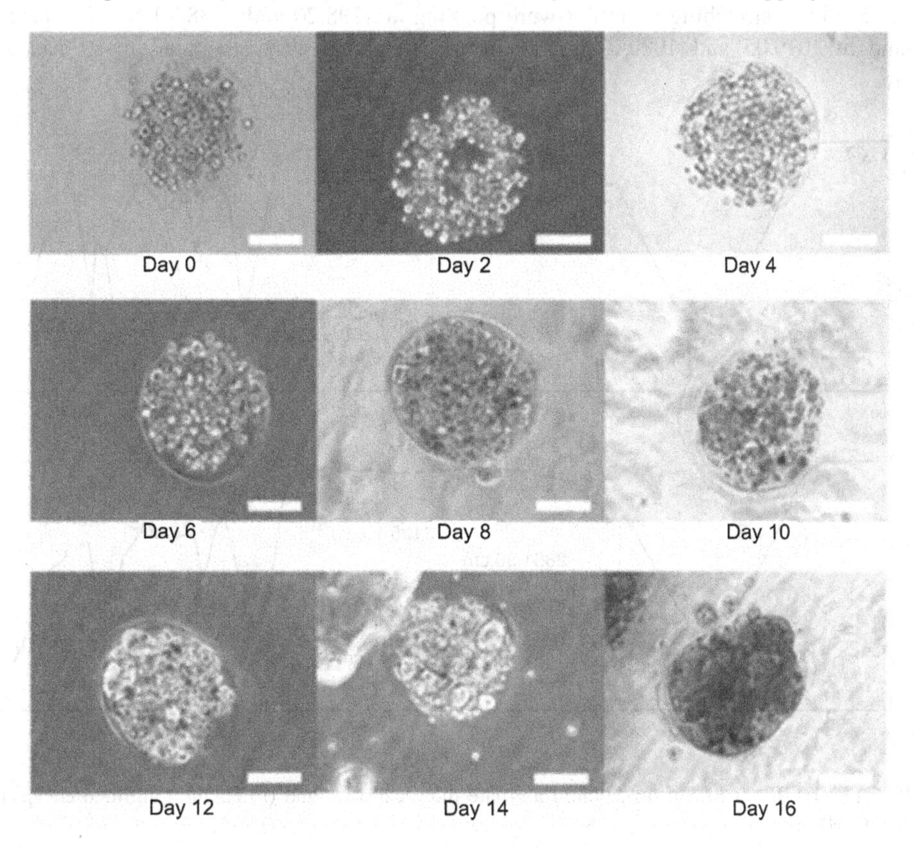

FIGURE 5.11 The phase contrast microscopic image of calcium alginate encapsulated 3D HaCaT for 16 days of culture at 100 × magnification (scale bar: 100 μm). Adapted from Leong et al. 2017, by permission of the *Bioengineering* under the Creative Commons Attribution License.

5.3.2 FTIR Spectrum of Calcium Alginate Encapsulated Cells

The FTIR spectra of sodium alginate, calcium alginate and cell laden calcium alginate in the wavelength of 4000 to 600 cm^{-1} are as shown in Fig. 5.12. In the spectrum of calcium alginate (Fig. 5.12b), the broad band at 3253.86 cm^{-1} corresponded to the C-H hydroxyl groups, while the dip at 2124.02 cm^{-1} was attributed to –OH stretching and the bending vibrations of N-H (amide II band) occurred at 1594.52 cm^{-1} (Daemi and Barikani 2012). The peaks at 1415.60 and 1297.81 cm^{-1} are associated with N-H stretching of amide and ether bonds and N-H stretching (amide II band), respectively. The peaks observed at 1078.02 and 1028.39 cm^{-1} are the secondary hydroxyl group (characteristic peak of CHOH in cyclic alcohols, C-O stretch) and primary OH (characteristic peak of $-CH_2$-OH in primary alcohol, C-O stretch) (Dianawati et al. 2012).

Figure 5.12b shows the FTIR spectrum of calcium alginate encapsulated HaCaT cells in the wavelength ranges of 4000-600 cm^{-1}. The broad band at 3274.31 cm^{-1} exhibited characteristics absorption band for C-H hydroxyl groups in the spectrum of calcium alginate encapsulated HaCaT cells. The FTIR spectrum displays strong phospholipid terminal $-CH_3$ stretching vibrations (both symmetric in the region of 2851.20 cm^{-1} and asymmetric in the region of 2963.10 cm^{-1}). Then, the dip at 2125.10 cm^{-1} was caused by –OH stretching. Moving to the fingerprint region, the FTIR spectra exhibited strong Amide II peaking at 1547.30 cm^{-1}. In addition, the bands at 1598.42 cm^{-1} and 1412.57 cm^{-1} are assigned to asymmetric and symmetric stretching peaks of carboxylate salt groups of carboxylate and carbonyl (Nagpal et al. 2013). Next, the amide III band and the asymmetric $-PO_2$- stretching vibration were peaking at 1298.20 and 1238.20 cm^{-1}, respectively. The band at 1079.05 and 1029.10 cm^{-1} (C-O-C stretching) was attributed to its saccharide structure.

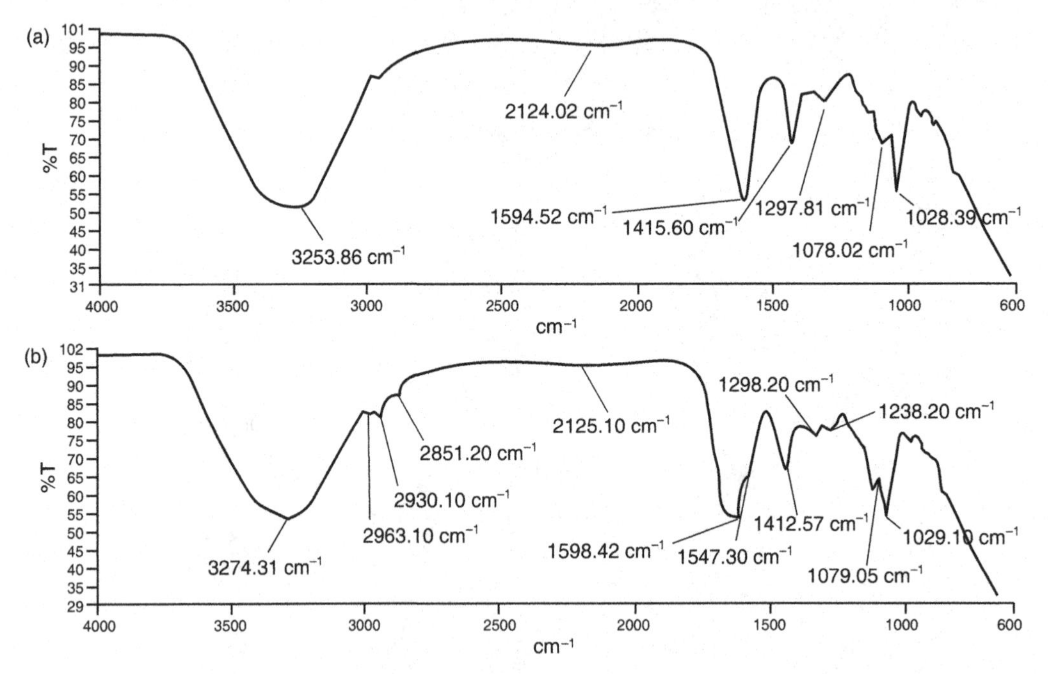

FIGURE 5.12 FTIR spectra of (a) calcium alginate microcapsules and (b) calcium alginate encapsulated with HaCaT cells.

The micrograph of DAPI staining in Fig. 5.13a indicates that the distributions of nuclei for HaCaT cells are highly concentrated in the microcapsules after 16 days of culture. Figure 5.13b shows the results of live and dead cell stainings for the HaCaT microtissues following 16 days of culture, respectively. The calcium alginate microcapsules membrane is semipermeable which allowed permeation of live/dead stains into the core of the encapsulated microtissues. The green (Calcein-AM) and red (Ethidium homodimer-1) fluorescence staining indicated live and dead cells within the microtissues, respectively. Green-fluorescent Calcein-AM (Invitrogen, Paisley, UK) indicates intracellular esterase activity, while red-fluorescent Ethidium homodimer-1 (EthD-1) (Invitrogen, Paisley, UK) indicates loss of plasma membrane integrity. Both microtissues were revealed with high viability of cells and only a few dead cells were observed.

Although the microtissues were confined in the spherical structure of the microcapsules, it is interesting to determine if the microtissues formed were also spherical. Figure 5.14a and 5.14b show the HaCaT microtissues before and after being removed from the alginate capsules, respectively. The calcium alginate microcapsules membrane dissolved completely within 2 minutes in the media containing 0.2 mg/ml of alginate lyase. The extracted microtissues of HaCaT were in arbitrary shape

(Fig. 5.14b) but indicated good cell-cell integrity and adhesions. No loose pieces of microtissues were observed.

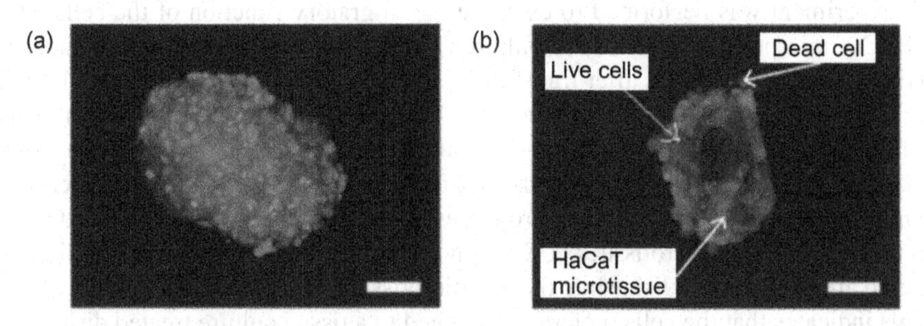

FIGURE 5.13 DAPI staining of cells in the microtissues of (a) HaCaT after 16 days of culture at 100 × magnification (scale bar: 100 μm). (b) Live and dead staining fluorescence microscopic images of HaCaT microtissues after 16 days of culture at 100 × magnification. Majority of the cells were stained green (scale bar: 100 μm). Adapted from Leong et al. 2017, by permission of the *Bioengineering* under the Creative Commons Attribution License.

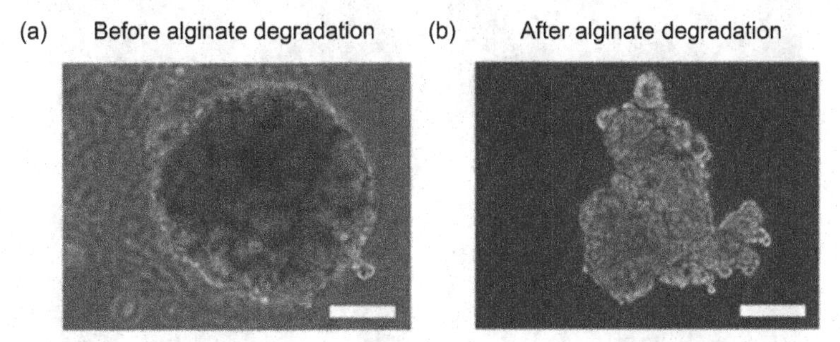

FIGURE 5.14 Phase contrast microscopic images of the calcium alginate encapsulated with HaCaT microtissues before alginate degradation, and the extracted HaCaT microtissues after alginate degradation at 100 × magnification (scale bar: 100 μm). Adapted from Leong et al. 2017, by permission of the *Bioengineering* under the Creative Commons Attribution License.

FE-SEM was applied to investigate the overall and surface structure of the microtissues. Figure 5.15 shows the FE-SEM micrographs of HaCaT microtissue after 15 days of culture at 300 × magnifications. FE-SEM images revealed good integrity of cells in the extracellular matrix which is in good agreement with the phase contrast micrographs of HaCaTs microtissues. The cells reconstructed the cell-cell junctions and formed aggregates by self-derived extracellular matrix proteins.

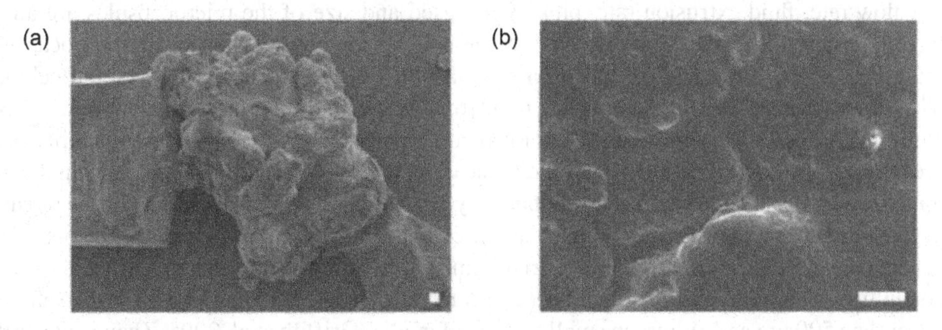

FIGURE 5.15 Field emission-scanning electron micrographs of HaCaT microtissue after 15 days of culture at (a) 300× and (b) 1500× magnifications, respectively (scale bar: 10 μm). Adapted from Leong et al. 2017, by permission of the *Bioengineering* under the Creative Commons Attribution License.

Although the live and dead cell staining suggested that the cells in the microtissues were mostly viable, the technique could not reveal the basic functionalities of the cells, such as motility. The re-seeding experiment was performed to examine the migratory function of the cells. Hence, the extracted HaCaT microtissues were then re-cultured in a petri dish to investigate their morphological changes (Fig. 5.16). Several hours after transferring the microtissues of HaCaT on a tissue culture-treated dish, the adhesion of the microtissues was examined by a mild perturbation of media by gently shaking the media. The microtissues were found to be well adhered to the surface of the petri dish. After 24 hr of re-seeding, individual cells were found to be migrating out of the microtissues. As more and more cells migrated out of the microtissues after 48 h of culture, this induced a monolayer of cells to form around the microtissues. The 2D monolayer of cells continued to proliferate with a larger covered area in the petri dish while the 3D microtissues gradually disintegrated after 72 hr of culture. This indicates that the cells preferably attached to a tissue culture treated dish.

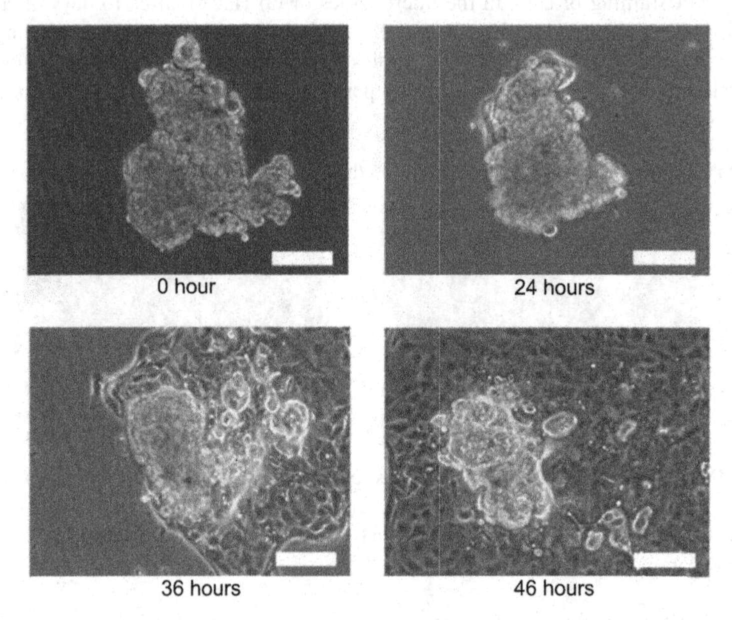

FIGURE 5.16 Phase contrast microscopic images of re-seeding the HaCaT microtissues at 100 × magnification (scale bar: 100 μm). Adapted from Leong et al. 2017, by permission of the *Bioengineering* under the Creative Commons Attribution License.

5.4 FACTORS INFLUENCING THE SUCCESS OF MICROENCAPSULATION

The air flow rate, fluid extrusion rate, biopolymer used and size of the microcapsules are among the factors in determining the success of growing cells in the confinement of the microcapsules. At a flow rate of 0.3 l/min and an extrusion rate of 20 μl/min, the aerosol atomization technique produced microcapsules of cells at 234.51 ± 18.52 μm and the microcapsules were cultured in a bath of culture media. Microcapsules within this range are recommended for supporting viability of the cells residing at the core of the microcapsules. The diameter of microcapsules in a few hundreds of micrometres reduces the mass transfer resistance by providing greater surface area for the diffusion of oxygen and nutrients to the cells (Wilson and McDevitt 2013). Hence, the cells at the outer layer and inner core would receive substantial nutrients and gases which allow them to stay viable after 16 days of culture. There is a chance for cell necrosis and nutrients deprivation of cells if the spheroids are larger than 500 μm and culture in small volume of media (Griffith et al. 2005, Zhang et al. 2016).

Under the strategy of microencapsulation, the cells were confined in small spatial volume, enabling the cells to reconstruct cell-cell adhesion in three dimensions which is important to stimulate the formation of microtissues (Wong et al. 2016). With the limited size and growth volume of microcapsules, the proliferation of cells seemed to have growth limit and they remained quiescent upon reaching the dormant phase of the microtissues. This is similar to culturing monolayer of cells to full confluency, in which the proliferation is limited by the size of the culture flask. During the culture, the phenol red of the culture media changed from red to yellow indicating the excretion of waste product and catabolites produced by the cells, hence decreasing the pH of the media. Other factors that could influence the growth of microtissues are such as the type of cells (Leong et al. 2017), type of biopolymer, viscoelasticity of the biopolymer, degradability of the biopolymer or soluble factors in the culture media. However, the cell density is also a crucial factor for different cell types to grow into microtissues. This could be due to the variation of cell growth rates for different cell types (Achilli et al. 2012).

The live/dead cell staining result indicated that the microtissues were viable after 16 days of culture since only a few HaCaT cells were stained in red (Fig. 5.13b). This result showed that the cells encapsulated in the hydrogel-like microcapsules in a large volume of bath media received sufficient nutrients and gases which allow them to stay alive even after long period of culture (Griffith et al. 2005).

The results showed that the microtissues may appear as spheroids according to the shape of the microcapsules. To reveal the actual appearance of the microtissues without disrupting the structure of the microtissues, alginate degradation was performed within a short period of time (< 2 min) to degrade the alginate capsule. The microtissues after alginate degradation were found in spheroids or non-spheroids depending on the self-organisation ability of cells to form microtissues as shown in Fig. 5.15. The integrity of the microtissues is also influenced by the ability of the HaCaT cells to produce extracellular matrix proteins that could support cell-cell adhesion as observed in the FE-SEM microscopy.

Growing cell spheroids based on the microencapsulation technique has the advantages of spheroid size control and re-association of cells based on self-generated extracellular matrix proteins. HaCaTs were demonstrated with the ability to self-assemble into microtissues if they are provided with suitable and biocompatible micro-environment such as alginate based microcapsules. The multicellular structure serves a protective layer for diffusion of molecules from outer surface to the core of microtissues. In regenerated 3D cells or microtissues, the morphology of cells, cell-cell adhesion, cell-matrix interactions and volumetric growth of microtissues are distinctly different from those in 2D culture, thus making the microtissue a translational and fundamental model for studying the efficacy of drug in the *in vivo* animal studies. For implementation of microencapsulation cells in tissue transplantation, the microcapsular membrane is a layer which protects the microtissues from the attack of the immune system of the recipients (Vaithilingam and Tuch 2011). Plausibly, the alginate membrane is non-toxic and can be easily removed if it is not needed as demonstrated in this work.

Without cell staining, the re-seeding is a simple experiment to confirm that the microtissues were composed of living cells that could adapt and migrate in the culture dish. The surface of tissue culture treated dishes is usually modified to be more hydrophilic by generating negatively charged hydroxyl group on the surface. The negative charges on the surface of the culture plastic changed the surface properties of the dish (Ramsey et al. 1984). This created a surface with high affinity to attract the migration of cells. Interestingly, this is in contrast with a previous study which showed that the microtissues could migrate in colony on a soft liquid crystal surface (Soon et al. 2018). The stiffness of a surface clearly influences the behaviour of the cells in the microtissues.

5.5 CONCLUSIONS

Aerosol atomization technique is an efficient technique to microencapsulate cells in calcium alginate leading to the growth of microtissues within 16 days. The encapsulated cells have the ability to proliferate, self-assemble and grow into microtissues in the microcapsules of calcium alginate. The size of the microtissues within the range of a few hundreds of micrometres is controllable by varying the extrusion rate of the syringe pump and air flow rate. In this technique, the drop distance is a factor to determine if spherical microcapsules can be produced. Degradation of calcium alginate membrane using alginate lyase confirmed the formation of 3D microtissues model. In addition to the live and dead cell staining, the microtissues re-seeding experiment is a simple test to examine the viability and basic cell functionality of the microtissues.

5.6 ACKNOWLEDGEMENTS

The authors disclose no conflicts of interest. The authors are thankful to the Ministry of Education (MOE), Malaysia for funding this project through Fundamental Research Grant Schemes (FRGS), Grant Vot No. K106.

REFERENCES

Achilli, T.M., J. Meyer and J.R. Morgan. 2012. Advances in the formation, use and understanding of multi-cellular spheroids. Expert Opinion on Biological Therapy 12: 1347-1360.

Ahn, S., H. Lee, L.J. Bonassar and G. Kim. 2012. Cells (MC3T3-E1)-laden alginate scaffolds fabricated by a modified solid-freeform fabrication process supplemented with an aerosol spraying. Biomacromolecules 13: 2997-3003.

Ahn, S. and G. Kim. 2015. Cell-encapsulating alginate microsized beads using an air-assisted atomization process to obtain a cell-laden hybrid scaffold. Journal of Materials Chemistry B 3: 9132-9139.

Aly, H.S.M., M.N.M. Jaafar and T.M. Lazim. 2013. Spray and Atomization Characterictics of Liquid. Malaysia: Penerbit UTM Press.

Antoni, D., H. Burckel, E. Josset and G. Noel. 2015. Three-dimensional cell culture: a breakthrough *in vivo*. International Journal of Molecular Sciences 16: 5517-5527.

Chan, E.S., B.B. Lee, P. Ravindra and D. Poncelet. 2009. Prediction models for shape and size of ca-alginate macrobeads produced through extrusion-dripping method. Journal of Colloid and Interface Science 338: 63-72.

Chicheportiche, D. and G. Reach. 1988. *In vitro* kinetics of insulin release by microencapsulated rat islets: effect of the size of the microcapsules. Diabetologia 31: 54-57.

Cui, J.H., J.S. Goh, S.Y. Park, P.H. Kim and B.J. Le. 2001. Preparation and physical characterization of alginate microparticles using air atomization method. Drug Development and Industrial Pharmacy 27: 309-319.

Daemi, H. and M. Barikani. 2012. Synthesis and characterization of calcium alginate nanoparticles, sodium homopolymannuronate salt and its calcium nanoparticles. Scientia Iranica 19: 2023-2028.

da Rocha, E.L., L.M. Porto and C.R. Rambo. 2014. Nanotechnology meets 3D *in vitro* models: tissue engineered tumors and cancer therapies. Materials Science and Engineering: C 34: 270-279.

de Vos, P., C.G. van Hoogmoed, J. van Zanten, S. Netter, J.H. Strubbe and H.J. Busscher. 2003. Long-term biocompatibility, chemistry, and function of microencapsulated pancreatic islets. Biomaterials 24: 305-312.

de Vos, P., H.A. Lazarjani, D. Poncelet and M.M. Faas. 2014. Polymers in cell encapsulation from an enveloped cell perspective. Advanced Drug Delivery Reviews 67-68: 15-34.

Dianawati, D., V. Mishra and N.P. Shah. 2012. Role of calcium alginate and mannitol in protecting Bifidobacterium. Applied and Environmental Microbiology 78: 6914-6921.

Edmondson, R., J.J. Broglie, A.F. Adcock and L. Yang. 2014. Three-dimensional cell culture systems and their applications in drug discovery and cell-based biosensors. Assay and Drug Development Technologies 12: 207-218.

Elger, D.F., B.C. Williams, C.T. Crowe and J.A. Roberson. 2014. Engineering Fluid Mechanics. Singapore: John Wiley & Sons, Singapore.

Gao, Q., Y. He, J.-z. Fu, J.-j. Qiu and Y.-a. Jin. 2015. Fabrication of shape controllable alginate microparticles based on drop-on-demand jetting. Journal of Sol-Gel Science and Technology 77: 610-619.

Gasperini, L., J.F. Mano and R.L. Reis. 2014. Natural polymers for the microencapsulation of cells. Journal of the Royal Society Interface 11: 20140817.

Gautier, A., B. Carpentier, M. Dufresne, Q. Vu Dinh, P. Paullier and C. Legallais. 2011. Impact of alginate type and bead diameter on mass transfers and the metabolic activities of encapsulated C3A cells in bioartificial liver applications. European Cells and Materials 21: 94-106.

Ghidoni, I., T. Chlapanidas, M. Bucco, F. Crovato, M. Marazzi, D. Vigo, M.L. Torre and M. Faustini. 2008. Alginate cell encapsulation: new advances in reproduction and cartilage regenerative medicine. Cytotechnology 58: 49-56.

Griffith, C.K., C. Miller, R.C. Sainson, J.W. Calvert, N.L. Jeon, C.C. Hughes and S.C. George. 2005. Diffusion limits of an *in vitro* thick prevascularized tissue. Tissue Engineering 11(1): 257-266.

Herran, C.L. and Y. Huang. 2012. Alginate microsphere fabrication using bipolar wave-based drop-on-demand jetting. Journal of Manufacturing Processes 14: 98-106.

Herrero, E.P., E.M. Mart´ın Del Valle and M.A. Galan. 2006. Development of a new technology for the production of microcapsules based in atomization processes. Chemical Engineering Journal 117: 137-142.

Hu, Y., Q. Wang, J. Wang, J. Zhu, H. Wang and Y. Yang. 2012. Shape controllable microgel particles prepared by microfluidic combining external ionic crosslinking. Biomicrofluidics 6: 26502-265029.

Huang, K.S., M.K., Liu, C.H. Wu, Y.T. Yen and Y.C. Lin. 2007. Calcium alginate microcapsule generation on a microfluidic system fabricated using the optical disk process. Journal of Micromechanics and Microengineering 17: 1428-1434.

Hunt, N.C. and L.M. Grover. 2010. Cell encapsulation using biopolymer gels for regenerative medicine. Biotechnology Letters 32: 733-742.

Kang, A., J. Park, J. Ju, G.S. Jeong and S.H. Lee. 2014. Cell encapsulation via microtechnologies. Biomaterials 35: 2651-2663.

Khademhosseini, A., G. Eng, J. Yeh, J. Fukuda, J. Blumling, 3rd, R. Langer and J.A. Burdick. 2006. Micromolding of photocrosslinkable hyaluronic acid for cell encapsulation and entrapment. Journal of Biomedical Materials Research Part A 79: 522-532.

Koh, W.G., A. Revzin and M.V. Pishko. 2002. Poly(ethylene glycol) hydrogel microstructures encapsulating living cells. Langmuir 18: 2459-2462.

Kunz-Schughart, L., J.P. Freyer, F. Hofstaedter and R. Ebner. 2004. The use of 3-D cultures for high throughput screening: the multicellular spheroid model. Journal of Biomolecular Screening 9: 273-285.

Lee, K.Y. and D.J. Mooney. 2012. Alginate: properties and biomedical applications. Progress in Polymer Science 37: 106-126.

Leong, W.Y., C.F. Soon, S.C. Wong, K.S. Tee, S.C. Cheong, S.H. Gan and M. Youseffi. 2017. *In vitro* growth of human keratinocytes and oral cancer cells into microtissues: an aerosol-based microencapsulation technique. Bioengineering 4: 1-14.

Lewinska, D., J. Bukowski, M. Kozuchowski, A. Kinasiewicz and A. Werynski. 2008. Electrostatic microencapsulation of living cells. Biocybernetics and Biomedical Engineering 28: 69-84.

Li, N., X.X. Xu, G.W. Sun, X. Guo, Y. Liu, S.J. Wang, Y. Zhang, W.T. Yu, W. Wang and X.J. Ma. 2013. The effect of electrostatic microencapsulation process on biological properties of tumour cells. Journal of Microencapsulation 30: 530-537.

Martin-Banderas. L., A.M. Ganan-Calvo and M. Fernandez-Arevalo. 2010. Making drops in microencapsulation processes. Letters in Drug Design & Discovery 7: 300-309.

Martinez, C.J., J.W. Kim, C. Ye, I. Ortiz, A.C. Rowat, M. Marquez and D. Weitz. 2012. A microfluidic approach to encapsulate living cells in uniform alginate hydrogel microparticles. Macromolecular Bioscience 12: 946-951.

Nagpal, M., S.K. Singh and D. Mishra. 2013. Synthesis characterization and *in vitro* drug release from acrylamide and sodium alginate based superporous hydrogel devices. International Journal of Pharmaceutical Investigation 3: 131-140.

Notingher, I., J.R. Jones, S. Verrier, I. Bisson, P. Embanga, P. Edwards, J.M. Polak and L.L. Hench. 2003. Application of FTIR and raman spectroscopy to characterisation of bioactive materials and living cells. Spectroscopy 17: 275-288.

Paredes Juarez, G.A., M. Spasojevic, M.M. Faas and P. de Vos. 2014. Immunological and technical considerations in application of alginate-based microencapsulation systems. Frontiers in Bioengineering and Biotechnology 2: 26.

Prüsse, U., L. Bilancetti, M. Bučko, B. Bugarski, J. Bukowski, P. Gemeiner, D. Lewińska, V. Manojlovic, B. Massart, C. Nastruzzi, V. Nedovic, D. Poncelet, S. Siebenhaar, L. Tobler, A. Tosi, A. Vikartovská and K.D. Vorlop 2008. Comparison of different technologies for alginate beads production. Chemical Papers 62: 364-374.

Rabanel, J.M., X. Banquy, H. Zouaoui, M. Mokhtar and P. Hildgen. 2009. Progress technology in microencapsulation methods for cell therapy. Biotechnology Progress 25: 946-963.

Ramsey, W.S., W. Hertl, E.D. Nowlan and N.J. Binkowski. 1984. Surface treatments and cell attachment. *In Vitro* 20: 802-808.

Schwinger, C., S. Koch, U. Jahnz, P. Wittlich, N.G. Rainov and J. Kressler. 2002. High throughput encapsulation of murine fibroblasts in alginate using the JetCutter technology. Journal of Microencapsulation 19: 273-280.

Shin, C.S., B. Kwak, B. Han and K. Park. 2013. Development of an *in vitro* 3D tumor model to study therapeutic efficiency of an anticancer drug. Molecular Pharmacology 10: 2167-2175.

Sohail, A., M.S. Turner, A. Coombes, T. Bostrom and B. Bhandari. 2011. Survivability of probiotics encapsulated in alginate gel microbeads using a novel impinging aerosols method. International Journal of Food Microbiology 145: 162-168.

Soon, C.F., K.T. Thong, K.S. Tee, A.B. Ismail, M. Denyer, M.K. Ahmad, Y.H. Kong, P. Vyomesh and S.C. Cheong. 2016. A scaffoldless technique for self-generation of three-dimensional keratinospheroids on liquid crystal surfaces. Biotechnic & Histochemistry 91(4): 1-13.

Soon, C.F., K.S. Tee, S.C. Wong, N. Nayan, S. Sundra, M.K. Ahmad, F. Sefat, N. Sultana and M. Youseffi. 2018. Comparison of biophysical properties characterized for microtissues cultured using microencapsulation and liquid crystal based 3D cell culture techniques. Cytotechnology 70: 13-29.

Souza, G.R., J.R. Molina, R.M. Raphael, M.G. Ozawa, D.J. Stark, C.S. Levin, L.F. Bronk, J.S. Ananta, J. Mandelin, M.M. Georgescu, J.A. Bankson, J.G. Gelovani, T.C. Killian, W. Arap and R. Pasqualini 2010. Three-dimensional tissue culture based on magnetic cell levitation. Nat Nanotechnol 5: 291-296.

Stevens, M.M., H.F. Qanadilo, R. Langer and V. Prasad Shastri. 2004. A rapid-curing alginate gel system: utility in periosteum-derived cartilage tissue engineering. Biomaterials 25: 887-894.

Sugiura, S., T. Oda, Y. Izumida, Y. Aoyagi, M. Satake, A. Ochiai, N. Ohkohchi and M. Nakajima. 2005. Size control of calcium alginate beads containing living cells using micro-nozzle array. Biomaterials 26: 3327-3331.

Sugiura S., T. Oda, Y. Aoyagi, R. Matsuo, T. Enomoto, K. Matsumoto, T. Nakamura, M. Satake, A. Ochiai, N. Ohkohchi and M. Nakajima 2007. Microfabricated airflow nozzle for microencapsulation of living cells into 150 micrometer microcapsules. Biomed Microdevices 9: 91-99.

Sun, A.M. 1997. Microencapsulation of cells: medical applications. The Annals of the New York Academy of Sciences 831: 271-279.

Swioklo, S., P. Ding, A.W. Pacek and C.J. Connon. 2016. Process parameters for the high-scale production of alginate-encapsulated stem cells for storage and distribution throughout the cell therapy supply chain. Process Biochemistry 59(B): 289-296.

Tendulkar, S., S.H. Mirmalek-Sani, C. Childers, J. Saul, E.C. Opara and M.K. Ramasubramanian. 2012. A three-dimensional microfluidic approach to scaling up microencapsulation of cells. Biomed Microdevices 14: 461-469.

Vaithilingam, V. and B.E. Tuch. 2011. Islet transplantation and encapsulation: an update on recent developments. The Review of Diabetic Studies 8: 51-67.

Wan, J. 2012. Microfluidic-based synthesis of hydrogel particles for cell microencapsulation and cell-based drug delivery. Polymers 4(2): 1084-1108.

Watson, B. 2010. Mobile Equipment Hydraulics: A Systems and Troubleshooting Approach. USA: Delmar Cengage Learning.

Whelehan, M. and I.W. Marison. 2011. Microencapsulation using vibrating technology. Journal of Microencapsulation 28: 669-688.

Wilson, J.L. and T.C. McDevitt. 2013. Stem cell microencapsulation for phenotypic control, bioprocessing, and transplantation. Biotechnology and Bioengineering 110: 667-682.

Wong, S.C., C.F. Soon, W.Y. Leong and K.S. Tee. 2016. Flicking technique for microencapsulation of cells in calcium alginate leading to the microtissue formation. Journal of Microencapsulation 33: 162-171.

Zhang, W. and X. He. 2009. Encapsulation of living cells in small (approximately 100 microm) alginate microcapsules by electrostatic spraying: a parametric study. Journal of Biomechanical Engineering 131: 074515.

Zhang, W., C. Li, B.C. Baguley, F. Zhou, W. Zhou, J.P. Shaw, Z. Wang, Z. Wu and J. Liu. 2016. Optimization of the formation of embedded multicellular spheroids of MCF-7 cells: how to reliably produce a biomimetic 3D model. Analytical Biochemistry 515: 47-54.

Zhao, G., X. Liu, K. Zhu and X. He. 2017. Hydrogel encapsulation facilitates rapid-cooling cryopreservation of stem cell-laden core-shell microcapsules as cell-biomaterial constructs. Advanced Healthcare Materials 6(23): 1-26.

Zhao, J. and L. Yang. 2012. Simulation and experimental study on the atomization character of the pressure-swirl nozzle. Open Journal of Fluid Dynamics 2: 271-277.

Zhu, B. and H. Yin. 2015. Alginate lyase: review of major sources and classification, properties, structure-function analysis and applications. Bioengineered 6(3): 125-131.

6

BioMEMS Devices for Tissue Engineering

Tao Sun[1] and Chao Liu[2,*]

6.1 INTRODUCTION

Over the past few decades, bio microelectromechanical systems (BioMEMS) has been a fast-growing discipline, and has emerged as a powerful tool in tissue engineering for the study of regulating cell behaviors with specific microstructures, modulating drug delivery profile, monitoring cellular activities and engineered tissue functions, and promoting tissue restoration and regeneration. The purpose of this chapter is to summarize the recent progress in design and integration of bioMEMS devices for tissue engineering applications. In the first section of this chapter, a detailed overview is given of regulating cellular behaviors via bioMEMS technologies to pattern surface micro-structure. In the following section, organs-on-chips to mimic organs' functions are introduced to demonstrate wide applications of bioMEMS technologies in tissue engineering. Subsequently, micro-electronic devices integrated with conventional scaffolds are presented to illustrate the next-generation, active and smart scaffolds for tissue restoration and regeneration. Future development of bioMEMS in tissue engineering is discussed in the final section.

6.2 REGULATING CELL BEHAVIORS VIA BioMEMS DEVICES

BioMEMS technology is able to create a topography with a feature size of micrometer or even nanometer scale on silicon-based materials, polymers and metals. By precisely regulating the pattern and micro-texture on the surface of substrates to tune the physical and chemical properties of the extracellular matrix (ECM), cell behaviors such as cell adhesion, cell orientation, cell proliferation and differentiation were reported to be controlled. For example, micro-gratings with nanoscale features were fabricated on polystyrene plates via nanoimprint lithography (Fig. 6.1a) (Crouch et al. 2009). The orientation of human dermal fibroblasts was investigated by changing the aspect ratio (depth/width) of gratings. Both cell alignment and elongation increase with increasing aspect ratio, and even a shallow grating with the aspect ratio of ~0.05 is sufficient to induce 85% cell alignment (Fig. 6.1b). As cytoskeleton has limited flexibility and cells are unable to frequently traverse significant steps, contact guidance was thought to be the mechanism for cell polarization and alignment in the direction without obstacle.

[1] Media Lab, Massachusetts Institute of Technology, Cambridge, 02139, United States.
[2] School of Electronics Science, Northeast Petroleum University, Daqing 163318, China.
[*] Corresponding author: msm-liu@126.com

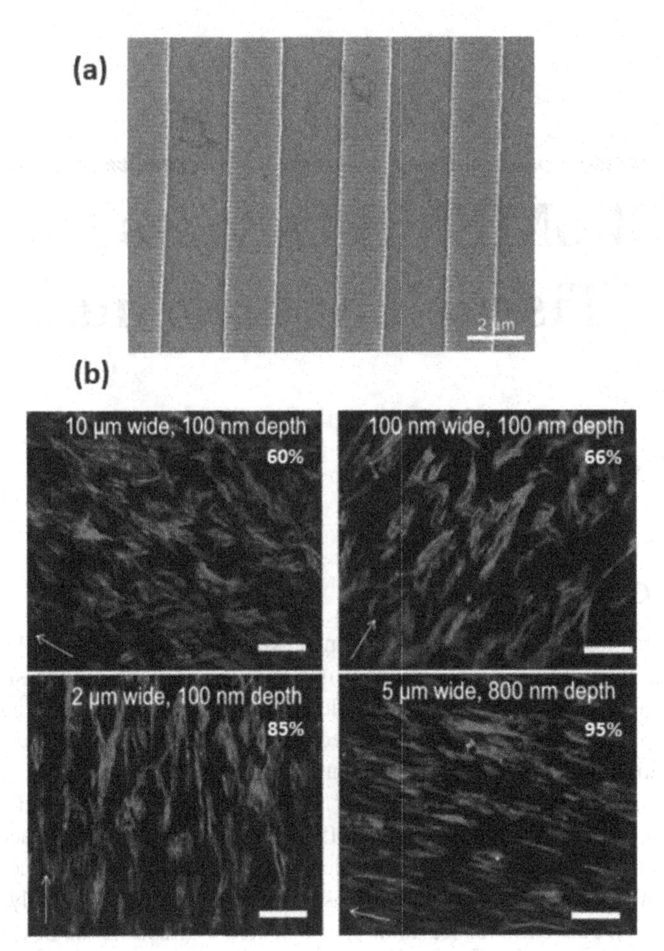

FIGURE 6.1 (a) SEM image of 2 µm lines embedded with 100 nm lines by double imprint at 95°C and 50 bar, (b) cells on lines of 10 µm wide, 100 nm depth; 100 nm wide, 100 nm depth; 2 µm wide, 100 nm depth; and 5 µm wide, 800 nm depth. Arrows indicate the grating direction. Cell alignment percentages are shown inside the images. The scale bar is 30 µm. (Reprinted with permission from (Crouch et al. 2009)).

Although controversial results were reported for cell proliferation, most of studies found that micro-textured surface led to a decrease in proliferation (Ni et al. 2009). The patterned surfaces also influence cell differentiation. Osteogenic responses of human mesenchymal stromal cells (hMSCs) were compared on square-patterned, inverse square-patterned, and planar titanium, chromium, diamond-like carbon (DLC), and tantalum (Kaivosoja et al. 2013). Early-marker alkaline phosphatase (ALP) reached highest values on both patterned titanium samples. Presence of hydroxyapatite showed that both types of patterning promoted ($p<0.001$) osteogenesis compared to planar samples. Therefore, micro-patterned interface is believed to provide physical differentiation cues to enhance stem cell differentiation.

6.3 MIMICKING ORGAN FUNCTIONS ON BioMEMS CHIPS

BioMEMS devices have been proposed to mimic the mechanical and biochemical microenvironment of tissues and organs for drug discovery and toxicity screening. Huh et al. described a biomimetic microsystem that reconstitutes the critical functional alveolar-capillary interface of the human lung (Huh et al. 2010). The micro-device was developed by micro-fabricating a microfluidic system containing two closely apposed micro-channels separated by thin (10 µm), porous, flexible membrane made of poly (dimethylsiloxane) (PDMS) (Fig. 6.2). Human alveolar epithelial cells and human pulmonary micro-vascular endothelial cells were cultured on opposite sides of the intervening membrane coated with ECM. After the cells were grown to confluence, air was pumped into the epithelial compartment to create an air-liquid interface and more precisely mimic the lining of the alveolar air space. The artificial lung micro-device prepared by soft lithography is able to reproduce complex integrated organ-level responses to bacteria and inflammatory cytokines introduced into the alveolar space.

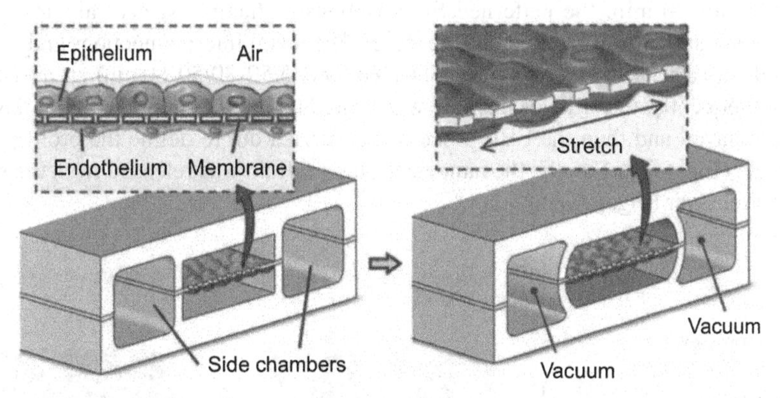

FIGURE 6.2 Schematic diagram of the micro-fabricated lung mimic device. (Reprinted with permission from (Huh et al. 2010)).

To investigate into basic mechanisms of organ physiology and disease, researchers have fabricated chips for the study of the liver, kidney, intestine, lung, heart, muscle, fat, bone, marrow, cornea, skin, blood vessels, nerve, and blood-brain barrier over the last decade (Bhatia and Ingber 2014). Compared to conventional 3D cell culture system, it is straightforward to integrate organ-on-chip with characterization methodology such as fluorescence confocal microscopy, micro-fluorimetry, multiple electrode arrays, etc. Accordingly, different types of cells can be precisely identified and characterized from others. Moreover, it is convenient to control fluid flow in organs-on-chips to regulate cell differentiation, function and longevity. Hence, organs-on-chips are expected to simulate the microenvironment of tissues and organs, and bring a deep insight into the design of scaffolds for better tissue regeneration. Although organs-on-chips still have not been widely accepted and recognized for drug assay, one should not ignore their amazing potential to replace animal testing someday.

6.4 DEVELOPING ELECTRONIC SCAFFOLDS FOR TISSUE ENGINEERING

Although both BioMEMS and tissue engineering are rapidly-growing fields, the research merging scaffolds with MEMS sensors or actuators for tissue engineering is still at the early stage, in part because traditional two-dimensional (2D) photolithography technology in MEMS is not compatible with the 3D fabrication process of porous scaffolds. Moreover, almost all materials widely used in the MEMS process are non-biodegradable, including metals (gold, platinum, titanium, etc), Si-based materials (Si, SiO_2, Si_3N_4) and polymers (polyimide, parylene, SU8), which are not suitable for the fabrication of biodegradable porous scaffold in tissue engineering. Therefore, the seamless integration of MEMS technology and tissue engineering remains a critical challenge in fabricating new-generation electronic scaffolds embedded with MEMS devices. The novel electronic scaffolds are expected to not only provide a mechanical support for cells to regenerate tissues, but also monitor cellular behaviors and control the drug release to surrounding tissues. Tian et al. designed a 3D scaffold containing flexible Si nanowire transistors to monitor electrical activities of cells cultured within the scaffold (Tian et al. 2012). The micro-fabrication of the nanowire nanoelectronic scaffold is based on a MEMS process and is shown in Fig. 6.3. Briefly, electron beam lithography (EBL) was applied to pattern a double layer resist, on which 100 nm nickel metal was deposited as the final relief layer for the nanoelectronic scaffolds (Fig. 6.3a and b). A 300-500 nm layer of SU-8 photoresist was then coated over the entire chip (Fig. 6.3c), followed by dropping an isopropanol solution containing the n^+-n-n^+ kinked Si nanowires on the SU-8 layer (Fig. 6.3d). Afterwards, EBL was used to pattern the overall SU-8 scaffold structure including a ring structure underneath the selected kink nanowire. Then the SU-8 developer was used to develop the pattern, and after curing at 180°C for 20 min, the patterned SU-8 ribbons as flexible structural support for metal interconnects and nanowires were created (Fig. 6.3e). The metal interconnections and bonding pads were produced with a stacked layer structure of Cr/Pd/Cr (1.5/50-80/50-80 nm) via a lift-off process (Fig. 6.3f). Subsequently, the silicon substrate was coated by a uniform 300-400 nm layer of SU-8 as a passivation layer, and then the EBL process was carried out to define the profile of the SU-8 passivation layer (Fig. 6.3g). Finally, the nanoelectronic scaffold was released from the substrate by etching the nickel layer (Fig. 6.3h).

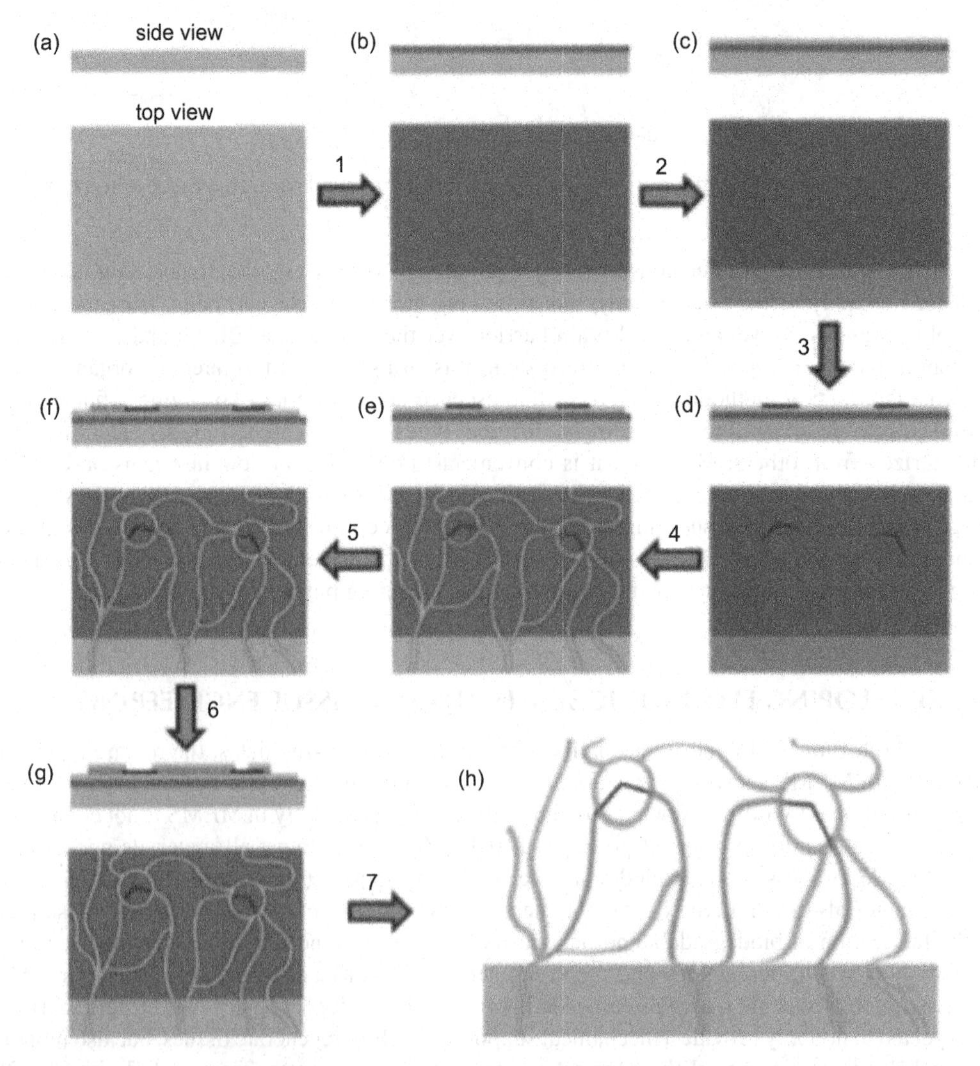

FIGURE 6.3 Schematic diagram of the nanoelectronic scaffold fabrication process. Materials involved in the fabrication process include silicon (cyan), nickel relief layer (blue), polymer ribbons (green), metal interconnects (gold) and silicon nanowires (black). (Reprinted with permission from (Tian et al. 2012)).

To monitor cellular behaviors or tissue functions, the nanoelectronic scaffold can be integrated with conventional macroporous biomaterials. Fig. 6.4a displays a confocal fluorescence micrograph of a nanoelectronic scaffold/collagen, indicating that collagen fibers (green) are fully entangled with the electronic scaffold (yellow). Scanning electron microscopy images (Fig. 6.4b) of the hybrid nanoelectronic scaffold/alginate prepared by lyophilization shows that the nanoelectronic scaffold intimately contacts with alginate framework. Similarly, the nanoelectro scaffold can be integrated with PLGA fibers on both sides via an electrospinning process (Fig. 6.4c). Moreover, the stability of the nanoelectronic scaffold was characterized in neural culture media over a nine-week period (Fig. 6.4d). The sensitivity variation (ΔS/S) was less than ± 11%.

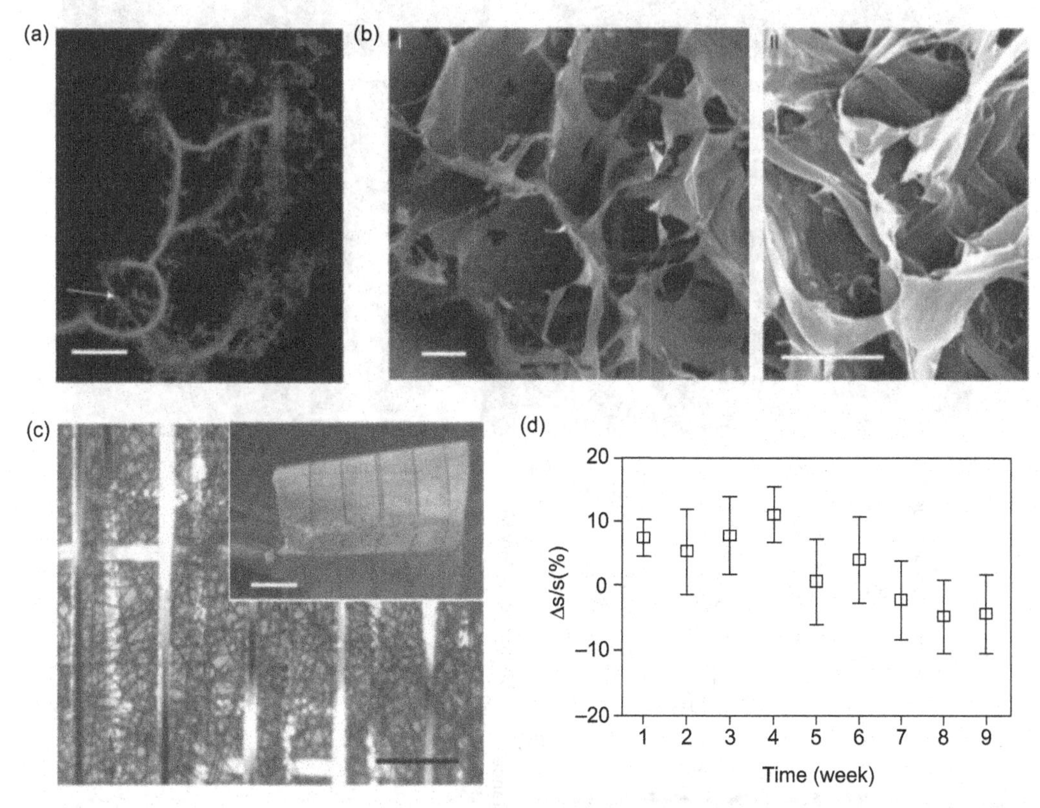

FIGURE 6.4 The integration of the nanoelectronic scaffolds with conventional macroporous biomaterials. (a) confocal fluorescence micrograph of nanoelectronic scaffold/collagen matrix, scale bar: 10 μm, (b) SEM images of nanoelectronic scaffold/alginate matrix, top (I) and side (II) views, scale bar: 200 μm (I) and 100 μm (II), (c) optical image of nanoelectronic/PLGA matrix, scale bar: 200 μm and 5 μm (insert), (d) Relative sensitivity variation for the nanoelectronic scaffold over time in culture. (Reprinted with permission from (Tian et al. 2012)).

The function of the nanoelectronic scaffold was evaluated *in vitro* by culturing several types of cells. After two weeks of culture into the hybrid nanoelectronic scaffold/matrigel, high-density neurons (red) with spatially interconnected neurites were present in Fig. 6.5a and b, and the neurits penetrated the scaffold and often passed through the ring structures. Similarly, confocal fluorescence microscopy of a cardiac 3D culture indicated that high density cardiomyocytes intimately attached on the surface of the nanoelectronic scaffold (Fig. 6.5c), and epifluorescence micrographys of cultured cardiac cells showed striations characteristic of cardiac tissue (Fig. 6.5d). Live/dead cytotoxicity assessment and metabolic activity assay were performed for neurons and cardiac cells cultured on the nanoelectronic scaffolds, respectively (Fig. 6.5e and Fig. 6.5f, respectively), revealing that the nanoelectronic scaffold had little effect on the cell viability and

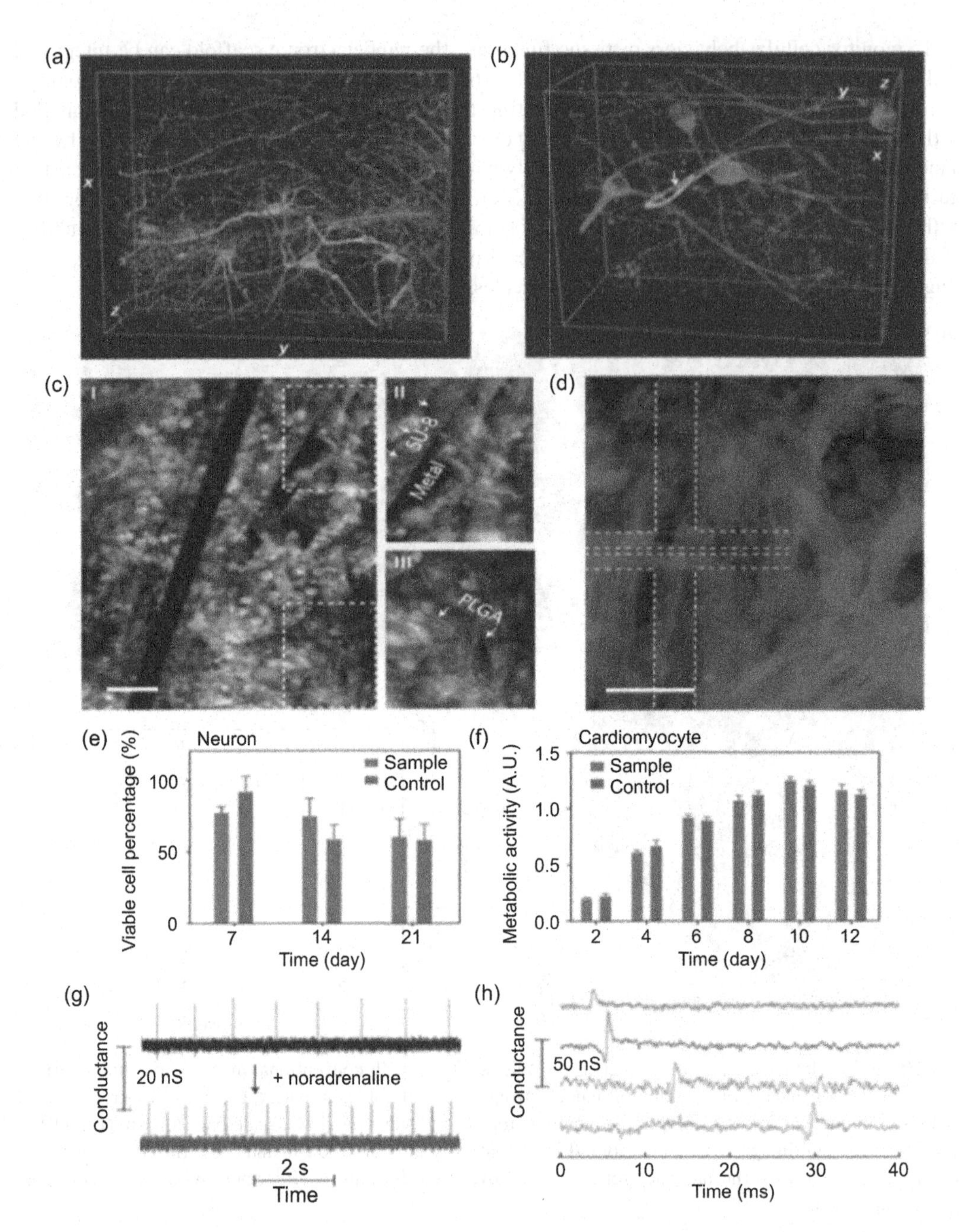

FIGURE 6.5 3D cell culture and electrical sensing from the nanoelectronic scaffold. (a) and (b) confocal images of rat hippocampal neurons after a two-week culture in nanoelectronic scaffold/matrigel matrix, (c) confocal fluorescence micrographs of a cardiac 3D culture on nanoelectronic scaffold, scale bar: 40 μm, (d) epifluorescence micrograph of the surface of the cardiac patch in Fig. 3c, scale bar: 40 μm, (e) cell viability of neurons cultured on nanoelectronic scaffolds and control samples, (f) metabolic activity assay for cardiomyocytes cultured on nanoelectronic scaffolds and control samples, (g) Electrical output from the nanoelectronic scaffold before (black) and after (blue) applying noradrenaline, (h) simultaneous electrical recording from different channels of the nanoelectronic scaffold. (Reprinted with permission from (Tian et al. 2012)).

Color version at the end of the book

thus can be utilized for a number of *in vitro* studies, including drug screening assays for synthetic neural and cardiac tissues. The monitoring capability of the nanoelectronic scaffold was verified by recording cellular electrical activities. As shown in Fig. 6.5g, the electrical output from the nanoelectronic scaffold with cardiomyocytes showed regularly spaced spikes with a frequency of ~1 Hz, a calibrated potential change of ~3 mV, and a ~2 ms width, which were consistent with extracellular recordings from cardiomyocytes. Moreover, a twofold increase in the contraction frequency was detected by the nanoelectronic scaffold, after a drug application of noradrenaline which stimulates cardiac contraction through β_1-adrenergic receptors. In addition, simultaneous recordings from four different channels of the nanoelectronic scaffold was demonstrated in Fig. 6.5h, with sub millisecond time resolution, indicating that the nanoelectronic scaffold was able to map cardiac and other synthetic tissue electrical activities over the entire constructs at high density in three dimensions.

In contrast to the nondegradable electronic scaffolds developed by Tian et al., Feiner et al. developed a biodegradable, functional, cardiac scaffold composed of electrospun albumin fibers and evaporated gold electrodes (Tian et al. 2012, Feiner et al. 2018). In this approach (Fig. 6.6), bovine serum albumin fibers were fist fabricated to form the bottom layer via electrospinning process, and then gold electrodes, interconnection and bonding pads were directly deposited on the bottom electronspun layer covered by a plastic sheet shadow mask. Afterwards, an ECM-based hydrogel was used as an adhesive, and another electron spun layer was placed on top of the patterned bottom layer as a passivation layer. Subsequently, a layer of polypyrrole loaded with dexamethasone was selectively deposited onto the specific electrodes. Additionally, neonatal rat ventricular cardiomyocytes were seeded onto the device. The seeded cardiomyocytes into the electronic

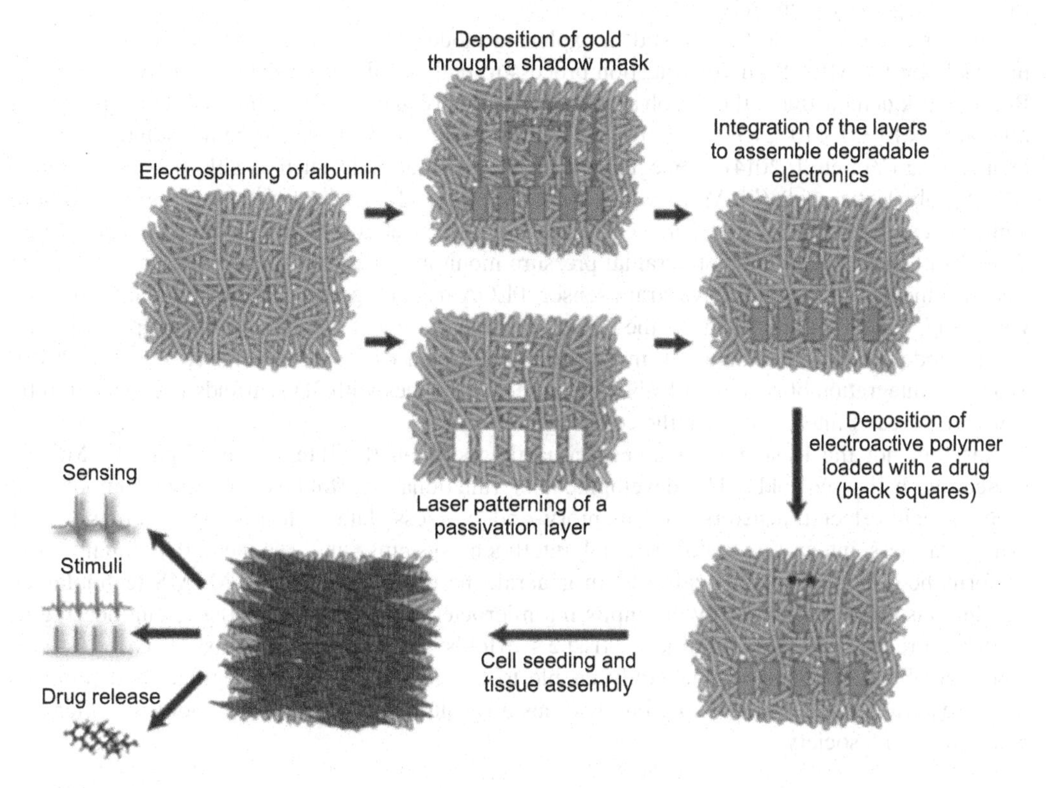

FIGURE 6.6 Schematic diagram of the fabrication and functions of the electronic scaffold developed by Feiner et al. (Reprinted with permission from (Feiner et al. 2018)).

scaffolds were able to organize into a confluent cardiac tissue. By contacting the electronic scaffold to external devices, extracellular signals can be recorded from the engineered tissue and stimuli may be delivered to the scaffold for pacing and drug release. After implantation, the inorganic materials within the scaffold were dissocated with the degradation of electronic scaffolds, indicating that the electronic scaffold can be used as transient electronic implants. Integrating flexible micro-fabricated electronic components or devices within 3D scaffolds has advanced the field of tissue engineering due to continuous monitoring and actuation of engineered tissues in real time, bringing us one step closer to creating fully functional engineered tissues.

6.5 CONCLUSIONS AND FUTURE OUTLOOK

Although both BioMEMS and tissue engineering are rapidly-growing fields, the research involving microfabricated electronic devices is still at a rudimentary stage, and is rarely reported for engineering tissues in three dimensional structures. Several fundamental and technical challenges need to be addressed to make further breakthrough in the device fabrication for tissue engineering. For instance, one of the primary challenges with BioMEMS based electronic devices is the intrinsically planar nature of the lithography technique used in the microfabrication process. Ning et al. proposed a micro-fabrication and packaging process to transform 2D piezoelectric devices to a 3D mesostructure (Ning et al. 2018). The 3D microscale mechanical frameworks containing multiple, independently addressable piezoelectric thin-film actuators have wide range applications from biosensing, vibratory excitation, precise control, mechanobiology, and energy harvesting. Fig. 6.7 illustrates various 3D architectures integrated with piezoelectric micro-actuators and their finite element analysis models.

Further attempts could be engaged in exploring biodegradable conductive and nonconductive materials for the MEMS microfabrication process to realize fully degradable 3D MEMS devices. Recent work demonstrated the dissolvability of Mg, Zn, W and Mo thin film, and their dissolution rates are 7×10^{-2}, 7×10^{-3}, $1.7\text{~}0.3 \times 10^{-3}$ and 3×10^{-4} $\mu m \cdot h^{-1}$, respectively, in Hank's solution at room temperature (Yin et al. 2014). Biodegradable polymeric materials, such as silk, polycaprolactone (PCL), poly(lactic acid) (PLA), poly(lactic-co-glycolic acid) (PLGA), and bio-organics (sodium alginate) can serve as a water barrier or encapsulation layer for electrical interconnects (Kang et al. 2018). A biodegradable intracranial pressure monitoring device was designed to incorporate a Si nanomembrane piezoresistive strain sensor, PLGA insulating layer, Mg interconnects and Mo wires (Fig. 6.8a). Figure 6.8b shows the intracranial measurement of temperature and pressure in a rodent model, and wireless data transmission unit connected with the Mo wires (Kang et al. 2016). Seamless integration of such dissolvable microelectric devices with 3D scaffolds is expected to be a growing area of interest to scientific community.

Last but not the least, most studies primarily focus on the integration of passive MEMS devices with the scaffolds. The development of functional scaffolds with active components, such as field-effect transistors, and memory and wireless data transmission modules, could bring seamless integration with artificial intelligent systems, and exponentially enhance the performance of conventional scaffolds. In general, the recent advances in MEMS technologies has demonstrated the feasibility of employing microelectronic devices for tissue engineering to render unprecedented functionalities to tissue scaffolds. The functional 3D tissue scaffolds with biodegradable MEMS devices not only enable to continuously monitor, regulate and promote tissue restoration and regeneration, but also has a revolutionary influence on both the scientific community and society.

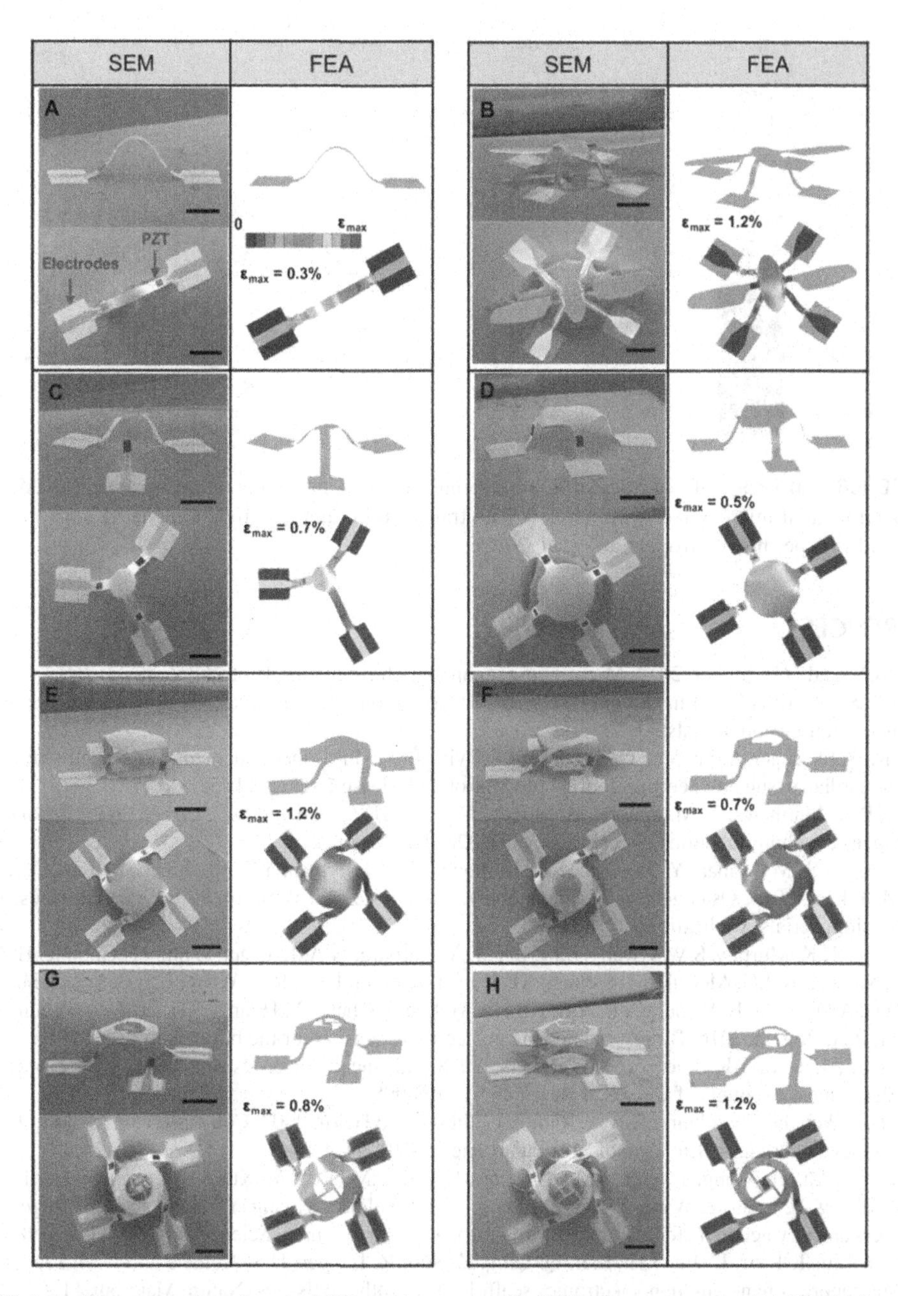

FIGURE 6.7 Demonstration of typical 3D architectures with PZT micro-actuators. (A) "Bridge" structure with two PZT micro-actuators. (B) "Fly" structure with a pair of actuators on the wings. (C) "Tilted pyramid truss" structure with three actuators. (D) "Four-leg table" structure with an actuator on each leg. (E) "Rotated table" structure with an actuator on each leg. (F) "Rotated table" with a central hole on top and four actuators. (G) "Double-floor rotated table" structure that consists of a large rotated table and a small one on the top, with four actuators. (H) "Double-floor rotated table" structure with five actuators (the additional one on top). Each panel includes a side and a top view. The yellow and blue regions correspond to the electrodes and PZT microactuators, respectively. Scale bars, 500 μm. The contour plots show results of FEA modeling for the maximum principal strain in the electrodes and PZT microactuators (Ning et al. 2018). (Reprinted with permission from (Ning et al. 2018)).

Color version at the end of the book

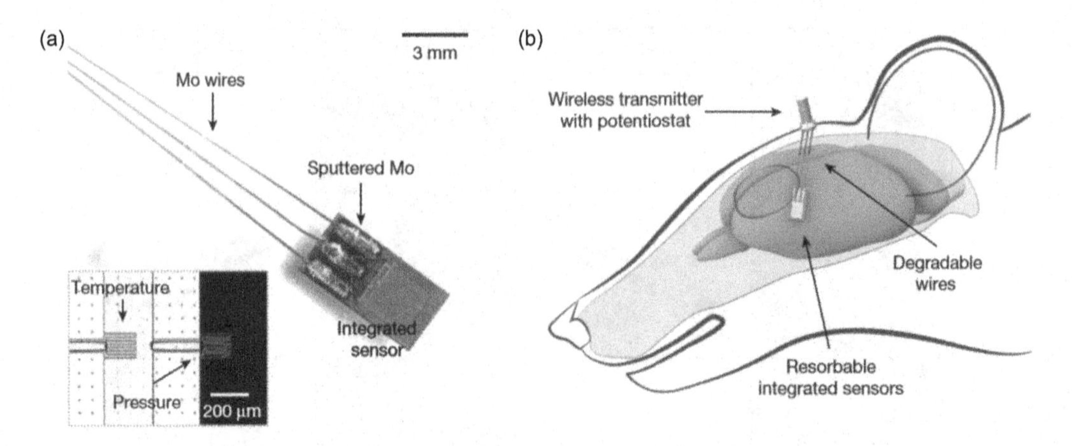

FIGURE 6.8 (a) Image of biodegradable intracranial pressure and temperature sensors with Mo wires, (b) the intracranial microsystem with wireless data transmission unit in a rodent model (Kang et al. 2016). (Reprinted with permission from (Kang et al. 2016)).

REFERENCES

Bhatia, S.N. and D.E. Ingber. 2014. Microfluidic organs-on-chips. Nature Biotechnology 32: 760.

Crouch, A.S., D. Miller, K.J. Luebke and W. Hu. 2009. Correlation of anisotropic cell behaviors with topographic aspect ratio. Biomaterials 30: 1560-1567.

Feiner, R., S. Fleischer, A. Shapira, O. Kalish and T. Dvir. 2018. Multifunctional degradable electronic scaffolds for cardiac tissue engineering. Journal of Controlled Release 281: 189-195.

Huh, D., B.D. Matthews, A. Mammoto, M. Montoya-Zavala, H.Y. Hsin and D.E. Ingber. 2010. Reconstituting organ-level lung functions on a chip. Science 328: 1662-1668.

Kaivosoja, E., S. Myllymaa, Y. Takakubo, H. Korhonen, K. Myllymaa, Y.T. Konttinen, R. Lappalainen and M. Takagi. 2013. Osteogenesis of human mesenchymal stem cells on micro-patterned surfaces. Journal of Biomaterials Applications 27: 862-871.

Kang, S.-K., R.K. Murphy, S.W. Hwang, S.M. Lee, D.V. Harburg, N.A. Krueger, J. Shin, P. Gamble, H. Cheng, S. Yu, Z. Liu, J.G. McCall, M. Stephen, H. Ying, J. Kim, G. Park, R.C. Webb, C.H. Lee, S. Chung, D.S. Wie, A.D. Gujar, B. Vemulapalli, A.H. Kim, K.M. Lee, J. Cheng, Y. Huang, S.H. Lee, P.V. Braun, W. Ray and J.A. Rogers. 2016. Bioresorbable silicon electronic sensors for the brain. Nature 530: 71.

Kang, S.-K., J. Koo, Y.K. Lee and J.A. Rogers. 2018. Advanced materials and devices for bioresorbable Electronics. Accounts of Chemical Research 51: 988-998.

Ni, M., H.T. Wen, D. Choudhury, N.A.A. Rahim, C. lliescu and H. Yu. 2009. Cell culture on MEMS platforms: a review. International Journal of Molecular Sciences 10: 5411-5441.

Ning, X., X. Yu, H. Wang, R. Sun, R.E. Corman, H. Li, C.M. Lee, Y. Xue, A. Chempakasseril, Y. Yao, Z. Zhang, H. Luan, Z. Wang, W. Xia, X. Feng, R.H. Ewoldt, Y. Huang, Y. Zhang and J.A. Rogers. 2018. Mechanically active materials in three-dimensional mesostructures. Science Advances 14: eaat8313.

Tian, B. J. Liu, T. Dvir, L. Jin, J.H. Tsui, Q. Qing, Z. Suo, R. Langer, D.S. Kohane and C.M. Lieber. 2012. Macroporous nanowire nanoelectronics scaffolds for synthetic tissues. Nature Materials 11: 986-994.

Yin, L., H. Cheng, S. Mao, R. Haasch, Y. Liu, X. Xie, S. Hwang, H. Jain, S. Kang, Y. Su, R. Li, Y. Huang and J.A. Rogers. 2014. Dissolvable metals for transient electronics. Advanced Functional Materials 24: 645-658.

7

Injectable Scaffolds for Bone Tissue Repair and Augmentation

Subrata Bandhu Ghosh[1], Kapender Phogat[1,2] and Sanchita Bandyopadhyay-Ghosh[1*]

7.1 INTRODUCTION

Every year, around millions of people suffer from bone fractures or other orthopedic-related injuries resulting from a variety of surgical, degenerative, and traumatic causes (Migliaresi et al. 2007, Lee et al. 2009, Bai et al. 2018). These result in increasing medical cost of treating bone related trauma, infection, and defects (Lee et al. 2009, Vo et al. 2012, Bai et al. 2018). Current clinical treatments for bone injuries such as autografts and allografts suffer from the potential risks of disease transmission, infection, and host rejection. Besides, there are additional problems related to insufficient bone cell availability and donor site morbidity (Lee et al. 2009, Bai et al. 2018, Gómez-Barrena et al. 2015, Kumawat et al. 2019). In this regard, synthetic bone scaffolds are gaining increased attention for repairing and regenerating defective bone tissues (Rose and Oreffo 2002, Winkler et al. 2018). These prefabricated scaffolds (preformed) aim to provide structural support and create appropriate environment for cell adhesion, migration, proliferation, and differentiation, while recapitulating the functional activity of the bone defects (Simeonov et al. 2016, Ghassemi et al. 2018, Bai et al. 2018). The various types of scaffold materials commonly used for bone scaffolding are based on metallic, polymeric, ceramic, and composite materials (Bandyopadhyay-Ghosh et al. 2008). Metallic materials although, have excellent mechanical properties, they usually cause stress-shielding due to their non-similar modulus of elasticity to that of bone (Ryan et al. 2006, Alvarez and Nakajima 2009). Besides, most of the metallic scaffolds are prone to corrosion, possess poor biological recognition on the material surface and are often known to release toxic ions (Ryan et al. 2006, Alvarez and Nakajima 2009, Navarro et al. 2008). Polymeric scaffolds on the other hand are easy to shape and light in weight, with the possibility to achieve tailored properties. However, they have insufficient mechanical properties and are not bioactive (Navarro et al. 2008, Bai et al. 2018). Bioceramic scaffold materials have the advantage that they can mimic the inorganic bone composition, and are usually bioactive (Bandyopadhyay-Ghosh 2008, Bandyopadhyay-Ghosh et al. 2008, Gerhardt and Boccaccini 2010). However, due to the difficulty in casting complex shapes and their poor fracture toughness, use of bioceramics as scaffold materials is restricted. Composite scaffold materials, in this regard, can play a vital role. Development of an interconnected polymer-inorganic (ceramics,

[1] Department of Mechanical Engineering, School of Automobile, Mechanical & Mechatronics Engineering, Manipal University Jaipur, Dehmi Kalan, Jaipur, Rajasthan, India-303007.

[2] Department of Mechanical Engineering, Faculty of Engineering, JECRC University, Ramchandrapura Industrial Area, Jaipur, Rajasthan, India-303905.

[*] Corresponding author: sanchitab.ghosh@jaipur.manipal.edu

glass, glass-ceramic) composite scaffold takes advantages of both polymer and inorganics to meet mechanical and physiological requirements of the host tissue (Fernandez-Yague et al. 2015). The new generation composite scaffolds contain bioactive components, which are released from the implant providing a supply of calcium and phosphate ions or other biologically active moieties, thereby accelerating the healing process. From a biological perspective, it also makes sense to combine biopolymers and inorganics to fabricate composite scaffolds for bone tissue engineering because native bone is the combination of a naturally occurring polymer and biological apatite (Bendtsen and Wei 2015).

The use of preformed scaffolds for replacing or treating the defective bone tissues although, have been on the rise, because of their obvious benefits described above, they always need invasive surgery for implantations, which may lead to medical complications associated with harvested tissue (Wu et al. 2018, Ghassemi et al. 2018, Kumawat et al. 2019). These kinds of surgical procedures are often painful and may lead to excessive bleeding, infection, nerve damage, and increased healing time. Other disadvantage of preformed scaffolds stems from their predefined shapes which may not match the broken area, with the possibility of insufficiently filling the defect area of the bone, resulting in poor integration with surrounding bone tissues (Wu et al. 2014, Bai et al. 2018). There is an urgent need therefore, to minimize or even avoid the invasive surgical procedures for such orthopedic treatments. In this regard, use of noninvasive, injectable scaffolds can provide promising alternatives, which can overcome the aforementioned limitations (Phogat and Bandyopadhyay-Ghosh 2018, Wu et al. 2018).

Injectable scaffolds for bone tissue regeneration are particularly interesting, not only because the use of these materials can potentially avoid the surgical interventions to implant the scaffolds at defective/fractured areas, but they can also reassemble within a short interval and can fill any irregularly shaped defects once inside the body (Hou et al. 2004, Jin et al. 2009, Naahidi et al. 2017). An injectable system also takes the advantage of a more homogeneous distribution of bioactive molecules within the matrix, that can readily be obtained by virtue of the scaffold components being in suspension or solution before solidification *in vivo*. The promising injectable materials that can be used as bone scaffolds are primarily based on hydrogel (Temenoff and Mikos 2000, Hou et al. 2004, Nourmohammadi et al. 2016) and paste (Migliaresi et al. 2007). Hydrogels are soft materials, having cross-linked hydrophilic networks that can absorb large quantities of water (or biological fluids) while maintaining their original structure (Ahmed and Aggor 2010, Jung et al. 2017, Portnov et al. 2017). When hydrogels are injected, they can readily wet all surfaces of the injured site and create a low-density aqueous cavity that contains the components necessary for bone tissue regeneration. Paste based injectable scaffolds are usually developed from calcium phosphate cements (CPC) and are also considered to be clinically important owing to their biocompatibility, bioactivity, ease of handling, moldability, and injectability (Li et al. 2009).

With this background, this chapter focusses on synthesis, characterization, and properties of injectable bone scaffolds. Motivations, benefits, and property requirements of such scaffolds have also been discussed in detail. Both hydrogel and paste based injectable scaffolds and their setting mechanisms have been elaborated. Emphasis has been given to injectable composite scaffold systems, with special reference to injectable nanocomposites. Finally, the challenges and future directions of injectable scaffolds for bone regeneration and repair have been discussed.

7.2 BONE

7.2.1 Ultrastructure of Bone

Bone is the primary structural and supportive connective tissue of the body which protects various organs and enable mobility. It comprises of relatively hard and lightweight composite material (Dinopoulos et al. 2012, Boskey 2007). From the material point of view, bone could be simplified as a three-phase material formed by organic phase, an inorganic nanocrystalline phase and a bone matrix (Boskey 2007, Boskey and Coleman 2010, Bigham et al. 2008). Bone has a hierarchical structure, where each level is designed to perform a range of mechanical, biological, and chemical functions (Wang et al. 2016). The hierarchical structure of bone consists of macroscale, microscale, sub-microscale, nanoscale, and sub-nanoscale (Fig. 7.1). In the macrostructure of bone, a smooth, dense, and continuous external layer forms the compact or cortical bone, while, the interior layer of bone is known as spongy or trabecular bone (Zhao 2010, Bigham et al. 2008, Dinopoulos et al. 2012). Compact bone consists of closely packed osteons or haversian systems and surrounds the medullary cavity, or bone marrow. It provides strength and protection to bones. Spongy (cancellous) bone on the other hand, is lighter and less dense than compact bone (Merolli and Leali 2012). Spongy bone consists of trabeculae, which are lamellae that are arranged as rods or plates and are adjacent to irregular cavities that contain red bone marrow. In bone, osteoblasts are bone-forming cell, osteoclasts resorb or break down bone, and osteocytes are mature bone cells. Osteoblasts produce a matrix that is eventually mineralized, or hardened, to become bone (Florencio-Silva et al. 2015, Buckwalter et al. 1996, SEER Training Modules 2019). At the micron- and nano-scales, reinforced collagen mainly consisting of aggregated type-I collagen and hydroxyapatite (HAp) form the building blocks for both compact and trabecular bones (Buckwalter et al. 1996, Dinopoulos et al. 2012, Bigham et al. 2008, Wang et al. 2016).

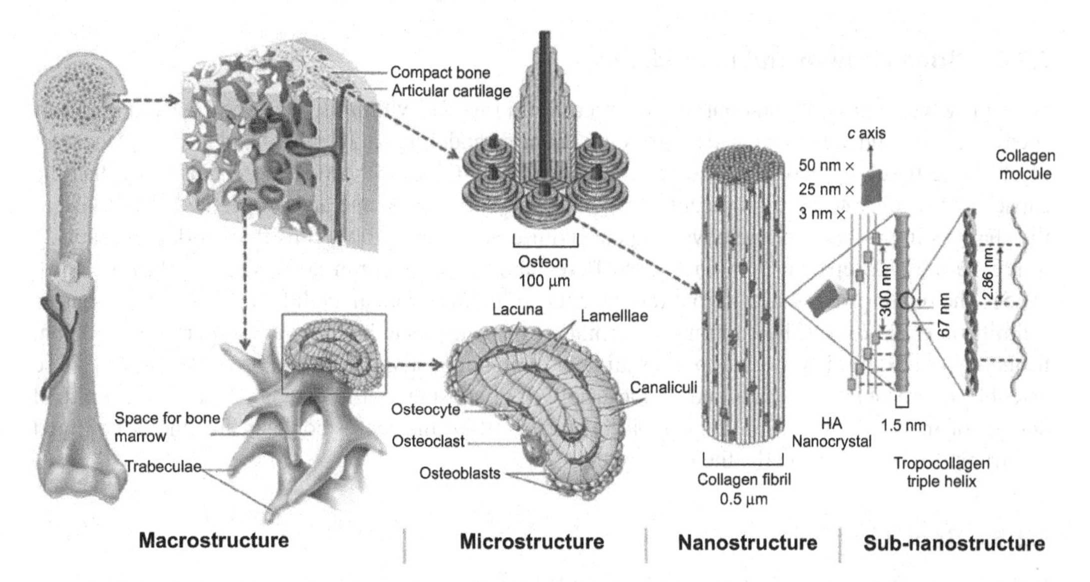

FIGURE 7.1 Hierarchical structure of bone. (Reprinted with permission from Wang, H., S. Xu, S. Zhou, W. Xu, M. Leary, P. Choong, M. Qian, M. Brandt and Y. M. Xie. "Topological design and additive manufacturing of porous metals for bone scaffolds and orthopaedic implants: A review," Biomaterials 83 (2016): 127-41).

Color version at the end of the book

7.2.2 Bone Composition

Bone is an inorganic-organic biocomposite material. The relative proportions of various components present in bone may vary with age, site, gender, disease, and treatment. In general, approximately 30% of bone is made up of organic matrix, while remaining 70% is mostly inorganic mineral hydroxyapatite ($Ca_{10}(PO_4)_6(OH)_2$) and water (Bigham et al. 2008, Leslie 2012, Boskey 2007, Boskey and Coleman 2010). The organic phase primarily includes collagen type I fibres along with non-collagenous proteins and numerous proteoglycans, while the inorganic phase comprises of nanocrystalline solid with apatite structure (hydroxyapatite) consisting of a variety of calcium phosphates (Feng 2009, Zhao 2010). The cross-linking ability of collagen stems from the unique chemistry of its hydroxylated and glycosylated molecules, leading to spontaneous formation of oriented fibrilar protein structure, on which HAp crystals of approximately 10-50 nm can deposit and grow (Taton 2001, Zhao 2010, Olszta et al. 2007, Leslie 2012, Boskey 2007). The organic matrix (reinforced by HAp crystal deposits) embeds osteocytes (osteoblasts and osteoclasts) which are responsible for organization and remodeling of bone (Zhao 2010, Olszta et al. 2007, Taton 2001). The interfacial compatibility, interactions, and structural relationships between collagen and HAp is crucial to impart bone's resilience and strength.

7.2.3 Bone Damages

Increasing incidence of bone fracture, owing to ageing population, trauma, degenerative diseases, or by normal wear and tear lead to significant damage to the skeletal structure and quality of life (Correlo et al. 2011, Acton 2013, Bigham et al. 2008). These kinds of defects can also result from an intrinsic abnormality of bone. Osteopenia (abnormally low bone density), osteomalacia (abnormal, deficient mineralization of bone), tumours (primary or metastatic) and tumour-like conditions of bone are few such examples of bone damages (King 2007, Dinopoulos et al. 2012, Bigham et al. 2008).

7.2.4 Bone Growth and Remodeling

Bone growth is a natural and continuous process and involves vitamins (D, C, and A) and minerals such as calcium, phosphorous, and magnesium. Bone remodeling on the other hand is the replacement of old bone tissue by new bone tissue (Bandyopadhyay-Ghosh S, 2008). The bone tissue has the capacity to regenerate small bone defects by adapting its mass, shape and properties in response to the changes in mechanical and physiological requirements through its cells that work continuously to regenerate and repair it (Einhorn 1998, Bates and Ramachandran 2007, Gómez-Barrena et al. 2015). The repair process stems from the interplay of biomechanical, cellular and molecular factors (Dimitriou et al. 2011). The continuous formation of bone tissue happens at the peripheral region, formed by an external crust and an internal layer with connective tissue and osteoblast cells. These osteoblasts are phosphate-rich and exude a jelly-like substance, the osteoid. Due to the gradual deposit of inorganic material, this osteoid becomes stiffer and the osteoblasts are finally confined and transformed in bone cells, the osteocytes.

7.2.5 Bone Grafting

Bone injury often requires bone grafting, where bone grafts are used for augmentation and formation of new bone, to repair damaged bone, or to expedite the bone healing and regeneration process. Primarily, three main types of bone grafts are used: autograft, allograft and synthetic bone graft or bone scaffold (Jia et al. 2015, Sheikh et al. 2015, O'Keeffe et al. 2008, Polo-Corrales et al. 2014). Autograft (autologous or autogenous bone graft) is considered to be the 'first choice' of material for the surgeons for reconstructive surgery, where the bone is obtained from the same patient, receiving the graft. However, this approach often leads to complications such as infection, pain, scarring, severe bleeding, and donor-site morbidity (Polo-Corrales et al. 2014, Zhao 2010, Feng 2009, O'Keeffe et al. 2008, Baroli 2009, Barriga et al. 2004, McCann et al. 2004). On the other hand, allografting technique relies on materials stored in regulated bone banks and thereby can overcome some of the challenges encountered with autografts. However, allografts lack the osteogenic capacity and risks of infections or immune rejection are often experienced (Zhao 2010, Barriga et al. 2004, Baroli 2009, Polo-Corrales et al. 2014). Although, both types of bone grafts discussed above (autografts, allografts) have been widely used, their limitations have driven the research efforts to develop other alternatives, referred to as bone graft substitutes (Kheirallah and Almeshaly 2016). These type of synthetic bone grafts, also known as bone scaffolds, are made from various natural or manmade materials and can be derived from metallic, polymeric, ceramic materials or composites. Usually, bioactive factors are incorporated into a scaffold to stimulate cells and to mimic the bone microenvironment (Polo-Corrales et al. 2014). The advent of bone tissue engineering has led to the development of engineered bone scaffolds which can enhance the osteogenic potential of bone, in addition to improving their osteoconductive and osteoinductive properties (Jia et al. 2015, Munting et al. 1993, Gotman and Fuchs 2011, Bandyopadhyay-Ghosh et al. 2010).

7.3 BONE SCAFFOLDS

Bone graft substitutes or bone scaffolds are gaining popularity owing to the advancement in tissue engineering which facilitates the formation of new bone and its integration with surrounding host bone (Lienemann et al. 2012, Chen et al. 2006). Based on the technique of implantation, the scaffolds are divided into three groups: preformed scaffold, injectable scaffold and injectable preformed scaffold.

7.3.1 Preformed Scaffolds

Preformed scaffolds are designed to have a predetermined shape. Invasive surgery is therefore, essential to implant these scaffolds into bony/skeletal defects (Bueno and Glowacki, 2011). Preformed scaffolds have the freedom to design targeted parameters that can be strictly controlled. They are intended for use in fracture sites with requirements for load bearing or for targeting specific shape (tumour resection or spinal fusion) by gathering information from diagnostic imaging.

7.3.2 Injectable Scaffolds

Injectable scaffolds eliminate the need for, or reduce the size of, any incisions required to implant the material. The injectable matrix materials are liquid before and during the injection and can pass through a small gauge-needle or a catheter; however, they undergo gelation or hardening *in situ*, allowing them to form a temporary three-dimensional template on which the cells can adhere to form a functional new tissue (Portnov et al. 2017). Such scaffolds are able to occupy the defective bone cavity of any shape. Their implantation may not involve invasive surgery; however, owing to their low mechanical strengths, they are indicated for non-load bearing sites or defects where the structural bone is largely intact (Bueno and Glowacki, 2011). Injectable scaffolds must allow *in vivo* application and hardening within a clinically acceptable time frame. Injectable scaffold materials, investigated for bone tissue engineering, include polyanhydride monomers (Burkoth and Anseth 2000), poly(propylene fumarates) (PPF) (Behravesh et al. 2003), lactic acid-based oligomers (Burdick et al. 2003), cellulosic nanocomposite (Phogat and Bandyopadhyay-Ghosh 2018) and calcium phosphate cements, among others (Seeherman et al. 2006).

7.3.3 Injectable Preformed Scaffolds

Another type of scaffold has recently been investigated which possesses characteristics of both preformed and injectable scaffolds (Bencherif et al. 2012, 2014). It has been reported that on application of appropriate shear stress, such scaffolds are reversibly compressed (up to 90% of its volume), resulting in injectable macroporous preformed scaffolds. Injectable preformed 3D scaffolds with shape memory properties, in the form of elastic sponge-like matrices are prepared by cryotropic gelation (at –20°C) of a methacrylated (MA)-alginate, using free-radical cross-linking mechanism (Bencherif et al. 2012, 2014). The cryogels shear-collapse upon application of appropriate mechanical force, thereby, assisting the biomaterial to pass through a conventional-gauge needle. These macroporous alginate-based scaffolds obtained in predefined geometrical sizes and shapes, could be passed through a surgical needle without mechanical distortion. However, on removal of shear force, the scaffolds quickly recover their original shapes, as shown in Fig. 7.2.

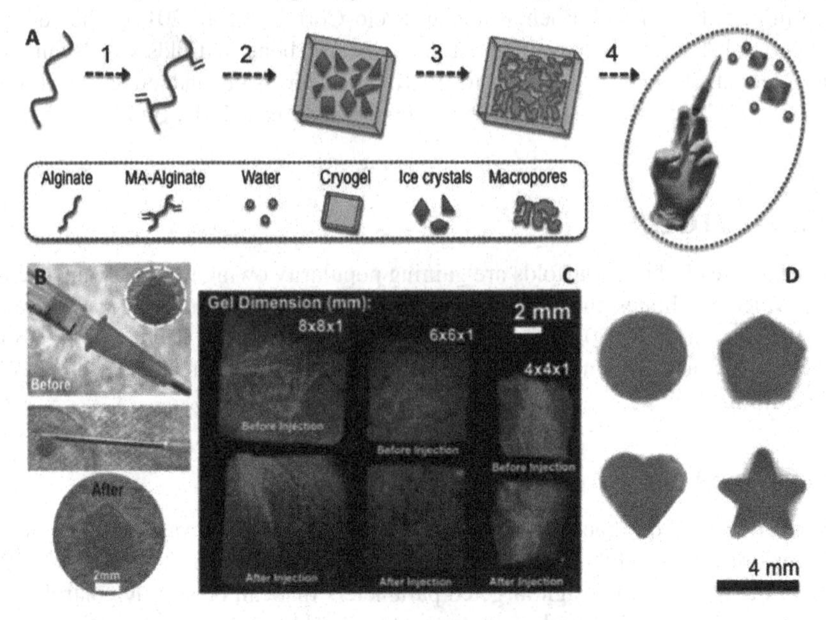

FIGURE 7.2 (A) Cryogelation process to synthesize injectable preformed scaffold with shape memory properties. (B) Photographs showing placement of a cryogel in 1-mL syringe (before injection) and gel recovery (after injection) via a conventional 16-gauge needle. (C) Rhodamine-labeled 1% MA-alginate gels with various sizes and shapes, prepared by cryogenic polymerization. Fluorescent square-shaped gels were syringe injected with a complete geometric restoration. (D) Photographs showing cryogels prepared with different geometric shapes retained their original shapes after syringe injection. (Reprinted with permission from Sidi A. Bencherif, R. Warren Sands, Deen Bhatta, Praveen Arany, Catia S. Verbeke, David A. Edwards and David J. Mooney. "Injectable preformed scaffolds with shape-memory properties," PNAS Early Edition (2012): 1-6).

7.4 INJECTABLE SCAFFOLDS

7.4.1 Motivations

Early tissue engineering approaches often relied on scaffolds such as preformed meshes or rigid scaffolds, that required surgical intervention for implantation (Peter and Elisseeff, 2005, Chan and Leong, 2008). Surgical implantation would need incisions, anesthesia, and can result in surgical and post-surgical complications. There is therefore, a need to develop strategies where the scaffolds can be implanted in a less invasive manner. In contrast to traditional open surgery, minimally invasive surgery uses laparoscopic devices, inserted through small incisions, to carry out the intervention, thereby decreasing surgery-related complications and morbidity. The advent of minimally invasive surgical techniques has also allowed for improvement in both post-operative convalescence and important clinical outcomes (Skovrlj et al. 2015). Additionally, such systems allow co-injection of cell suspension and scaffold materials, thereby producing a cell-scaffold construct which can fill irregular shaped cavity (Hou et al. 2004, Amini and Nair 2012, Portnov et al. 2017).

7.4.2 Benefits

An injectable scaffold offers several advantages as compared to preformed scaffolds used in tissue engineering approaches as described below (Peter and Elisseeff 2005, Williams and Elisseeff 2005, Laurencin and Nair 2008, Gutowska et al. 2001, Amirthalingam et al. 2018):

- Injectable scaffolds do not require invasive surgical procedures for implantation, thereby, limiting surgery related complications, unwanted scarring and expediting post-operative recovery process.
- An injectable scaffold can fill any defect shape.
- Once injected, such scaffolds have the potential to polymerize *in situ*.
- Therapeutic agents (e.g., growth factors) along with bioactive and biodegradable components may be readily incorporated.
- It is possible to deliver encapsulated cells along with the injectable scaffold materials into the target site in a predictable manner.

7.4.3 Types of Injectable Scaffolds

Depending upon the type of materials used, injectable scaffolds can either be in the hydrogel form or in the form of a cementitious paste.

7.4.3.1 Hydrogel Based

Injectable hydrogel scaffolds for bone tissue repair offer several advantages, as they can be injected into any irregularly shaped defect, can readily wet surfaces and can act as a low-density aqueous reservoir, incorporating the necessary components for bone tissue repair and augmentation as shown in Fig. 7.3 (Portnov et al. 2017, Migliaresi et al. 2007, Tommasi et al. 2016, Amirthalingam et al. 2018, Phogat and Bandyopadhyay-Ghosh 2018). Besides, hydrogels can be useful in protecting the damage sites from undesirable cellular elements that stems from soft tissue. Cells and dissolved minerals can also enter the scaffold, in a similar manner as the non-mineralized extra cellular matrix (ECM). Besides, the cell proliferation can initiate generation of collagenous ECM within the hydrogel scaffold, leading to formation of mineralized new bone throughout, as opposed to their formation at the surface only in traditional scaffolds (Migliaresi et al. 2007).

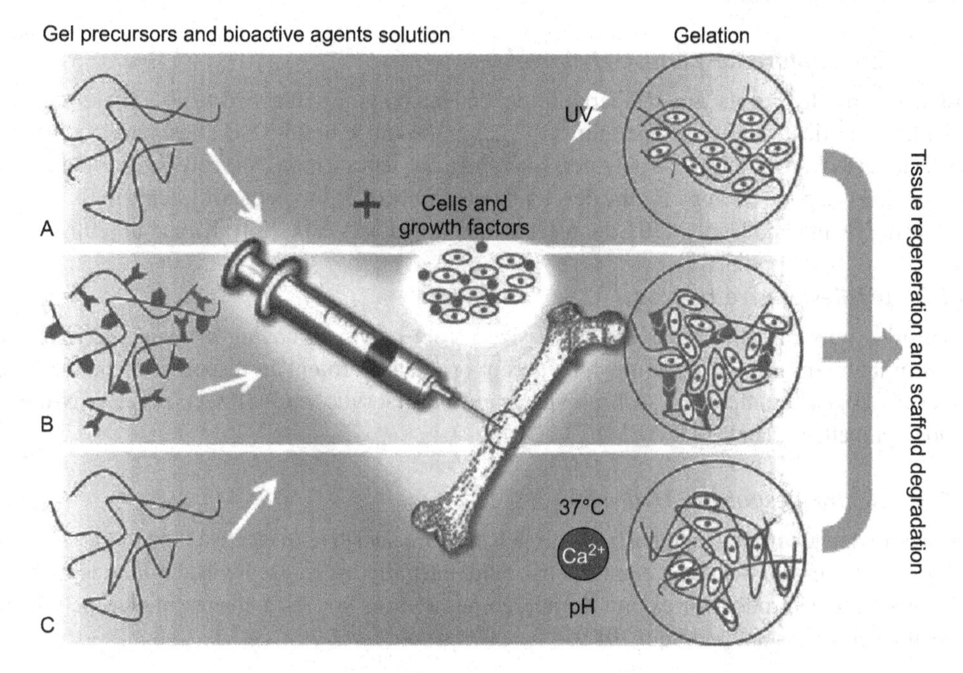

FIGURE 7.3 A schematic representation of injectable *in situ* gelling hydrogel scaffolds for bone tissue engineering. (Reprinted with permission from Portnov, T., T.R. Shulimzon and M. Zilberman. "Injectable hydrogel-based scaffolds for tissue engineering applications," Rev. Chem. Eng. 33(1) (2017): 91-107).

7.4.3.2 Cementitious Paste Based

Injectable cements are obtained by combining a ceramic (or glass or glass-ceramic) powder with a polymer or an organic compound. These ceramic based pastes are known for their favourable bone response. It is expected that dissolution and bioresorption of such injectable scaffolds provide an appropriate biological environment to stimulate bone regeneration process (Migliaresi et al. 2007). The cement material breaks down into ionic components in the body fluid during the bioresorption process, while, simultaneous remodeling of bone, generates new bone-ceramic interfaces, thereby controlling growth rate of new bone (Migliaresi et al. 2007).

7.5 INJECTABLE HYDROGEL SCAFFOLDS

In recent days, significant emphasis is being given to the design of "injectable" or "minimally invasive" hydrogel scaffolds that solidify after injection *in vivo* (Amirthalingam et al. 2018). The hydrogel precursor can be introduced in a minimally invasive manner directly to the target defect site, thereby, reducing patient discomfort, risk of infection, recovery time, and treatment cost (Bakaic et al. 2015). The following section reviews the different types of injectable hydrogel scaffold formation techniques based on stimuli responses and through the use of physical cross-linking under the physiological conditions of the target tissue, or by chemical cross-linking, exploiting various polymerization techniques.

7.5.1 Classification based on Stimuli Responses

Different types of external stimuli can be applied to induce injectable hydrogel formation *in situ* and to provide a means for controlling 'on-off' release of the encapsulated therapeutic reagents within the scaffolds (Kona et al. 2011). The hydrogels that can be classified based on the stimuli (Qiu and Park 2001) as listed below:

7.5.1.1 Temperature-Responsive Hydrogel

Most injectable hydrogels are temperature-dependent polymerizing hydrogels. These materials have a lower critical solution temperature (LCST) in the range of 25-32°C (Yoshizawa et al. 2004, Kona et al. 2011). At LCST, these polymers undergo a phase transition. The reverse thermal gelation phenomenon is a good strategy for the development of injectable biomedical systems for bone tissue engineering (Kang and Huang 2011, Jeong et al. 1997, Qiu and Park 2001, Kim et al. 2009).

7.5.1.2 pH-Responsive Hydrogel

pH-sensitive injectable hydrogels are synthesized from polymers having pendant acidic or basic groups in its main backbone chain, which accept or release protons in response to changes in pH (Liu et al. 2017). Examples of such polymers are poly(acrylic acid) (PAA), poly(N, N-diethylaminoethyl methacrylate) (PDEAEM).

7.5.1.3 Electro-Responsive Hydrogel

Polyelectrolytes are used to synthesize hydrogels that are sensitive to electric field or current. They swell or shrink upon application of an electric field. Partially hydrolyzed polyacrylamide hydrogels in contact with electrodes, for example, undergo phase transition by a change in electric potential across the gel (Gupta and Siddiqui 2012).

7.5.1.4 Photo-Responsive Hydrogel

Photo-responsive hydrogels can be either UV-sensitive or visible light-sensitive. Presence of UV irradiation swells the hydrogels owing to an increase in osmotic pressure within the gel. UV-degradable injectable self-healing hydrogel has been synthesized based on biocompatible polymer, poly (ethylene glycol)-b-poly (γ-o-nitrobenzyl-L-glutamate) (Zhao et al., 2018). Visible light-sensitive hydrogels contain a light-sensitive chromophore such as trisodium salt of copper chlorophyllin (Maharjan et al. 2008). The light absorbed by the chromophore is dissipated locally as heat thereby altering the swelling behavior of the polymer hydrogel (Qiu and Park, 2001). The excitation wavelength has long been known as a challenging issue for photo responsive hydrogels. Near-infrared (NIR) light (wavelengths roughly 700-1000 nm) is preferred in that respect because of its deeper tissue penetration and less detrimental effect on healthy cells as compared to ultraviolet (UV) light (Zhang et al. 2017). Synthesis of NIR-responsive hydrogels, however, possesses challenge, due to the difficulty in triggering appropriate photochemical reactions, inducing photo-cleavage.

7.5.1.5 Other Stimuli-Responsive Hydrogels

Some hydrogels respond to specific ions and molecules. Poly diallyl dimethyl ammonium chloride hydrogels, for instance, are sensitive to a critical concentration of sodium chloride in aqueous solution. Further, some hydrogels shrink by cross-linking interactions upon antigen-antibody binding and swell in the presence of free antigens reducing the cross-linking density (Miyata et al. 1999).

7.5.2 Classification based on Cross-linking Mechanisms

Hydrogels, if covalently (chemically) cross-linked, are called 'permanent' or 'chemical' gels (Hoffman 2012, Hennink and Nostrum 2002), whereas, if the networks are held together by molecular interactions, and/or secondary forces including H-bonding, Van der Waals forces or hydrophobic forces (physical cross-linking), they are known as 'reversible', or 'physical' gels (Campoccia et al. 1998, Prestwich et al. 1998, Colla et al. 2016). Appropriate precursor formulation and selection of suitable cross-linking methods play vital roles in developing suitable injectable bone scaffolds. Following sections will review some of the techniques used for formation of such injectable hydrogels.

7.5.2.1 Chemically Cross-linked Hydrogels

In this type of hydrogel, a chemical initiator forms free radicals that react with the functional groups (usually unsaturated bonds) in the monomers or macromers (Kona et al. 2011). This cross-linking reaction proceeds until all the monomer is cross-linked. Several chemical cross-linking methods have been reported for synthesis of hydrogels, as listed in Table 7.1. This method of cross-linking improves the mechanical properties of the injectable scaffolds. By varying the chemical initiator concentrations and the type of cross-linkers, the density of cross-linking network can be altered, which in turn varies the properties of the formed scaffolds (Timmer et al. 2003a, b). Though this system has the benefit of being activated with change in amount or type of chemical initiators, a major limitation is the control of the gelling time (Kona et al. 2011). An example of this type of material is oligo (poly (ethylene glycol) fumarate (OPF). In addition to OPF, hyaluronic acid (HA) derivatives, unsaturated ultra-low molecular weight poly (L-lactide) (ULMW PLA), etc., are also examples of such chemically cross-linked hydrogels.

TABLE 7.1 Mechanisms, processes, and examples of chemically cross-linked hydrogels

Mechanisms	Processes	Examples
Radical polymerization	Chemically cross-linked gels can be obtained by radical polymerization of low molecular weight monomers in the presence of cross-linking agents	Poly (2-hydroxyethyl methacrylate) (pHEMA) based hydrogel has been synthesized by using polymerization of HEMA in the presence of ethylene glycol dimethacrylate (Hennink and Nostrum 2002, Achilias and Siafaka 2017)
Michael addition	Michael addition induced cross-linking by nucleophilic addition of a carbanion to an α, β-unsaturated carbonyl compound	Hyaluronic acid, chitosan, and PEG are used for injectable hydrogel preparation via the Michael addition under physiological conditions (Chen et al. 2013, Bakaic et al. 2015, Rodell et al. 2015)
Schiff base reaction	Schiff base reaction induced cross-linking by forming imine bond between amino and aldehyde groups	Injectable chitosan-based polysaccharide hydrogel, cross-linked by Schiff base reaction between amino functionalities of chitosan and the aldehyde groups of dextran aldehyde (Cheng et al. 2013, Bakaic et al. 2015)
Click chemistry	Cross-linking by 'click chemistry' based reaction, including Diels–Alder reaction, the thiol-ene reaction, and thiol-maleimide coupling etc	Dendron–polymer–dendron conjugate-based injectable hydrogel through radical thiol-ene "click" reactions (Kaga et al. 2016)
Photo-cross-linking	Photo-cross-linking via initiation, propagation, and termination steps, induced by electromagnetic radiation in the visible and ultraviolet regions	Poly (ethylene glycol) dimethacrylate copolymer, chitosan based injectable hydrogel by photo-cross-linking (Zhao et al. 2018, Bakaic et al. 2015)
Enzymatic	Suitable enzyme can be used as catalysts for *in situ* gelling	Chitosan-based injectable hydrogels have been reported where water-soluble chitosan-glycolic acid/tyrosine conjugates were prepared by using tyrosinase enzyme (Jin et al. 2009)

7.5.2.2 Physically Cross-linked Hydrogels

There is a growing trend to develop the injectable hydrogel scaffolds using physical cross-linking techniques (Huh et al. 2012). This is because of the fact that chemical cross-linking involves use of cross-linkers which are often toxic in nature and the unreacted cross-linkers if present, have to be removed/extracted, before application inside the body. Besides, the chemical cross-linkers can adversely affect the immobilized substances (e.g. proteins, cells) within the hydrogel. Physical cross-linking can surpass the above limitations; however, the polymeric network, obtained primarily from inter-chain secondary bonding usually results into limited mechanical strength and stability (Hoffman 2012, Hou et al. 2004, Huh et al. 2012). Physical cross-linking may be initiated by different techniques. Examples of such techniques include ionic interactions, crystallization, use of amphiphilic block and graft copolymers, hydrogen bonding, protein interactions, *in situ* precipitation, etc (Jeong et al. 2016). The physically cross-linked gels have been used as injectable delivery vehicles for bone scaffold applications (Paige et al. 1995). Table 7.2 lists some of the useful techniques for synthesis of physical cross-linked hydrogels.

TABLE 7.2 Mechanism, process, and examples of physically cross-linked hydrogels

Mechanisms	Processes	Examples
Ionic interaction	Cross-linking takes place by ionic interactions of charged polymers. Charged polymers that are soluble in aqueous solvents or water are used and form gels when they react with di- or tri-valent counter ions. The process is a reversible gelation and upon removal of the cations, the gel becomes liquid	Alginate gels have been synthesized by reacting with di-valent calcium ions ($CaCO_3$) (Hennink and Nostrum 2002) A shear-thinning and self-healing collagen-Au hybrid hydrogel system was synthesized by the electrostatic complexation of positively charged collagen proteins and anionic precursor ions of gold nanoparticles (Ruirui et al. 2016)
Crystallization	Crystallization induced hydrogel formation, where the crystallites act as physical cross-linking sites in the network	Poly (vinyl alcohol) (PVA), dextran polysaccharide based hydrogels have been prepared using repeated freeze-thaw cycles (Thangprasert et al. 2019)
Use of amphiphilic block and graft copolymers	Hydrogel formation by use of amphiphilic copolymers in which the hydrophobic segments of polymers are aggregated by self-assembly	Triblock polymeric hydrogel with hydrophobic poly (lactic-co-glycolic acid) (PL(G)A) segment in the middle have been prepared by coupling of two PEG-PL(G)A diblock copolymers (Hennink and Nostrum 2002)
In situ precipitation	*In situ* precipitation injectable materials are polymer solutions that precipitate when they come in contact with a non-solvent (physiological fluids) when injected into the body. The polymer precipitates out to form the gel as it is not soluble in water	Hydrophobic polymer hydrogels are prepared in physiologically tolerant solvents like dimethyl sulfoxide (DMSO) that are also miscible with water (Coombes et al. 2004, Eliaz and Kost 2000, Kona et al. 2011)

7.6 INJECTABLE PASTE BASED SCAFFOLDS

Cementitious pastes have been used increasingly in minimally invasive bone tissue engineering applications. Calcium phosphate cements (CPCs), mineral trioxide aggregates, in this regard are gaining attention as bone substitutes, due to their injectability and self-setting ability in physiological conditions. CPCs also possess structures similar to biological apatites and often demonstrate excellent bioactivity. Injectable calcium phosphate cements (CPCs) consist of aqueous mixtures of calcium phosphates and other soluble calcium salts. This material exhibit injectability, short and long-term mechanical strength as well as increased compressive strength with an increase in the aging time. These materials also show significant *in vitro* bioactivity in simulated body fluid (SBF). The intrinsic micropores present in such bone cements also help in passage of nutrients and wastes, which in turn facilitates bone cell proliferation and regeneration (Ginebra 2008). Their ability to stimulate osteoblast proliferation and promote osteoblastic differentiation of the bone marrow stromal cells, make them promising candidates as injectable bone scaffolds (Huan and Chang 2009). Along with injectabilty, the setting parameters (such as setting time, dimensional integrity, resultant toxicity, if any) of such injectable pastes are equally crucial. The selection of setting mechanism, therefore, play vital role in designing the appropriate injectable paste based scaffold. Although, several routes have been attempted for preparing the paste based scaffolds, some of these routes are applicable for pre-set cements only, and therefore, cannot be used for injectable applications. The following section reviews the approaches towards developing paste based injectable scaffolds and their hardening mechanisms.

7.6.1 Self-Setting

Reinforcing bone by injecting a cementitious bone substitute such as poly (methyl methacrylate) (PMMA) is a common technique and is found to be effective for vertebral fractures. However, during the setting process, PMMA based cement has been reported to generate monomer toxicity and excessive heat that may lead to bone necrosis (Schoenfeld et al. 1979, Karlsson et al. 1995, Galibert et al. 1987, Deramond et al. 1997, Bohner et al. 2006). An alternative to PMMA is calcium phosphate cements (CPC) which are made of calcium phosphate particles dispersed in an aqueous solution. They are usually considered to be injectable, although their injectability is often found to be poor. Self-setting of calcium phosphate cements can happen *in vivo* assisted by the aqueous medium present at the biological target site (Hou et al. 2004, Bohner et al. 2006).

7.6.2 Thermoplastic Hardening

Thermoplastic polymer pastes have also been used as injectable scaffolds. With their low melting point (usually lower than 65°C) they can be injected into the targeted body site readily. The polymer solution cools rapidly to body temperature forming a partially crystallized structure (Hatefi and Amsden 2002). The moderate degree of crystallinity facilitates cell growth within the scaffold and release of growth factors or drugs during the scaffold formation process (Kona et al. 2011).

7.6.3 Dual Setting

Non-reactive polymers such as collagen, chitosan, hyaluronic acid etc. have been used in conjunction with ceramic powder to develop polymer–cement composite pastes. In such cases, the aqueous solution of the polymer is reacted (solid state reaction) with the cement powder at a suitable temperature/pressure (Geffers et al. 2015, Geffers 2016). In some cases, simultaneous gelation–polymerisation process is carried out by adding reactive monomer systems in the cement liquid, thereby developing an interconnected matrix within the porous cement structure as shown in Fig. 7.4.

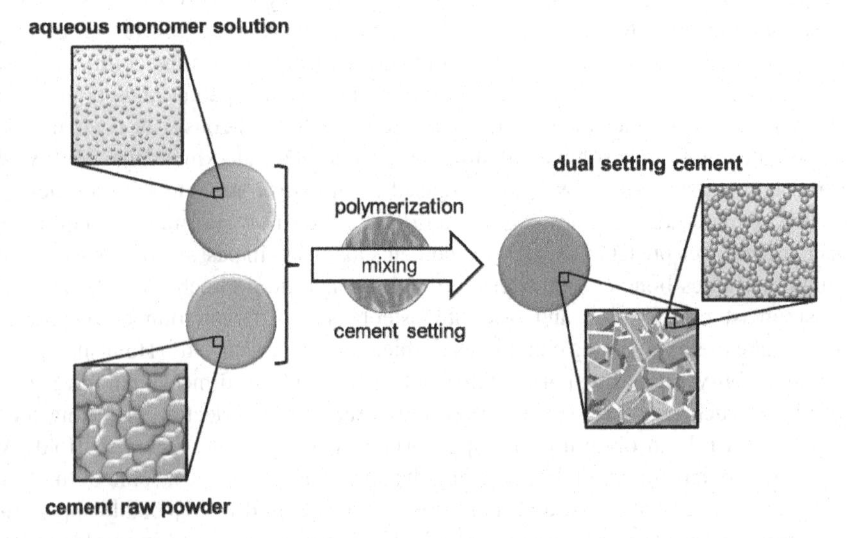

FIGURE 7.4 Hardening mechanism of dual-setting cements with the formation of interconnected matrices of hydrogel and precipitated cement crystals. (Reprinted with permission from Geffers, M., J. Groll and U. Gbureck. "Reinforcement Strategies for Load-Bearing Calcium Phosphate Biocements: A review," Materials 8 (2015): 2700-17).

7.6.4 Visible Light activated Setting

Injectable paste systems (single or dual) have been reported, which polymerize when being placed under normal ambient light conditions (Hasel 2001). Such systems have been found to be capable of being syringed through a hypodermic syringe from 25 to 16-gauge needles. The unpolymerized pastes contain chemicals that will generate initiating free radicals on exposure to visible light of wavelength 470 ± 480 nm (Douglas et al. 1979).

7.6.5 Ultraviolet Light activated Setting

Another way to activate the hardening effect in the injectable paste at targeted site is use of ultraviolet (UV) light. Si doped HAp nanopowder incorporated photocurable aligate paste was prepared for orthopaedic application. Si doped HAp nanopowder incorporated alginate-based paste showed good injectability and hardened after UV exposure to generate a sparsely porous matrix (Gupta 2018). However, the chromosomal and genetic instability of exposed cells possess concern while using the UV lights for photo-polymerization (Dahle et al. 2005, Kappes et al. 2006). Care should also be taken with use of toxic photoinitiator or intense UV light irradiation along with these cell-encapsulated pastes as they can lead to cell damage.

7.7 PROPERTY REQUIREMENTS OF INJECTABLE SCAFFOLDS

Injectable scaffolds are expected to meet certain property requirements, as described below, which help in designing and fabricating an efficient system:

7.7.1 Rheological Properties

Injectable scaffold material should be fluid in the form of stable dispersion (liquid/solid) or gel and should be injected through a needle of appropriate size (Hou et al. 2004). Such scaffold material upon injection, must be able to be transformed into elastic or viscoelastic solid, thereby ensuring its retention at the defect site and the capability to support mechanical load (Hou et al. 2004, Wang et al. 2006, Migliaresi et al. 2007). The rheological properties of an injectable bone substitute material include: injectability, cohesion and viscosity (Hou et al. 2004, Khairoun et al. 1998, Mclemore 2011).

7.7.1.1 Injectability

Rheological properties of the formulations dictate the injectability of a scaffold. Besides, solidification rate and setting time of such scaffolds, injectability is also governed by scaffold structure, their formulation and processing conditions (Hou et al. 2004, Mclemore et al. 2006, Wang et al. 2006, Bohner and Baroud 2005, Mclemore 2011, Khairoun et al. 1998, Montufar et al. 2011). Malleability of polymeric hydrogel is also considered to be a key to injectability (Jahan et al. 2019). Highly viscous polymer solution or slightly cross-linked gel may be capable of undergoing shear-thinning behavior (reduction in viscosity) on removal of shear force generated by injection. Several hydrogels exhibit such a behavior and have been used for tissue engineering applications (Gutowska et al. 2001, Phogat and Bandyopadhyay-Ghosh 2018). This is particularly interesting for development of injectable system. It has also been found that presence of bioactive fillers, may also affect the shear thinning behavior of injectable hydrogel (Phogat and Bandyopadhyay-Ghosh 2018). For chemically cross-linked scaffolds, several factors, including chemistry of formulation and concentration of the monomers and the initiators strongly influence the injectability of the scaffold (Hu et al. 2019, Tan et al. 2009, Mclemore et al. 2006, Hou et al. 2004). Other factors that influence injectability of a system include diameter of the injection device and molecular interactions (Page et al. 2013).

7.7.1.2 Cohesion (Resistance against Washout)

Cohesion is the ability of the injectable material to keep its geometrical integrity in a body fluid. Poor cohesion may adversely affect setting and can result in negative *in vivo* response owing to the release of incorporated particulate components (Bohner 2010, Tan et al. 2009, Yan et al. 2018). Various approaches can be used to improve cohesion such as decrease of mean particle size, liquid-to-solid ratio (LSR), and enhancement of ionic character of the mixing solution (Bohner 2010, Montufar et al. 2009, 2011, 2012).

7.2.1.3 Viscosity

Generally, bone cement formulations are thixotropic (shear-thinning) (Baroud et al. 2005, Liu et al. 2006b). It is important to control the viscosity of injectable formulation. Decrease in viscosity may be achieved by increase in LSR and an increase in particle size. Additives, such as citrate ions or poly (acrylic acid) can decrease particle interactions and in turn can reduce the viscosity. Another approach to increase the viscosity of the mixing formulation is to use hydrogels (Bohner 2010, Hu et al. 2019, Bohner et al. 2006, Cherng et al. 1997).

7.2.2 Biocompatibility

Injectable scaffold must be biocompatible in nature, which also means that neither the components nor their degradation products should trigger an inflammatory response, immunogenicity or cytotoxicity (Hou et al. 2004, Tan et al. 2009, Kim et al. 2010). The effects of the components of injectable formulation, residual leachable byproducts or degradation products strongly influence cell viability and therefore warrant careful consideration.

7.7.3 Gelation/Solidification

The precursors of injectable scaffolds, containing cells and bioactive molecules, need to undergo mild solidification process, in the physiological conditions (Hu et al. 2019, Hou et al. 2004). Several parameters, including the solidification mechanism, formulation of scaffold composition and the related solidification process need careful consideration. The gelation time or solidification time should be appropriate to maintain the desired cohesiveness at the injection site, while, allowing sufficient time for proper insertion procedure. A working time in the range of 5-15 minutes is usually preferred for clinical use (Habib et al. 2008, Page et al. 2012, Sarvestani et al. 2007, Harmata 2015).

7.7.4 Mechanical Strength

Mechanical properties of injectable scaffolds primarily depend on chemical structure, cross-linking mechanism, cross-linking density, molecular weight, aqueous solution concentration, and swelling capacity under physiological conditions (Hou et al. 2004). Injectable scaffolds should possess good dimensional stability following injection and solidification, while, it should also possess sufficient mechanical strength to provide temporary support for the cells, while withstanding biomechanical loading (Hu et al. 2019, Lee and Mooney 2001, Elbert and Hubbell 2001, Balgude et al. 2001).

7.7.5 Degradability

Physiological degradability of injectable scaffolds should happen at a rate that matches with the rate of new tissue regeneration (Hou et al. 2004). While, slow degradation rate can affect the deposition and distribution of extracellular matrix along with cell migration, thereby impeding new tissue function, a faster degradation can detrimentally affect the support to cells (Hou et al. 2004, Tan et al. 2009).

7.7.6 Porous Micro-architecture

Porosity is one of the key requirements of scaffold materials designed for bone regeneration (Hutmacher 2000, Jones 2013, Kim et al. 2010). Injectable scaffolds, developed *in situ* ideally should possess interconnected porous microstructure to promote cell ingrowth, transport of nutrients and removal of cellular wastes. Furthermore, porosity enhances material resorption to allow complete regeneration of the bone tissue (Bohner and Baumgart 2004). Primary influencing parameters include pore size, pore distribution, overall porosity and interconnectivity of the porous micro-architecture (Hou et al. 2004). Several methods have been attempted to fabricate porous injectable scaffolds. They include particulate leaching, *in situ* pore formation, gas forming, air entraining etc. Presence of gradient porosity having micropores (pores smaller than 100 μm) and macropores (pores larger than 100 μm) have been found to favour growth of fibroblasts and formation of both non-mineralized and mineralized bone (Ginebra and Montufar 2014, Green et al. 2002, Yang et al. 2001).

7.7.7 Ability to Carry Drugs or Other Bioactive Substances

Injectable scaffolds should be able to incorporate growth factors and cell adhesion promoter towards stimulation of cell function and tissue growth (Hou et al. 2004). They have the advantage to act as reservoir and release the molecules at the targeted defect site, thereby creating an environment, conducive to tissue regeneration (Whitaker et al. 2001).

7.8 INJECTABLE COMPOSITE SCAFFOLDS

Composites are made of two or more distinctly different materials (or phases) such as ceramics and polymers. The composite scaffolds are designed to improve both mechanical and osteoconductive properties by combining biodegradable polymers with bioactive ceramics (Phogat and Bandyopadhyay-Ghosh 2018, Huang and Chu 2019). Considering that bone is a biocomposite of natural polymer matrix and inorganic bioapatite filler, there has been a growing trend to fabricate composite scaffolds for bone tissue engineering by combining biopolymers and inorganic fillers. Furthermore, the composites scaffolds can be tailored to have patient specific bone properties, such as similar stiffness, fractural strength and bending strength, etc., which can lead to better biological responses in the form of reduced stress shielding (Hu et al. 2019, Tanner 2017, Huang and Chu 2019).

An injectable composite scaffold can utilize the benefits from composite and injectable systems (Phogat and Bandyopadhyay-Ghosh 2018). In injectable composite scaffolds, bioactive molecules can be homogeneously distributed within the matrix by remaining in suspension or solution before solidification i*n vivo*, making them promising materials for bone tissue induction or regeneration (Hou et al. 2004). Injectable composite scaffolds for bone regeneration can be classified into two main categories: flowable polymer-ceramic paste with the ability to be set *in situ* and composite hydrogels with the ability to co-inject scaffold suspension and cells to the defect site (Migliaresi et al. 2007, Huang and Chu 2019).

7.8.1 Injectable Composite Hydrogels

Hydrogels with their aqueous matrix closely resemble ECM, can encapsulate the bioactive components, and provide a network structure that can bear mechanical stresses. Suitable ceramic components are often used in injectable hydrogel matrix to form composites (Migliaresi et al. 2007, Huang and Chu 2019). Based on the type of matrix, injectable composite hydrogels can be classified as follows:

7.8.1.1 Natural Polymer Matrix Based

Injectable composite hydrogels have been developed using natural polymeric materials like alginate, collagen and hyaluronic acid (HA) derivatives with inorganic materials such as calcium phosphate, hydroxyapatite (HAp) etc. Composite injectable material made of beta-tricalcium phosphate (β-TCP) beads as the solid phase and alginate as the gel phase (Fig. 7.5) exhibited the ability to support new bone formation (Hasnain et al. 2019, Matsuno et al. 2008). The 3D-formed composite was obtained as soon as alginate hydrogel was pushed out from the tip of the syringe owing to instantaneous cross-linking within the syringe (Jeznach et al. 2018). Hydrogels based on polysaccharides also have wide range applications due to their high swelling capacity and water uptake (Hasan and Abdel-Raouf 2019). An injectable multiphasic bone substitute (IBS) hydrogel composed of 2% aqueous solution of methyl hydroxypropyl-cellulose (MHPC) and biphasic calcium phosphate (BCP, 60% HAp and 40% β-TCP) was evaluated for percutaneous orthopedic surgery (Grimandi et al. 1998). A preliminary *in vivo* test in rabbit femoral epiphysis showed bone in-growth into the scaffold after one week. In anotherstudy, a composite of carboxymethyl-chitin (CM-chitin) with hydroxyapatite (HAp) was examined for its ability to repair bone in animals (Chang and Zhang 2011, Tokura and Tamura 2001).

FIGURE 7.5 Injectable 3D-formed composite of β-TCP beads and alginate: a light microscope photograph of the composite, b SEM photograph of the composite, c SEM photograph of the composite surface. Bar is 1,000 µm in (b) and 100 µm in (c). (Reprinted with permission from Matsuno, T., Y. Hashimoto, S. Adachi, K. Omata, Y. Yoshitaka, Y. Ozeki, Y. Umezu, Y. Tabata, M. Nakamura and T. Satoh. "Preparation of injectable 3D-formed β-tricalcium phosphate bead/alginate composite for bone tissue engineering," Dent. Mater. J. 27(6) (2008): 827-34).

7.8.1.2 Synthetic Polymer Matrix Based

Several synthetic (man-made) polymer based composite materials have also shown promise as injectable materials. Injectable formulations are often prepared from highly hydrated synthetic polymers which self-regulate to form hydrogels by physical or chemical cross-linking mechanisms (Hasnain et al. 2019). Synthetic polymers that have been employed as injectable carriers for tissue-engineering applications comprise of a variety of hydrophilic/hydrophobic copolymer combinations of poly(lactic acid) (PLA), poly(lactic-co-glycolic) acid (PLGA), poly(ethylene glycol) (PEG), poly(glycolic acid) (PGA), poly(vinyl alcohol) (PVA), poly(glutamic acid) (PGA) poly(N-isopropylacrylamide) (PNIPAM), poly(phosphazenes) (PPZ) etc. and copolymers of poly(ethylene oxide) (PEO) and poly(propylene oxide) (PPO), under the commercial names of Pluronics and Polyoxamer (Migliaresi et al. 2007, Tan and Marra 2010, Tian et al. 2012, Rojo et al. 2014). Polypropylene fumarate, poly(hydroxyethyl methacrylate) (PHEMA) have also been used as an injectable polymer to fill bone defects, particularly in cancellous bone. The material can be injected at the targeted site, degrade with respect to time and are cross-linkable *in situ*. In another example, porous PLGA/Ca-P cement composites have exhibited osteoinductive properties and bone-like mineralization, when used assubcutaneous implants (Migliaresi et al. 2007, Ruhé et al. 2006).

7.8.2 Injectable Composite Pastes

Several injectable composite pastes have been reported. Examples of these materials include bioactive composites of β-tricalcium phosphate, hydroxyapatite, calcium phosphates and bioglass etc. Cement pastes, for example calcium phosphate based cements have popularly been used as injectable scaffolds (Jeong et al. 2016). In case of a composite cement, the dry powder (for example, combination of monocalcium phosphate, tricalcium phosphate, and calcium carbonate) and a polymeric solution are mixed to form a paste (Jeong et al. 2016). The paste is malleable and injectable for a certain short period. The setting time can be adjusted by use of modifier. These cements usually harden without generating significant heat. Additionally, they can develop compressive strength, followed by slow remodeling *in vivo*. An ideal cement should be resorbed gradually with time and get replaced by host bone (Olson et al. 2007, Polo-Corrales et al. 2014).

7.8.2.1 Bioactive Ceramic Filler Based

Calcium phosphate based injectable composite paste developed in combination with gelatin and soybean-derived polymers have been tested as injectable bone grafts in a femoral diaphyseal critical size defect model in rabbits. The injected paste adopted the shape of the cavity, filling it completely. In one study, injectable bone scaffold was developed with the matrix phase based on chitosan, citric acid and glucose solution, reinforced with β-tricalcium phosphate powder (Liu et al. 2006a, b). The material showed good paste consistency, injectability and could be molded into the desired shape (Fig. 7.6). The cell culture studies demonstrated cytocompatibility of the composite scaffolds (Liu et al. 2006b). One of the challenges of working with CPCs is their rapid setting (often less than 10 seconds); and accordingly, use of less soluble reagents has been recommended to delay the setting process (only to a limited extent) and thereby providing additional time for safe CPC placement to the defect site (Zhao 2010, Bohner 2007).

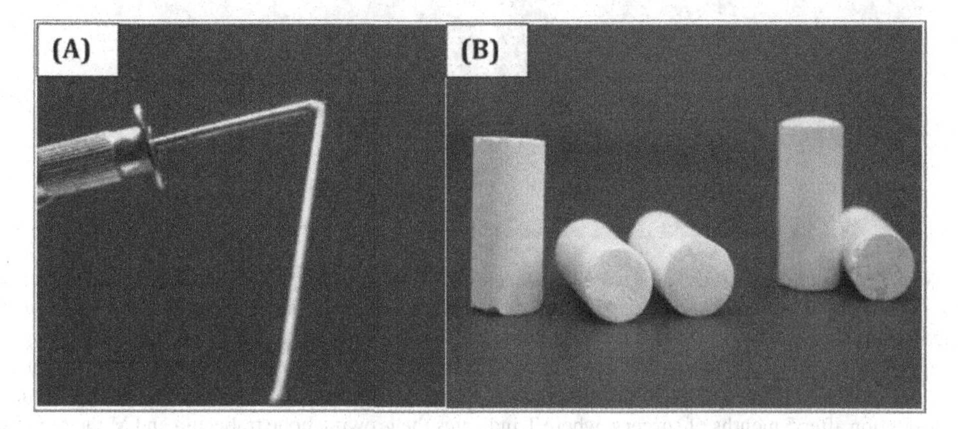

FIGURE 7.6 Photograph (A) shows the paste consistency of the composite biomaterial after mixing. The material was injectable and could be molded into the desired shape, as shown in the photograph (B). (Reprinted with permission from Liu, H., H. Lie, W. Cheng, Y. Yang, M. Zhu and C. Zhou. "Novel injectable calcium phosphate/chitosan composites for bone substitute materials", Acta Biomater., 2(2006b): 557-65).

In one report, when quick forming hydroxyapatite (HAp)/agarose gel composite was injected into the medial femoral condyle of rabbits, newly-formed bone was observed at the edge of the bone defect site two weeks postoperatively (Zhao 2010, Watanabe et al. 2007). At four weeks postoperatively, excellent bone regeneration was observed and the composite gradually degraded at eight weeks postoperatively. These results indicated that the composite dissolved rapidly and was replaced by newly formed bone, suggesting that the quick-forming HAp/agarose gel composite was a good candidate as an injectable scaffold (Fig. 7.7) (Watanabe et al. 2007, Ginebra and Montufar

2014). Injectable composite material composed of calcium phosphate and HA was reported to have significant osteoclastic and osteoblastic activities in the bone tissue in young adult rabbit knee (Liu et al. 2006b, Zhao 2010, Gao et al. 2002).

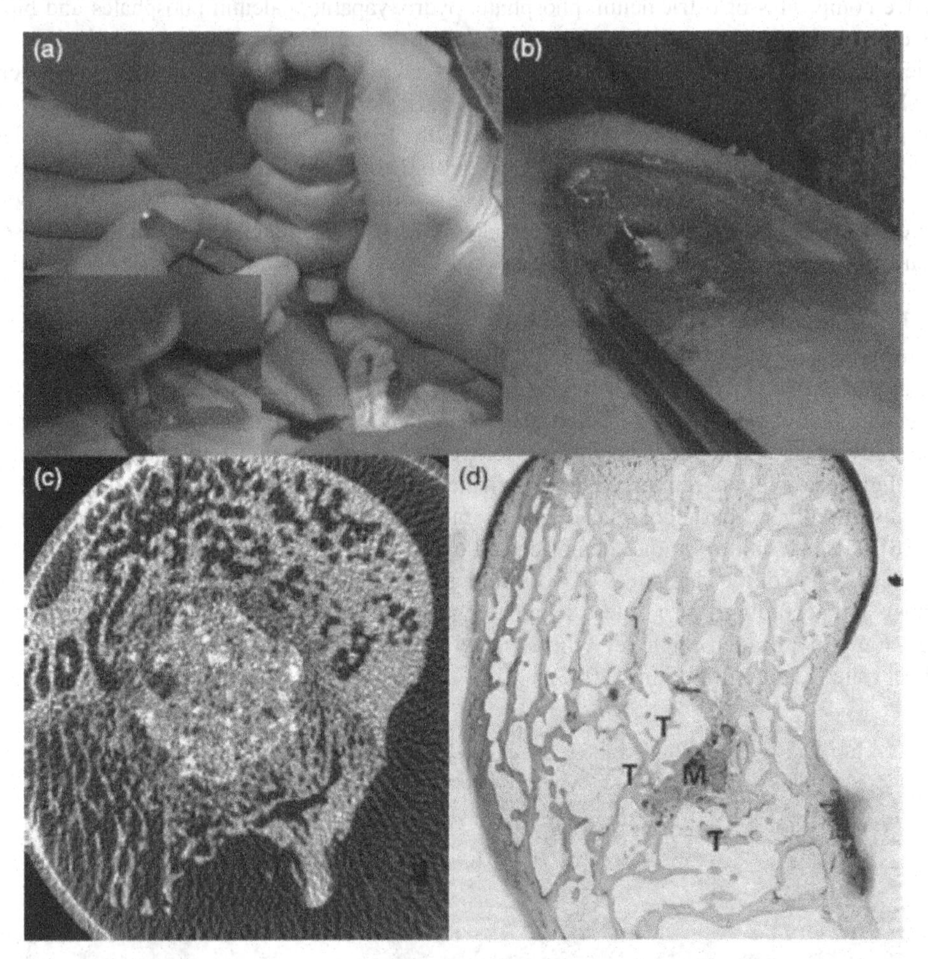

FIGURE 7.7 Preclinical study of a calcium-deficient hydroxyapatite porous scaffold. (a) The scaffold was implanted by injection in a critical size defect at the femoral diaphysis of New Zealand white rabbits. (b) Scaffold placed in the bone defect. Good cohesion was observed after injection, with no collapse or disintegration. (c) Microtomography of the site of implantation after 1 month of surgery. The bright zone corresponds to the implanted scaffold that perfectly adapts to the shape of the defect. (d) Histology of the site of implantation after 5 months of surgery, where T indicates the growing bone trabecula and M the remaining material. (Reprinted with permission from Ginebra, M.P. and E.B. Montufar. 2014. *Injectable biomedical foams for bone regeneration.* pp. 281-312. *In:* P.A. Netti [ed.]. Biomedical Foams for Tissue Engineering Applications. Woodhead Publishing, Kidlington, UK).

7.8.2.2 Bioactive Glass Filler Based

Injectable composite composed of bioactive glass (BAG) particulates and poly(ε-caprolactone-co-D, L-lactide) has been used as bone fillers in cancellous and cartilaginous subchondral bone defects in rabbits (Ginebra and Montufar 2014). Results showed that the incorporation of BAG particulates could lead to appreciable osteoconductivity and bone bonding. In addition, injectable self-curing systems based on phosphate-free bioactive glasses and poly(methyl methacrylate) (PMMA) were investigated *in vivo* by injecting a cement dough into a defect created in the femur of rabbits and curing the cement *in situ*. In contrast to control (PMMA), all bioactive formulations containing bioactive glass showed resorption of the PMMA cement. This could be attributed to the presence of the resorbable bioactive glass. Furthermore, cements formulated with bioactive glasses showed maximum neo-bone formation within two weeks and a more stable bone at the end of the eight weeks (Gonzalez Corchon et al. 2006). It was thus established that these composites possess substantial potential in bone tissue engineering (Kondiah et al. 2016).

7.9 INJECTABLE NANOCOMPOSITE SCAFFOLDS

A variety of nanocomposite materials designed to mimic the nanostructure and composition of natural bone have been proposed. Compared to micron-sized reinforcements, the incorporation of nanoreinforcement offers more biomimetic environment. It has been reported that use of such nanoreinforcement generally results into an improved cell-mediated resorption during scaffold biodegradation. Nanocomposite fibrous scaffolds obtained from a polymer and a nanosized ceramic phase have been reported. Several techniques such as molecular self-assembly, phase separation, and electrospinning fabrication have been employed to develop such nanocomposite scaffolds. The studies have confirmed that incorporation of bioactive agents to nanofibers can lead to biomimetic scaffolds. This was attributed to presence of chemical cues affecting cell-substrate interaction leading to cell adhesion, proliferation and differentiation (Pina et al. 2015, Stankus et al. 2008). Nanocomposites containing electro spun meshes primarily use wet lay-up process where the hydrogel solution is allowed to infiltrate the mesh, often assisted by mechanical pressure, gentle agitation, or vacuum-assisted infiltration (Pina et al. 2015, Butcher et al. 2014, Moutos et al. 2010, Han et al. 2013, Tonsomboon and Oyen 2013, Strange et al. 2014, Moutos and Guilak 2010, Jonoobi et al. 2014). In some cases, near net shape nanofibrous hydrogel composite have been prepared by using electrospinning in conjunction with electro spraying technique (Butcher et al. 2014, Hong et al. 2011, Thorvaldsson et al. 2012, Ekaputra et al. 2008).

In one such example of injectable nanocomposite, aqueous dispersions of poly (L-lactide-co-ethylene oxide-co-fumarate) terpolymer reinforced with nanosized HAp particles have been proposed as injectable multiphasic polymer/ceramic nanocomposites. The terpolymer was reported to interact with the surface of the apatite nanoparticles by polar interactions and hydrogen bonding. This nanocomposite was found to support attachment and migration of marrow stromal cells (Sarvestani and Jabbari 2006). A team from Texas, A&M University embedded ultrathin nanosilicate particles into a collagen-based biocompatible hydrogel in order to promote bone growth through a complex signaling mechanism (Xavier et al. 2015). The injectable hydrogel was engineered to remain at the defect site for a specific amount of time by controlling interactions between the nanosilicates and the matrix. An important feature of the material is its ability to work without protein-based growth factors, in contrast to conventional treatments, where significant amount of growth factor is used to stimulate bone cell growth, which can lead to adverse side effects. In one study, Au-based 4-arms thiol terminated poly (ethylene glycol) [Au-(PEGSH)4] based hydrogel nanocomposites were synthesized by reinforcing bioactive glass (BAG) nanoparticles as depicted in Fig. 7.8. *In vitro* studies were carried out using human osteosarcoma (HOS) cells to understand the cytotoxicity behavior. It was reported that the slow diffusion of the BAG inside the 3D matrix allowed controlled degradation of the BAG nanoparticles while maintaining the pH around physiological values and resulted in reduced cytotoxicity of the nanocomposites. *In vitro* studies also confirmed formation of hydroxyapatite due to the degradation of BAG nanoparticles embedded in the nanocomposite hydrogel. The authors reported that such injectable and self-healing Au-(PEGSH)4-BAG hydrogel nanocomposite biomaterial can be considered as potential scaffold to induce and promote bone self-repair (Gantar et al. 2016).

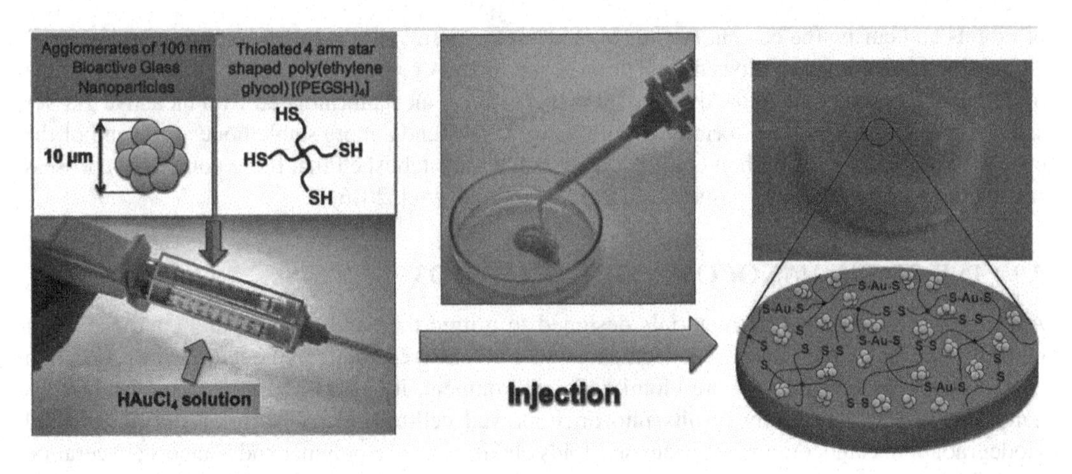

FIGURE 7.8 Schematic representation of the formation of dynamic hydrogel nanocomposites based on 10 wt% $(PEGSH)_4$ homopolymer with 20 mol.% of $HAuCl_4$ containing 10 wt% of 100 nm BAGs nanoparticles (agglomerated in clusters of 10 microns). (Reprinted with permission from Ana Gantar, Nataša Drnovšek, Pablo Casuso, Adrián Pérez-San Vicente, Javier Rodriguez, Damien Dupin, Saša Novak, and Iraida Loinaz. "Injectable and self-healing dynamic hydrogel containing bioactive glass nanoparticles as a potential biomaterial for bone regeneration," RSC Adv., 6(2016): 69156-66).

An injectable and *in situ* forming gel composite (GC) comprised of calcium alginate hydrogel and nano-hydroxyapatite/collagen (nHApC) has been reported to have controllable initial and final setting time. The injectability of GC was reported to be tunable, suggesting its suitability for bone repair and bone tissue engineering (Parameswaran-Thankam et al. 2018, Tan et al. 2009). In another study, an injectable hydrogel system was developed from chitin and poly(butylene succinate) (PBSu) incorporated with nano-sized fibrin nanoparticles (FNPs) and magnesium-doped bioglass (MBG). The hydrophilic nature of PBSu could support good cellular adhesion and growth (Ngamviriyavong et al. 2014). The addition of MBG was found to increase the elastic modulus of the hydrogels. Nanocomposites containing 5% MBG and 2% FNPs showed good rheological properties, injectability, temperature stability, biomineralization, and protein adsorption (Boccaccini and Blaker 2005).

In another study, injectable hyaluronic acid (HA)-based hydrogels were developed using cellulose nanocrystals (CNCs) as a reinforcement (Domingues et al. 2015). The injectable hydrogels were composed of adipic acid dihydrazide-modified HA (ADH-HA) and aldehyde-modified HA (a-HA) matrix reinforced with aldehyde-modified CNCs (a-CNCs) (Parameswaran-Thankam et al. 2018, Domingues et al. 2015). The authors reported that the CNCs could initiate both physical cross-linking (through secondary bonding and interactions) and chemical cross-linking (through covalent bonding) with the HA matrix. The resulting nanocomposites were reported to have high storage modulus which could be attributed to a compact network structure assisted by smaller pore dimensions. It was demonstrated that the biocompatibility of HA-CNCs nanocomposites varied with the concentration of CNCs and could be directly correlated with cell viability, metabolic activity and proliferation rate (Domingues et al. 2015). Development of injectable hydrogel nanocomposites based on carboxymethyl cellulose/dextran matrix, reinforced with cellulose nanocrystals (CNCs) and aldehyde-functionalized CNCs (CHO-CNCs) have also been reported (Yang et al. 2013). The researchers reported that the gelation occurred immediately. Unique physical and chemical properties and low cytotoxicity/ecotoxicity of organic cellulose nanocrystals were utilized in such nanocomposites. Besides, the favorable mechanical properties CNCs could make them ideal as load-bearing component in the bone scaffolds.

Bacterial nanocellulose (BC) has also been employed as a template for the ordered formation of calcium-deficient hydroxyapatite (CdHAP) (Chang and Zhang 2011). The BC-CdHAP nanocomposites were produced by introducing the mineral phase into the bacteria culture media. The results indicated formation of CdHAP spherical clusters, composed of nanosized crystallites, which could be attributed to the similarity of natural bone apatite with CdHAP precursor (Chang and Zhang 2011). The injectable nanocomposites could exploit the bioactivity of CdHAP and biocompatibility of the BC hydrogel for potential orthopedic applications. It was reported that use of carboxymethyl cellulose (CMC) reduced the average diameter of cellulose fibers significantly (almost 50% lower) than that of unmodified fibers (Chang and Zhang 2011, Polo-Corrales et al. 2014). In another attempt, BC-HAp nanocomposite membranes were tested in noncritical bone defects in rat tibiae for upto16 weeks. *In vivo* tests showed absence of any inflammatory reaction after 1 week, while, all defects were found to be completely filled in by regenerated bone tissue (Ginebra and Montufar 2014). Phogat and Bandyopadhyay-Ghosh have reported development of nanocellulose mediated bio-nanocomposite injectable hydrogel as bone graft substitute (Phogat and Bandyopadhyay-Ghosh 2018). They incorporated ultrafine-flurocanasite glass ceramic particulates within wheat straw derived nanocellulose matrix and the resulting bio-nanocomposite hydrogel demonstarted tunable vicoelasticity, self-standing property, and injectability.

7.10 CHARACTERIZATIONS AND PROPERTIES

7.10.1 Pore Microstructure

Several techniques are used for the study of porous systems, with their distinct advantages and disadvantages. Although, microscopy gives information about pore morphology, homogeneity, pore size of the injectable scaffolds, the bi-dimensional images may lead to erroneous estimations. In general, conventional optical microscopes have resolutions around 1-5 μm, while, for smaller pore sizes, scanning electron microscopy (SEM) having larger depth of field is a better alternative. In one study, porosity of poly (acrylamide-co-acrylic acid) based hydrogel was characterized using SEM and the authors reported that the micrograph clearly indicated presence of interconnected pores and capillary channels (Chavda and Patel 2011). Other techniques include mercury pycnometry, mercury intrusion porosimetry (MIP) to obtain information about the pore entrance size distribution, pore shape and tortuosity. Microcomputed tomography is an alternative method to determine porosity, pore morphology, and pore size of injectable scaffolds. The advantages of this technique over MIP are that it is a non-destructive method and it can be very useful for the analysis and finite element modeling of the pore structure through three dimensional reconstructions of porous materials.

7.10.2 Radiopacity

Radiopacity (image density using angiographic instrumentation) is another critical aspect of the injectable material to ensure that the end user can track and control the exact placement of the resulting implant. Radiographic materials can significantly affect implant properties including viscosity, strength etc., when compared to the original material. Liquid contrast agents used in such measurement can also diffuse out from solidified hydrogels over time. Therefore, the timeframe that the injectable implant is considered radiopaque must be determined.

7.10.3 Rheological Properties

7.10.3.1 Viscoelastic Properties and Injectability

An injectable scaffold system should be easily injected and rapidly cross-linked in a reasonable time frame to ensure uniform mixing of the matrix with fillers such as HA prior to injection (Jiao et al. 2012). An injectable hydrogel, for example, ideally should remain as low-viscosity liquid during implantation, while, upon implantation, a fast phase transition should ensure its transformation into elastic solid. Gelation (phase transition) time can be estimated by understanding the variation of the dynamic storage modulus (G′) and the dynamic loss modulus (G″), the real and imaginary components of the complex shear modulus respectively. A thorough rheological study with different sweeps like amplitude, angular frequency and temperature sweep of such viscoelastic materials is therefore important (Gantar et al. 2016).

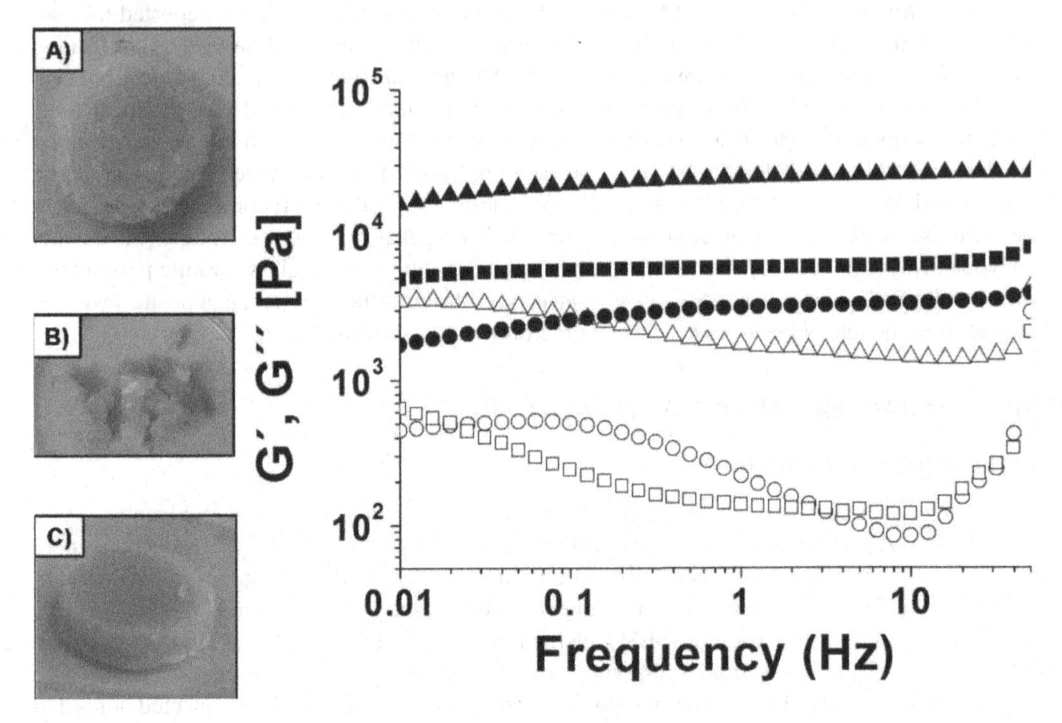

FIGURE 7.9 (Left) Digital pictures of Au-(PEGSH)4-BAG10 hydrogels nanocomposite at pH 7: (A) as-prepared, (B) cut into pieces with a scalpel and (C) self-healed overnight in a disc shaped mould; (Right) Frequency sweep studies for (triangles) Au-(PEGSH)4-BAG20, (squares) Au-(PEGSH)4-BAG10, and (circles) Au-(PEGSH)4-BAG5 hydrogel nanocomposites adjusted at pH 7 after being cut with a scalpel and self-healed in disc shape mould with light pressure overnight. Note that filled and open symbols represent G′ and G″, respectively. (Reprinted with permission from Ana Gantar, Nataša Drnovšek, Pablo Casuso, Adrián Pérez-San Vicente, Javier Rodriguez, Damien Dupin, Saša Novak, and Iraida Loinaz. "Injectable and self-healing dynamic hydrogel containing bioactive glass nanoparticles as a potential biomaterial for bone regeneration," RSC Adv., 6(2016): 69156-66).

Jiao et al. studied the rheological properties via time sweep oscillatory experiments of poly (ethylene glycol) maleate citrate (PEGMC)/hydroxyapatite (HAp) based injectable composite, cross-linked with citric acid-based biodegradable cross-linker. Initially, G″ was found to be larger than G′, as the viscous properties of the liquid state dominated. With time, the composite precursor gained elastic properties resulting in a gel-like state of the hydrogel composite (Jiao et al. 2012). Although, both of the moduli increased with time, it was observed that the slope of G′ was higher than that of G″. HA fillers did not affect the gelation kinetics; however, the elastic moduli improved through interactions between HAp and the polymer. In one study, chitosan nanoparticle loaded gellan xanthan injectable hydrogels were developed and their rheological properties were evaluated. Rheological measurements were performed using frequency (0.1-100 Hz) and amplitude (0.09-50 strain %) sweeps of the composite gels at 37°C. The samples were also tested for viscosity versus temperature and viscosity versus shear stress profiles (Dyondi et al. 2013). Gantar and his co-workers have reported rheological behavior of Au-$(PEGSH)_4$-BAG hydrogels nanocomposites (Gantar et al. 2016). Gelation time of such hydrogel nanocomposites were monitored at 1% deformation and frequency of 0.1 Hz. As shown in Fig. 7.9, G′ was greater than G″ at all frequencies for all samples.

At higher frequencies, conformational change of polymer chains resulted in an increase of G″; although, G′ decreased slightly (Gantar et al. 2016). Tan et al. studied the rheological properties of injectable and *in situ*-forming gel composite comprised of nano-hydroxyapatite/collagen reinforced calcium alginate hydrogel. Rheological results indicated occurrence of pre-gel, sol-gel phase and post-gel transformation in the process of gelation (Tan et al. 2009, 2010). Yang et al. studied the viscoelasticity behavior of cellulose nanocrystals (CNC) reinforced polysaccharide hydrogel based injectable nanocomposites. Parallel plate rheometry was used to measure G′ and G″ (Yang et al. 2013). It was reported that G′ values for the hydrogel nanocomposites exceeded 2000 Pa, while, the G″ values were found to be below 800 Pa, indicating highly elastic nature of the hydrogel nanocomposites (Yang et al. 2013). Phogat and Bandyopadhyay-Ghosh studied the rheological properties of cellulosic nanocomposite injectable hydrogels at various amplitudes, angular frequency and temperatures and reported that both G′ and G″ increased with increase in the loading level of reinforcing fluorcanasite glass-ceramic particulates, indicating enhanced dynamic stability of the nanocomposite hydrogels. With increased loading of glass-ceramic particulates, the gel points of the samples shifted towards the low strain range, ensuring enhanced injectability (Phogat and Bandyopadhyay-Ghosh 2018). Injectable hydrogel nanocomposites made of carboxymethyl cellulose (CMC) and halloysite nanotubes (HNT) have been reported for local treatment of bone defects (Del Buffa et al. 2016). The dynamic viscosity of the composite hydrogel sample was reported as a function of the shear rate. At very low shear rates, the samples demonstrated very short Newtonian region (up to about 0.1 s^{-1}), followed by an extended shear-thinning region due to disentanglement of the polymeric network. However, at higher shear rates (about 50 s^{-1}), a hysteresis effect was observed on application of a decreasing shear rate ramp, typical of CMC concentrated solutions (Del Buffa et al. 2016).

Injectability of paste is defined as its ability to be extruded through an aperture without the separation of the liquid and the powder that compose it (Ginebra and Montufar 2014, Bohner and Baroud 2005). In one study, the injectability of cement paste was measured by determining the percentage of paste that could be extruded from a syringe (Khairoun et al. 1998). Bohner and Baroud proposed a theoretical approach to correlate the effect of geometry and paste formulation on injectability, although, the model only considered the flow of Newtonian fluid through a porous medium (Bohner and Baroud 2005). Weiss and Fatimi have taken into account the rheological behavior of the injectable CaP formulation along with the geometrical components of the injection system to develop model for the paste injectability (Weiss and Fatimi 2010).

7.10.3.2 Gelation/Setting Time

The intermediate period between mixing and gelation (working time) is an important parameter to characterizing settability. Working time is usually calculated from the cure profile (rheological) obtained under constant strain conditions (Page et al. 2013). Gelation time can be evaluated by 'inverted tube test', as proposed by Gupta and co-workers (Gupta et al. 2006). In this test, if a tube is tilted and the injectable solution flows, it is defined as a sol phase, while it is considered to be in gel phase if the solution cannot flow. The time at which the gel does not flow is defined as the gelation time. International standard ISO 9917 is used to determine the setting times of the bone cements. A Gilmore needle method is usually used to find out the setting time. The cement specimen is considered to be set when a tip needle of 1 mm diameter, loaded with 400-g mass does not make perceptible circular indentation on the surface of the cement (Li et al. 2008).

7.10.4 Mechanical Stability

The mechanical properties of injectable hydrogels (e.g., rigidity and brittleness) are important parameters to maintain the gel structure against the compression from surrounding tissues. In addition, the mechanical stiffness can potentially be utilized to regulate the gene expression of the cells distributed in hydrogels (Ishikawa et al. 1999).

In one study, compressive elastic modulus (E) and shear modulus (G) of nanocomposite hydrogels comprised of calcium alginate hydrogel and nano-hydroxyapatite/collagen was found to be dependent on alginate concentration, although, the variations were not linear (Parameswaran-Thankam et al. 2018). This could be attributed to change in cross-linking density with increasing concentration of alginate. However, it was also reported that the presence of nHAC and $Ca_3(PO_4)_2$ particles could result in stress concentration and reduction of mechanical strength (Tan et al. 2009). Biphasic calcium phosphate mixed with silated hydroxypropylmethyl cellulose (Si-HMPC) has been used as an injectable bone substitute. The yield strength of the hydrogel composite (16.4 ± 7.2 MPa) was reported to be significantly higher than for the host trabecular bone tissue (2.7 ± 0.4 MPa) (Chang and Zhang 2011). Liu et al. developed an injectable bone scaffold composed of chitosan, citric acid and glucose solution based matrix and tetracalcium phosphate powder as the reinforcement. When the concentration of citric acid was increased, the compressive strength of the specimen was found to be increased. Addition of the carbon nanotubes (CNTs) to chitosan hydrogel also has been reported to increase the compressibility of the materials and Young's modulus, which could be attributed to higher cross-linking density, reinforcing effect of CNTs and their uniform dispersion in the polymer matrix. Similarly, reinforcement of β-TCP and $CaCO_3$ have been reported to increase both the compressive modulus and the maximum compressive strength of injectable composites of poly [D, L-lactide-co-(ε-caprolactone)]. The compressive strain at break was also found to decrease with increasing filler loading (Liu et al. 2009, López et al. 2010).

7.10.5 Biocompatibility

Biocompatibility tests of injectable scaffolds must be performed as they come in contact with the patient (Igwe et al. 2011). Several biocompatibility tests are required to assess an injectable biomaterial's purity, toxicity profile, irritation and sensitization profile, tissue and blood reactivity and mutagenic profile.

7.10.5.1 In Vitro Biocompatibility

A battery of *in vitro* tests should be performed to assess the biocompatibility of an injectable scaffold. Chemical compatibility of the injectable biomaterial is determined by leachable testing which helps in understanding whether the injectable biomaterial may act as a solvent and extract impurities from the delivery device during implantation (Igwe et al. 2011). Cytotoxicity tests evaluates the biological reactivity, while, genotoxicity tests assess the indication of any bacterial mutations, chromosomal damage or increased metabolic activation. Finally, hemocompatibility tests are used to evaluate if the injectable biomaterial leads to blood cell hemolysis, changes in blood count and blood clotting time.

CNC-reinforced nanocomposite based on injectable polysaccharide hydrogels were studied for *in vitro* cytotoxicity by using a cell viability MTT (3-(4,5-dimethylthiazol-2-yl)-2,5-diphenyltetrazolium bromide) assay with NIH 3T3 fibroblast cells. Minimal decreases in relative cell viability was observed after 1 day of cell exposure. It was also reported that none of the hydrogels had significant cytotoxicity, indicating promising potential of the hydrogels for bone tissue engineering (Yang et al. 2013, De France et al. 2017). Domingues et al. reported development of an injectable bionaocomposite hydrogels composed of adipic acid dihydrazide-modified hyaluronic acid (ADH-HA) reinforced with aldehyde-modified CNCs (a-CNCs). The surface of nanocomposites was examined after 1, 3 and 7 days of cell culture to investigate the ability of developed hydrogels to support the cell adhesion (Domingues et al. 2015). HA-CNC based nanocomposite hydrogels demonstrated cell supportive properties, possibly because of enhanced structural integrity and possible interactions of micro-environmental cues with CNC's sulphate groups (De France et al. 2017, Domingues et al. 2015). Besides, the hydrogels exhibited pronounced proliferative activity. The ability of developed nanocomposites to encapsulate cells and to control cellular behavior were investigated through studies on DNA content, metabolic activity and viability of encapsulated human adipose tissue derived mesenchymal stem cells (hASCs), following their culture at 1, 3 and 7 days. Mitochondrial activity of the cells within the nanocomposites was evaluated by Alamar Blue assay (Domingues et al. 2015). In another study, *in vitro* degradation behaviour of an injectable bone regeneration composites (IBRC) which comprised of nano-hydroxyapatite/collagen particles in alginate hydrogel carrier was investigated. *In vitro* results showed that immersion medium (SBF) had influence on degradation of alginate molecules (Petinakis et al. 2010).

7.10.5.2 *In Vivo* Biocompatibility

In vivo biocompatibility studies further characterize an injectable biomaterial's reactivity and immune response. *In vivo* study of CPC–alginate scaffold was carried out by Polo-Corrales et al., following implantation into rat calvarium. Formation of new bone tissue within the scaffold was revealed after 6 weeks of implantation (Polo-Corrales et al. 2014). In another study, *in vivo* osteogenic activity of the vancomycin-loaded hydroxyapatite/poly (amino acid) injectable composite scaffold was conducted on a rabbit model. The results showed an increase in bone growth as compared to the control group. Similarly, *in vivo* degradation study of nano-hydroxyapatite/collagen composite indicated no obvious decreasing of molecular weight of alginate up to 8 weeks. The scaffold was replaced by connective tissue following degradation after 24 weeks of implantation without any acute inflammatory reaction, indicating controllable degradability and biocompatibility. Biomimetic injectable hydrogel consisting of chitosan (CS) and mineralized collagen fibrils (nHAC) was studied for *in vivo* biocompatibility. A lateral arthrotomy of the knee joint was performed to introduce osseous critical-sized defect at the distal femoral end of rabbit (Fig. 7.10). The CS/nHAC composites with or without mesenchymal stem cells (MSCs) were then injected into the bone defects. The results confirmed formation of new bone and collagen fibrous tissues in the center of the defects after 12 weeks and established that the injectable bone graft substitute may be used to repair a large bone defect effectively.

FIGURE 7.10 Intraoperative view of a cylindrical 7 × 10 mm osseouscritical-sized defect at the distal femoral end of the rabbit. (Reprinted with permission from Huang, Z., Y. Chen, Q.-L. Feng, W. Zhao, B. Yu, J. Tian, S.-J. LI and B.-M. Lin. "*In vivo* bone regeneration with injectable chitosan/hydroxyapatite/collagen composites and mesenchymal stem cells", Front. Mater. Sci., 5(3) (2011): 301-10).

Color version at the end of the book

7.10.5.3 Antibacterial Activity

An injectable system composed of chitosan nanoparticles reinforced gellan xanthan gels has been reported to utilize the antimicrobial properties of chitosan. The components were tested for antibacterial efficacy against multiple antibiotic resistant strains of bacteria clinically prevalent in orthopedic implant infections, including pseudomonas aeruginosa, staphylococcus epidermidis, and staphylococcus aureus (Dyondi et al. 2013). The nanocomposite demonstrated excellent antibacterial activity with increasing concentration of chitosan nanoparticles.

7.11 CONCLUSIONS

Injectable bone scaffolds promise to offer exciting opportunities for bone replacement. While scaffold based bone tissue engineering has already been established, an injectable scaffold and its application with minimally invasive interventions represents clinically-appealing strategy. Development of such scaffolds are also driven by their ability to reach the target sites through limited exposures, tunability of microstructure, controlled distribution of therapeutic agents and cell suspension, resulting into a cell-scaffold construct which can fill any irregular shaped cavity. Many of the injectable biomaterials that have been reviewed in this chapter, by no means are exhaustive however, as new materials, new applications of known materials and new combinations of materials are evolving rapidly. It is important to highlight that the preclinical studies with injectable composites, although, are promising clinical implementation of most of these strategies thus far is based on only a few studies in a few clinical indications. A better understanding of bone biology in harmony with recent developments in tissue engineering, biomaterials, characterization techniques and biomedical engineering will make this truly interdisciplinary efforts more achievable and ambitious. Injectable composites, especially, the nanocomposites promise to furnish even superior properties as compared to unmodified and traditional injectable scaffolds. However, there is limited information on physical and biological effects of nanofillers. Although the potential benefits are clear, it is critically important to ensure that the injectable scaffolds do not pose any threat to the healing process. Process optimization, controlled solidification kinetics and mechanical compatibility with the target tissue are some of the important aspects of an efficient injectable scaffold. Besides, the ability to release incorporated bioactive agents and growth factors must also be considered. Successful implementation of injectable composite scaffolds for bone tissue engineering also depends on their ability to sustain and improve cellular growth as well as to create simulated environment for ECM complementation. A number of crucial challenges therefore remain, before the injectable bone scaffolds can be made clinically relevant. Detailed and concerted research efforts are therefore needed to unlock the potential of these unique bone scaffolds and produce tangible significant benefits in terms of improved life span and quality of lives for many people.

REFERENCES

Achilias, D.S. and P.I. Siafaka. 2017. Polymerization kinetics of poly (2-hydroxyethyl methacrylate) hydrogels and nanocomposite materials. Processes 5(4): 21.

Acton, Q.A. 2013. Bone Fractures: New Insights for the Healthcare Professional. Scholarly Editions, Atlanta, Geogaria.

Ahmed, E.M. and F.S. Aggor. 2010. Swelling kinetic study and characterization of cross-linked hydrogels containing silver nanoparticles. Journal of Applied Polymer Science 117: 2168-2174.

Alvarez, K. and H. Nakajima. 2009. Metallic scaffolds for bone regeneration. Materials 2(3): 790-832.

Amini, A.A. and L.S. Nair. 2012. Injectable hydrogels for bone and cartilage repair. Biomedical Materials 7: 1-13.

Amirthalingam, S., A. Ramesh, S.S. Lee, N.S. Hwang and R. Jayakumar. 2018. Injectable *in situ* shape forming osteogenic nanocomposite hydrogel for regenerating irregular bone defects. ACS Applied Bio Materials 1(4): 1037-1046.

Bai, X., M. Gao, S. Syed, J. Zhuang, X. Xu and X.-Q. Zhang. 2018. Bioactive hydrogels for bone regeneration. Bioactive Materials 3(4): 401-417.

Bakaic, E., N.M.B. Smeets and T. Hoare. 2015. Injectable hydrogels based on poly(ethylene glycol) and derivatives as functional biomaterials. RSC Advances 5(45): 35469-35486.

Balgude, A.P., X. Yu, A. Szymanski and R.V. Bellamkonda. 2001. Agarose gel stiffness determines rate of DRG neurite extension in 3D cultures. Biomaterials 22: 1077-1084.

Bandyopadhyay-Ghosh, S. 2008. Bone as a collagen-hydroxyapatite composite and its repair. Trends in Biomaterials & Artificial Organs 22: 116-124.

Bandyopadhyay-Ghosh, S., I.M. Reaney, A. Johnson, K. Hurrell-Gillingham, I.M. Brook and P.V. Hatton. 2008. The effect of investment materials on the surface of cast fluorcanasite glasses and glass-ceramics. Journal of Materials Science: Materials in Medicine 19: 839-846.

Bandyopadhyay-Ghosh, S., P.E.P. Faria, A. Johnson, D.N.B. Felipucci, I.M. Reaney, L.A. Salata, I.M. Brook and P.V. Hatton. 2010. Osteoconductivity of modified fluorcanasite glass–ceramics for bone tissue augmentation and repair. Journal of Biomedical Materials Research Part A 94A(3): 760-768.

Baroli, B. 2009. From natural bone grafts to tissue engineering therapeutics: brainstorming on pharmaceutical formulative requirements and challenges. Journal of Pharmaceutical Sciences 98(4): 1317-1375.

Baroud, G., E. Cayer and M. Bohner. 2005. Rheological characterization of concentrated aqueous beta-tricalcium phosphate suspensions: the effect of liquid-to-powder ratio, milling time, and additives. Acta Biomaterialia 1(3): 357-363.

Barriga, A., P. az-de-Rada, J.L. Barroso, M. Alfonso, M. Lamata, S. Hernaez, J.L. Beguiristain, M. San-Julian and C. Villas. 2004. Frozen cancellous bone allografts: positive cultures of implanted grafts in posterior fusions of the spine. European Spine Journal 13(2): 152-156.

Bates, P. and M. Ramachandran. 2007. Bone injury, healing and grafting. pp. 123-134. *In*: M. Ramachandran [ed.]. Basic Orthopaedic Sciences. The Stanmore Guide. CRC Press/Taylor & Francis Group, London, UK.

Behravesh, E., K. Zygourakis and A.G. Mikos. 2003. Adhesion and migration of marrow-derived osteoblasts on injectable *in situ* cross-linkable poly(propylene fumarate-co-ethylene glycol)-based hydrogels with a covalently linked RGDS peptide. Journal of Biomedical Materials Research 65A(2): 260-270.

Bencherif, S.A., R.W. Sands, D. Bhatta, P. Arany, C.S. Verbeke, D.A. Edwards and D.J. Mooney. 2012. Injectable preformed scaffolds with shape-memory properties. Proceedings of the National Academy of Sciences 109(48): 19590-19595.

Bencherif, S.A., D.J. Mooney, D. Edwards and R.W. Sands. 2014. Injectable preformed macroscopic 3-dimensional scaffolds for minimally invasive administration. US Patent: US 20140112990A1.

Bendtsen, S.T. and M. Wei. 2015. Synthesis and characterization of a novel injectable alginate-collagen-hydroxyapatite hydrogel for bone tissue regeneration. Journal of Materials Chemistry B 3(15): 3081-3090.

Bigham, A.S., S.N. Dehghani, Z. Shafiei and S. Torabi Nezhad. 2008. Xenogenic demineralized bone matrix and fresh autogenous cortical bone effects on experimental bone healing: radiological, histopathological and biomechanical evaluation. Journal of Orthopaedics and Traumatology 9(2): 73-80.

Boccaccini, A.R. and J.J. Blaker. 2005. Bioactive composite materials for tissue engineering scaffolds. Expert Review of Medical Devices 2(3): 303-317.

Bohner, M. and F. Baumgart. 2004. Theoretical model to determine the effects of geometrical factors on the resorption of calcium phosphate bone substitutes. Biomaterials 25(17): 3569-3582.

Bohner, M. and G. Baroud. 2005. Injectability of calcium phosphate pastes. Biomaterials 26: 1553-1563.

Bohner, M., N. Doebelin and G. Baroud. 2006. Theoretical and experimental approach to test the cohesion of calciumphosphate pastes. European Cells & Materials 12: 26-35.

Bohner, M. 2007. Reactivity of calcium phosphate cements. Journal of Materials Chemistry 17(38): 3980-3986.

Bohner, M. 2010. Design of ceramic-based cements and putties for bone graft substitution. European Cells & Materials 20: 1-12.

Boskey, A. 2007. Mineralization of bones and teeth. Elements Magazine 3: 385-392.

Boskey, A.L. and R. Coleman. 2010. Aging and bone. Journal of Dental Research 89: 1333-1348.

Buckwalter, J.A., M.J. Glimcher, R.R. Cooper and R.R. Recker. 1996. Bone biology I: structure, blood supply, cells, matrix, and mineralization. Instructional Course Lectures 45: 371-386.

Bueno, E.M. and J. Glowacki. 2011. Biologic foundations for skeletal tissue engineering. Synthesis Lectures on Tissue Engineering Vol. 3(1): 1-220.

Burdick, J.A., D. Frankel, W.S. Dernell and K.S. Anseth. 2003. An initial investigation of photocurable three-dimensional lactic acid based scaffolds in a critical-sized cranial defect. Biomaterials 24(9): 1613-1620.

Burkoth, A.K. and K. Anseth. 2000. A review of photo cross-linked polyanhydrides: *In situ* forming degradable networks. Biomaterials 21: 2395-2404.

Butcher, A.L., G.S. Offeddu and M.L. Oyen. 2014. Nanofibrous hydrogel composites as mechanically robust tissue engineering scaffolds. Trends in Biotechnology 32(11): 564-570.

Campoccia, D., P. Doherty, M. Radice, P. Brun, G. Abatangelo and D.F. Williams. 1998. Semisynthetic resorbable materials from hyaluronan esterification. Biomaterials 19: 2101-2127.

Chan, B.P. and K.W. Leong. 2008. Scaffolding in tissue engineering: general approaches and tissue-specific considerations. European Spine Journal 17(Suppl 4): 467-479.

Chang, C. and L. Zhang. 2011. Cellulose-based hydrogels: Present status and application prospects. Carbohydrate Polymers 84(1): 40-53.

Chavda, H.V. and C.N. Patel. 2011. Effect of cross-linker concentration on characteristics of superporous hydrogel. International Journal of Pharmaceutical Investigation 1(1): 17-21.

Chen, C., L. Wang, L. Deng, R. Hu and A. Dong. 2013. Performance optimization of injectable chitosan hydrogel by combining physical and chemical triple cross-linking structure. Journal of Biomedical Materials Research Part A 101(3): 684-693.

Chen, Y., A.F.T. Mak, M. Wang, J. Li and M.S. Wong. 2006. PLLA scaffolds with biomimetic apatite coating and biomimetic apatite/collagen composite coating to enhance osteoblast-like cells attachment and activity. Surface and Coatings Technology 201(3-4): 575-580.

Cheng, Y., A.A. Nada, C.M. Valmikinathan, P. Lee, D. Liang, X. Yu and S.G. Kumbar. 2013. *In situ* gelling polysaccharide-based hydrogel for cell and drug delivery in tissue engineering. Journal of Applied Polymer Science 131(4): 39934.

Cherng, A., S. Takagi and L.C. Chow. 1997. Effects of hydroxyl propyl methylcellulose and other gelling agents on the handling properties of calcium phosphate cement. Journal of Biomedical Materials Research 35: 273-277.

Colla, G., R.A.N. Pértile and L.M. Porto. 2016. Assembling a novel uniform bacterial nanocellulose hydrogel for tissue engineering. COBEQ 2016 - XXI Brazilian Congress of Chemical Engineering, ENBEQ 2016 - XVI Brazilian Meeting on the Teaching of Chemical Engineering, September 25-29, 2016.

Coombes, A.G., S.C. Rizzi, M. Williamson, J.E. Barralet, S. Downes and W.A. Wallace. 2004. Precipitation casting of polycaprolactone for applications in tissue engineering and drug delivery. Biomaterials 25: 315-325.

Correlo, V.M., J.M. Oliveira, J.F. Mano, N.M. Neves and R.L. Reis. 2011. Natural origin materials for bone tissue engineering – properties, processing, and performance. pp. 557-586. *In*: A. Atala, R. Lanza, J.A. Thomson and R. Nerem [eds.]. Principles of Regenerative Medicine (second edition). Academic Press, London, UK.

Dahle, J., E. Kvam and T. Stokke. 2005. Bystander effects in UV-induced genomic instability: antioxidants inhibit delayed mutagenesis induced by ultraviolet A and B radiation. Journal of Carcinogenesis 4(1): 11.

De France, K.J., T. Hoare and E.D. Cranston. 2017. Review of hydrogels and aerogels containing nanocellulose. Chemistry of Materials 29(11): 4609-4631.

Del Buffa, S., E. Rinaldi, E. Carretti, F. Ridi, M. Bonini and P. Baglioni. 2016. Injectable composites via functionalization of 1D nanoclays and biodegradable coupling with a polysaccharide hydrogel. Colloids and Surfaces B: Biointerfaces 145: 562-566.

Deramond, H., C. Depriester, P. Toussaint and P. Galibert. 1997. Percutaneous vertebroplasty. Seminars in Musculoskeletal Radiology 1: 285-296.

Dimitriou, R., E. Jones, D. McGonagle and P.V. Giaanoudi. 2011. Bone regeneration: current concepts and future directions. BMC Medicine 9: 66.

Dinopoulos, H., R. Dimitriou and P.V. Giannoudis. 2012. Bone graft substitutes: what are the options? Surgeon 10(4): 230-239.

Domingues, R.M.A., M. Silva, P. Gershovich, S. Betta, P. Babo, S.G. Caridade, J.F. Mano, A. Motta, R.L. Reis and M.E. Gomes. 2015. Development of injectable hyaluronic acid/cellulose nanocrystals bionanocomposite hydrogels for tissue engineering applications. Bioconjugate Chemistry 26(8): 1571-1581.

Douglas, W.H., R.G. Craig and C.J. Chen. 1979. A new composite restorative based on a hydrophobic matrix. Journal of Dental Research 58(10): 1981-1986.

Dyondi, D., T.J. Webster and R. Banerjee. 2013. A nanoparticulate injectable hydrogel as a tissue engineering scaffold for multiple growth factor delivery for bone regeneration. International Journal of Nanomedicine 8: 47-59.

Einhorn, T.A. 1998. The cell and molecular biology of fracture healing. Clinical Orthopaedics and Related Research 355(Suppl): S7-S21.

Ekaputra, A.K., G.D. Prestwich, S.M. Cool and D.W. Hutmacher. 2008. Combining electrospun scaffolds with electrosprayed hydrogels leads to three-dimensional cellularization of hybrid constructs. Biomacromolecules 9(8): 2097-2103.

Elbert, D.L. and J.A. Hubbell. 2001. Conjugate addition reactions combined with free-radical cross-linking for the design of materials for tissue engineering. Biomacromolecules 2(2): 430-441.

Eliaz, R.E. and J. Kost. 2000. Characterization of a polymeric PLGA-injectable implant delivery system for the controlled release of proteins. Journal of Biomedical Materials Research 50: 388-396.

Feng, X. 2009. Chemical and biochemical basis of cell-bone matrix interaction in health and disease. Current Chemical Biology 3(2): 189-196.

Fernandez-Yague, M.A., S.A. Abbah, L. McNamara, D.I. Zeugolis, A. Pandit and M.J. Biggs. 2015. Biomimetic approaches in bone tissue engineering: integrating biological and physicomechanical strategies. Advanced Drug Delivery Reviews 84: 1-29.

Florencio-Silva, R., G.R. da S. Sasso, E. Sasso-Cerri, M.J. Simões and P.S. Cerri. 2015. Biology of bone tissue: structure, function, and factors that influence bone cells. BioMed Research International 2015: 1-17.

Galibert, P., H. Deramond, P. Rosat and D. Le Gars. 1987. Preliminary note on the treatment of vertebral angioma by percutaneous acrylic vertebroplasty. Neurochirurgie 33(2): 166-168.

Gantar, A., N. Drnovšek, P. Casuso, A. Pérez-San Vicente, J. Rodriguez, D. Dupin, S. Novak and I. Loinaz. 2016. Injectable and self-healing dynamic hydrogel containing bioactive glass nanoparticles as a potential biomaterial for bone regeneration. RSC Advances 6(73): 69156-69166.

Gao, J., J.E. Dennis, L.A. Solchaga, V.M. Goldberg and A.I. Caplan. 2002. Repair of osteochondral defect with tissue-engineered two-phase composite material of injectable calcium phosphate and hyaluronan sponge. Tissue Engineering 8: 827-837.

Geffers, M., J. Groll and U. Gbureck. 2015. Reinforcement strategies for load-bearing calcium phosphate biocements: a review. Materials 8: 2700-2717.

Geffers, M. 2016. Novel Dual Setting Approaches for Mechanically Reinforced Mineral Biocements. Universität Würzburg, Switzerland.

Gerhardt, L.C. and A.R. Boccaccini. 2010. Bioactive glass and glass-ceramic scaffolds for bone tissue engineering. Materials 3: 3867-3910.

Ghassemi, T., A. Shahroodi, M.H. Ebrahimzadeh, A. Mousavian, J. Movaffagh and A. Moradi. 2018. Current concepts in scaffolding for bone tissue engineering. The Archives of Bone and Joint Surgery 6(2): 90-99.

Ginebra, M.P. 2008. Calcium phosphate bone cements. pp. 206-230. *In*: S. Deb [ed.]. Orthopaedic Bone Cements. Woodhead Publishing Series in Biomaterials. CRC Press, New York, Washington, DC.

Ginebra, M.P. and E.B. Montufar. 2014. Injectable biomedical foams for bone regeneration. pp. 281-312. *In*: P.A. Netti [ed.]. Biomedical Foams for Tissue Engineering Applications. Woodhead Publishing, Kidlington, UK.

Gonzalez Corchon, M.A., M. Salvado, B.J. de la Torre, F. Collia, J.A. de Pedro, B. Vazquez and J.S. Roman. 2006. Injectable and self-curing composites of acrylic/bioactive glass and drug systems. A histomorphometric analysis of the behaviour in rabbits. Biomaterials 27: 1778-1787.

Gotman, I. and S. Fuchs. 2011. Bio-inspired resorbable calcium phosphate-polymer nanocomposites for bone healing devices with controlled drug release. pp. 225-258. *In*: M. Zilberman [ed.]. Active Implants and Scaffolds for Tissue Regeneration. Studies in Mechanobiology, Tissue Engineering and Biomaterials. vol 8. Springer, Berlin, Germany.

Green, D., D. Walsh, S. Mann and R.O.C. Oreffo. 2002. The potential of biomimesis in bone tissue engineering: lessons from the design and synthesis of invertebrate skeletons. Bone 30: 810-815.

Grimandi, G., P. Weiss, F. Millot and G. Daculsi. 1998. *In vitro* evaluation of a new injectable calcium phosphate material. Journal of Biomedical Materials Research 39(4): 660-666.

Gupta, A.K. and A.W. Siddiqui. 2012. Environmental responsive hydrogels: a novel approach in drug delivery system. Journal of Drug Delivery and Therapeutics 2(1): 81-88.

Gupta, S.D. 2018. Preparation and characterization of alginate-hydroxyapatite paste for bone regeneration. International Journal of Advances in Science Engineering and Technology 6(3): 16-22.

Gutowska, A., B. Jeong and M. Jasionowski. 2001. Injectable gels for tissue engineering. The Anatomical Record 263: 342-349.

Gómez-Barrena, E., P. Rosset, D. Lozano, J. Stanovici, C. Ermthaller and F. Gerbhard. 2015. Bone fracture healing: cell therapy in delayed unions and nonunions. Bone 70: 93-101.

Habib, M., G. Baroud, F. Gitzhofer and M. Bohner. 2008. Mechanisms underlying the limited injectability of hydraulic calcium phosphate paste. Acta Biomaterialia 4(5): 1465-1471.

Han, N., P.A. Bradley, J. Johnson, K.S. Parikh, A. Hissong, M.A. Calhoun, J.J. Lannutti and J.O. Winter. 2013. Effects of hydrophobicity and mat thickness on release from hydrogel-electrospun fiber mat composites. Journal of Biomaterials Science, Polymer Edition 24(17): 2018-2030.

Harmata, A.J. 2015. Injectable, Settable Polyurethane Biocomposites for Bone Remodeling in Weight-Bearing and Contaminated Fractures. Ph.D. Thesis, Vanderbilt University, USA.

Hasan, A.M.A. and M.El-S. Abdel-Raouf. 2019. Cellulose-based superabsorbent hydrogels. pp. 245-267. *In*: Md.I.H. Mondal [ed.]. Cellulose-Based Superabsorbent Hydrogels. Springer International Publishing, Cham, Switzerland.

Hasel, R.W. 2001. Method of restoring a tooth. U.S. Patent # 6,315,567.

Hasnain, M.S., S.A. Ahmad, N. Chaudhary, M.N. Hoda and A.K. Nayak. 2019. Biodegradable polymer matrix nanocomposites for bone tissue engineering. pp. 1-37. *In*: A.M. Asiri Inamuddin and A. Mohammad [eds.]. Applications of Nanocomposite Materials in Orthopedics, Woodhead Publishing Series in Biomaterials. Kidlington, UK.

Hatefi, A. and B. Amsden. 2002. Biodegradable injectable *in situ* forming drug delivery systems. Journal of Controlled Release 80: 9-28.

Hennink, W.E. and C.F.V. Nostrum. 2002. Novel cross-linking methods to design hydrogels. Advanced Drug Delivery Reviews 54: 13-36.

Hoffman, A.S. 2012. Hydrogels for biomedical applications. Advanced Drug Delivery Reviews 64: 18-23.

Hong, Y., A. Huber, K. Takanari, N.J. Amoroso, R. Hashizume, S.F. Badylak and W.R. Wagner. 2011. Mechanical properties and *in vivo* behavior of a biodegradable synthetic polymer microfiber–extracellular matrix hydrogel biohybrid scaffold. Biomaterials 32(13): 3387-3394.

Hou, Q., P.A. De Bank and K.M. Shakesheff. 2004. Injectable scaffolds for tissue regeneration. Journal of Materials Chemistry 14: 1915-1923.

Hu, W., Z. Wang, Y. Xiao, S. Zhang and J. Wang. 2019. Advances in cross-linking strategies of biomedical hydrogels. Biomaterials Science 7: 843-855.

Huan, Z. and J. Chang. 2009. Novel bioactive composite bone cements based on the β-tricalcium phosphate–monocalcium phosphate monohydrate composite cement system. Acta Biomaterialia 5(4): 1253-1264.

Huang, W.-S. and I.-M. Chu. 2019. Injectable polypeptide hydrogel/inorganic nanoparticle composites for bone tissue engineering. PLOS ONE 14(1): e0210285.

Huh, K.M., H.C. Kang, Y.J. Lee and Y.H. Bae. 2012. pH-sensitive polymers for drug delivery. Macromolecular Research 20(3): 224-233.

Hutmacher, D.W. 2000. Scaffolds in tissue engineering bone and cartilage. Biomaterials 21: 2529-2543.

Igwe, J., A. Amini, P. Mikael, C. Laurencin and S. Nukavarapu. 2011. Nanostructured scaffolds for bone tissue engineering. pp. 169-192. *In*: M. Zilberman [ed.]. Active Implants and Scaffolds for Tissue Regeneration. Studies in Mechanobiology, Tissue Engineering and Biomaterials. vol 8. Springer, Berlin, Heidelberg.

Ishikawa, K., Y. Miyamoto, M. Takechi, Y. Ueyama, K. Suzuki, M. Nagayama and T. Matsumura. 1999. Effects of neutral sodium hydrogen phosphate on setting reaction and mechanical strength of hydroxyapatite putty. Journal of Biomedical Materials Research 44(3): 322-329.

Jahan, K., M. Mekhail and M. Tabrizian. 2019. One-step fabrication of apatite-chitosan scaffold as a potential injectable construct for bone tissue engineering. Carbohydrate Polymers 203: 60-70.

Jeong, B., Y.H. Bae, D.S. Lee and S.W. Kim. 1997. Biodegradable block copolymers as injectable drug-delivery systems. Nature 388(6645): 860-862.

Jeong, S.H., Y-H. Koh, S.W. Kim, J.U. Park, H.E. Kim and J. Song. 2016. Strong and biostable hyaluronic acid-calcium phosphate nanocomposite hydrogel via *in situ* precipitation process. Biomacromolecules 17(3): 841-851.

Jeznach, O., D. Kołbuk and P. Sajkiewicz. 2018. Injectable hydrogels and nanocomposite hydrogels for cartilage regeneration. Journal of Biomedical Materials Research Part A 106(10): 2762-2776.

Jia, T.Y., S. Gurmeet, A. Asni and R. Ramanathan. 2015. Proximal tibia bone graft: an alternative donor source especially for foot and ankle procedures. Malaysian Orthopaedic Journal 9(1): 14-17.

Jiao, Y., D. Gyawali, J.M. Stark, P. Akcora, P. Nair, R.T. Tran and J. Yang. 2012. A rheological study of biodegradable injectable PEGMC/HA composite scaffolds. Soft Matter 8(5): 1499-1507.

Jin, R., L.S. Moreira Teixeira, P.J. Dijkstra, M. Karperien, C.A. van Blitterswijk, Z.Y. Zhong and J. Feijen. 2009. Injectable chitosan-based hydrogels for cartilage tissue engineering. Biomaterials 30(13): 2544-2551.

Jones, J.R. 2013. Review of bioactive glass: from Hench to hybrids. Acta Biomaterialia 9: 4457-4486.

Jonoobi, M., Y. Aitomäki, A.P. Mathew and K. Oksman. 2014. Thermoplastic polymer impregnation of cellulose nanofibre networks: morphology, mechanical and optical properties. Composites Part A: Applied Science and Manufacturing 58: 30-35.

Jung, Y.S., W. Park, H. Park, D.K. Lee and K. Na. 2017. Thermo-sensitive injectable hydrogel based on the physical mixing of hyaluronic acid and pluronic F-127 for sustained NSAID delivery. Carbohydrate Polymers 156: 403-408.

Kaga, S., T.N. Gevrek, A. Sanyal and R. Sanyal. 2016. Synthesis and functionalization of dendron-polymer conjugate based hydrogels via sequential thiol-ene "click" reactions. Journal of Polymer Science Part A: Polymer Chemistry 54: 926-934.

Kang, M.K. and R. Huang. 2011. Swelling-induced instability of substrate-attached hydrogel lines. International Journal of Applied Mechanics 03(02): 219-233.

Kappes, U.P., D. Luo, M. Potter, K. Schulmeister and T.M. Rünger. 2006. Short- and long-wave UV light (UVB and UVA) induce similar mutations in human skin cells. Journal of Investigative Dermatology 126(3): 667-675.

Karlsson, J., W. Wendling, D. Chen, J. Zelinsky, V. Jeevanandam, S. Hellman and C. Carlsson. 1995. Methylmethacrylate monomer produces direct relaxation of vascular smooth muscle *in vitro*. Acta Anaesthesiologica Scandinavica 39: 685-689.

Khairoun, I., M.G. Boltong, F.C. Driessens and J.A. Planell. 1998. Some factors controlling the injectability of calcium phosphate bone cements. Journal of Materials Science: Materials in Medicine 9: 425-428.

Kheirallah, M. and H. Almeshaly. 2016. Bone graft substitutes for bone defect regeneration: a collective review. International Journal of Dentistry and Oral Science 03(5): 247-257.

Kim, D.D., D.H. Kim and Y.J. Son. 2010. Three-dimensional porous scaffold of hyaluronic acid for cartilage tissue engineering. pp. 329-349. *In*: M. Zilberman [ed.]. Active Implants and Scaffolds for Tissue Regeneration. Studies in Mechanobiology, Tissue Engineering and Biomaterials. vol 8. Springer, Berlin, Heidelberg.

Kim, H.K., W.S. Shim, S.E. Kim, K.-H. Lee, E. Kang, J.-H. Kim, K. Kim, I.C. Kwon and D.S. Lee. 2009. Injectable *in situ*–forming ph/thermo-sensitive hydrogel for bone tissue engineering. Tissue Engineering Part A 15(4): 923-933.

King, T.C. 2007. Elsevier's Integrated Pathology. Mosby Elsevier, Philadelphia, PA.

Kona, S., A.S. Wadajkar and K.T. Nguyen. 2011. Tissue engineering applications of injectable biomaterials. pp. 142-182. *In*: B. Vernon [ed.]. Injectable Biomaterials: Science and Applications. Woodhead Publication, Cambridge, UK.

Kondiah, P., Y. Choonara, P. Kondiah, T. Marimuthu, P. Kumar, L. du Toit and V. Pillay. 2016. A review of injectable polymeric hydrogel systems for application in bone tissue engineering. Molecules 21(11): 1580.

Kumawat, V.S., S.B. Ghosh and S. Bandyopadhyay-Ghosh. 2019. Microporous biocomposite scaffolds with tunable degradation and interconnected microarchitecture-a synergistic integration of bioactive chain silicate glass-ceramic and poly(ε-caprolactone). Polymer Degradation and Stability 165: 20-26.

Laurencin, C.T. and L.S. Nair. (eds) 2008. Nanotechnology and Tissue Engineering. The Scaffold. CRC Press, Taylor & Francis Group, Boca Raton-Florida.

Lee, K., C.K. Chan, N. Patil and S.B. Goodman. 2009. Cell therapy for bone regeneration-bench to bedside. Journal of Biomedical Materials Research Part B: Applied Biomaterials 89B(1): 252-263.

Lee, K.Y. and D.J. Mooney. 2001. Hydrogels for tissue engineering. Chemical Reviews 101(7): 1869-1879.

Leslie, W.D. 2012. Clinical review: ethnic differences in bone mass-clinical implications. The Journal of Clinical Endocrinology and Metabolism 97: 4329-4340.

Li, J., Z.Y. Qiu, L. Zhou, T. Lin, Y. Wan, S.Q. Wang and S.M. Zhang. 2008. Novel calcium silicate/calcium phosphate composites for potential applications as injectable bone cements. Biomedical Materials 3(4): 044102.

Li, X., Y. Sogo, A. Ito, H. Mutsuzaki, N. Ochiai, T. Kobayashi, S. Nakamura, K. Yamashita and R.Z. LeGeros. 2009. The optimum zinc content in set calcium phosphate cement for promoting bone formation *in vivo*. Materials Science and Engineering: C 29(3): 969-975.

Lienemann, P.S., M.P. Lutolf and M. Ehrbar. 2012. Biomimetic hydrogels for controlled biomolecule delivery to augment bone regeneration. Advanced Drug Delivery Reviews 64(12): 1078-1089.

Liu, C., H. Shao, F. Chen and H. Zheng. 2006a. Rheological properties of concentrated aqueous injectable calcium phosphate cement slurry. Biomaterials 27: 5003-5013.

Liu, H., H. Lie, W. Cheng, Y. Yang, M. Zhu and C. Zhou. 2006b. Novel injectable calcium phosphate/chitosan composites for bone substitute materials. Acta Biomaterialia 2: 557-565.

Liu, H., X. Chen, C. Zhou and H. Li. 2009. Basic properties of calcium phosphate cement containing chitosan in its liquid phase. *In*: Proceedings of 2nd International Conference on Biomedical Engineering and Informatics. October 17-19, 2009: 1-3.

Liu, M., X. Zeng, C. Ma, H. Yi, Z. Ali, X. Mou, S. Li, Y. Deng and N. He. 2017. Injectable hydrogels for cartilage and bone tissue engineering. Bone Research 5: 17014.

López, A., C. Persson, J. Hilborn and H. Engqvist. 2010. Synthesis and characterization of injectable composites of poly[D, L-lactide-co-(ε-caprolactone)] reinforced with β-TCP and $CaCO_3$ for intervertebral disk augmentation. Journal of Biomedical Materials Research Part B: Applied Biomaterials 95: 75-83.

Maharjan, P., B.W. Woonton, L.E. Bennett, G.W. Smithers, K. DeSilva and M.T.W. Hearn. 2008. Novel chromatographic separation-the potential of smart polymers. Innovative Food Science and Emerging Technologies 9(2): 232-242.

Matsuno, T., Y. Hashimoto, S. Adachi, K. Omata, Y. Yoshitaka, Y. Ozeki, Y. Umezu, Y. Tabata, M. Nakamura and T. Satoh. 2008. Preparation of injectable 3D-formed beta-tricalcium phosphate bead/alginate composite for bone tissue engineering. Dental Materials Journal 27(6): 827-834.

McCann, S., J.L. Byrne, M. Rovira, P. Shaw, P. Ribaud, S. Sica, L. Volin, E. Olavarria, S. Mackinnon and P. Trabasso. 2004. Outbreaks of infectious diseases in stem cell transplant units: a silent cause of death for patients and transplant programmes. Bone Marrow Transplant 33(5): 519-529.

McLemore, R., M.C. Preul and B.L. Vernon. 2006. Controlling delivery properties of a waterborne, *in situ*-forming biomaterial. Journal of Biomedical Materials Research Part B: Applied Biomaterials 79B(2): 398-410.

Mclemore, R. 2011. Rheological properties of injectable biomaterials. *In*: B. Vernon [ed.]. Injectable Biomaterials: Science and Applications. Oxford: Woodhead Publications.

Merolli, A. and P.T. Leali. 2012. Hard tissue structure and functionality. pp. 81-94. *In*: M. Santin and G. Phillips [eds.]. Biomimetic, Bioresponsive, and Bioactive Materials. John Wiley & Sons, New Jersey, USA.

Migliaresi, C., A. Motta and A.T. DiBenedetto. 2007. Injectable scaffolds for bone and cartilage regeneration. pp. 95-109. *In*: F. Bronner, M.C. Farach-Carson and A.G. Mikos [eds.]. Engineering of Functional Skeletal Tissues. Topics in Bone Biology, vol 3. Springer, London.

Miyata, T., N. Asami and T. Uragami. 1999. A reversibly antigen-responsive hydrogel. Nature 399(6738): 766-769.

Montufar, E.B., T. Traykova, C. Gil, I. Harr, A. Almirall, A. Aguirre, E. Engel, J.A. Planell and M.P. Ginebra. 2009. Foamed surfactant solution as a template for self-setting injectable hydroxyapatite scaffolds for bone regeneration. Acta Biomaterialia 6: 876-885.

Montufar, E.B., T. Traykova, J.A. Planell and M.P. Ginebra. 2011. Comparison of a low molecular weight and a macromolecular surfactant as foaming agents for injectable self-setting hydroxyapatite foams: polysorbate 80 versus gelatine. Materials Science and Engineering: C 31: 1498-1504.

Montufar, E.B., D. Ben-David, M. Espanol, E. Livne and M.P. Ginebra. 2012. BMP-2 release from low-temperature processed calcium phosphate foams. Journal of Tissue Engineering and Regenerative Medicine 6: 328.

Moutos, F.T. and F. Guilak. 2010. Functional properties of cell-seeded three-dimensionally woven poly(ε-caprolactone) scaffolds for cartilage tissue engineering. Tissue Engineering Part A 16(4): 1291-1301.

Moutos, F.T., B.T. Estes and F. Guilak 2010. Multifunctional hybrid three-dimensionally woven scaffolds for cartilage tissue engineering. Macromolecular Bioscience 10(11): 1355-1364.

Munting, E., A.A. Mirtchi and J. Lemaitre. 1993. Bone repair of defects filled with phosphoclacic hydraulic cemet-an *in vitro* study. Journal of Materials Science: Materials in Medicine 4: 337-344.

Naahidi, S., M. Jafari, M. Logan, Y. Wang, Y. Yuan, H. Bae, B. Dixon and P. Chen. 2017. Biocompatibility of hydrogel-based scaffolds for tissue engineering applications. Biotechnology Advances 35(5): 530-544.

Navarro, M., A. Michiardi, O. Castano and J.A. Planell. 2008. Biomaterials in orthopedics. Journal of the Royal Society Interface 5(27): 1137-1158.

Ngamviriyavong, P., S. Patntirapong, W. Janvikul, S. Arphavasin, P. Meesap and W. Singhatanadgit. 2014. Development of poly(butylene succinate)/calcium phosphate composites for bone engineering. Composite Interfaces 21: 431-441.

Nourmohammadi, J., A. Ghaee and S.H. Liavali. 2016. Preparation and characterization of bioactive composite scaffolds from polycaprolactone nanofibers-chitosan-oxidized starch for bone regeneration. Carbohydrate Polymers 138: 172-179.

Olson, S.A., M.W. Kadrmas, J.D. Hernandez, R.R. Glisson and J.L. West. 2007. Augmentation of posterior wall acetabular fracture fixation using calcium-phosphate cement: a biomechanical analysis. Journal of Orthopaedic Trauma 21(9): 608-616.

Olszta, M.J., X.G. Cheng, S.S. Jee, R. Kumar, Y.Y. Kim, M.J. Kaufman, E.P. Douglas and L.B. Gower. 2007. Bone structure and formation: a new perspective. Materials Science and Engineering R: Reports 58(3-5): 77-116.

O'Keeffe, R.M., B.L. Riemer and S.L. Butterfield. 2008. Harvesting of autologous cancellous bone graft from the proximal tibia metaphysis: a review of 230 cases. Journal of Orthopaedic Trauma 5(4): 469-474.

Page, J.M., E.M. Prieto, J.E. Dumas, K.J. Zienkiewicz, J.C. Wenke, P.B. Baer and S.A. Guelcher. 2012. Biocompatibility and chemical reaction kinetics of injectable, settable polyurethane/allograft bone biocomposites. Acta Biomaterialia 8(12): 4405-4416.

Page, J.M., A.J. Harmata and S.A. Guelcher. 2013. Design and development of reactive injectable and settable polymeric biomaterials. Journal of Biomedical Materials Research Part A 101(12): 3630-3645.

Paige, K.T., L.G. Cima, M.J. Yaremchuk, J.P. Vacanti and C.A. Vacanti. 1995. Injectable cartilage. Plastic and Reconstructive Surgery 96(6): 1390-1398.

Parameswaran-Thankam, A., C.M. Parnell, F. Watanabe, A.B. RanguMagar, B.P. Chhetri, P.K. Szwedo, A.S. Biris and A. Ghosh. 2018. Guar-based injectable thermoresponsive hydrogel as a scaffold for bone cell growth and controlled drug delivery. ACS Omega 3(11): 15158-15167.

Peter X.M. and J. Elisseeff (eds). 2005. Scaffolding in Tissue Engineering. Taylor & Francis Group – CRC Press, Boca Raton, Florida.

Petinakis, E., X. Liu, L. Yu, C. Way, P. Sangwan, K. Dean, S. Bateman and G. Edward. 2010. Biodegradation and thermal decomposition of poly (lactic acid)-based materials reinforced by hydrophilic fillers. Polymer Degradation and Stability 95(9): 1704-1707.

Phogat, K. and S. Bandyopadhyay-Ghosh. 2018. Nano-cellulose mediated injectable bio-nanocomposite hydrogel scaffold-microstructure and rheological properties. Cellulose 25(10): 5821-5830.

Pina, S., J.M. Oliveira and R.L. Reis. 2015. Natural-based nanocomposites for bone tissue engineering and regenerative medicine: a review. Advanced Materials 27(7): 1143-1169.

Polo-Corrales, L., M. Latorre-Esteves and J.E. Ramirez-Vick. 2014. Scaffold design for bone regeneration. Journal of Nanoscience and Nanotechnology 14(1): 15-56.

Portnov, T., T.R. Shulimzon and M. Zilberman. 2017. Injectable hydrogel-based scaffolds for tissue engineering applications. Reviews in Chemical Engineering 33(1): 91-107.

Prestwich, G.D., D.M. Marecak, J.F. Marecek, K.P. Vercruysse and M.R. Ziebell. 1998. Controlled chemical modification of hyaluronic acid: synthesis, applications, and biodegradation of hydrazide derivatives. Journal of Controlled Release 53: 93-103.

Qiu, Y. and K. Park. 2001. Environment-sensitive hydrogels for drug delivery. Advanced Drug Delivery Reviews 53: 321-339.

Rodell, C.B., J.W. MacArthur, S.M. Dorsey, R.J. Wade, L.L. Wang, Y.J. Woo and J.A. Burdick. 2015. Shear-thinning supramolecular hydrogels with secondary autonomous covalent cross-linking to modulate viscoelastic properties *in vivo*. Advanced Functional Materials 25: 636-644.

Rojo, L., B. Vázquez and J.S. Román. 2014. Biomaterials for scaffolds: synthetic polymers. pp. 263-300. *In*: C. Migliaresi and A. Motta [eds.]. Scaffolds for tissue engineering-biological design, materials, and fabrication. Pan Stanford Publishing, CRC Press. Singapore.

Rose, F.R.A.J. and R.O.C. Oreffo. 2002. Bone tissue engineering: hope vs hype. Biochemical and Biophysical Research Communications 292(1): 1-7.

Ruhé, P.Q., E.L. Hedberg-Dirk, N.T. Padron, P.H.M. Spauwen, J.A. Jansen and A.G. Mikos. 2006. Porous Poly (DL-lactic-co-glycolic acid)/calcium phosphate cement composite for reconstruction of bone defects. Tissue Engineering 12(4): 789-800.

Ruirui, X., L. Kai, J. Tifeng, Z. Ning, M. Kai, Z. Ruiyun, Z. Qianli, M. Guanghui and Y. Xuehai. 2016. An injectable self-assembling collagen–gold hybrid hydrogel for combinatorial antitumor Photothermal/photodynamic therapy. Advanced Materials 28(19): 3669-3676.

Ryan, G., A. Pandit and D.P. Apatsidis. 2006. Fabrication methods of porous metals for use in orthopaedic applications. Biomaterials 27: 2651-2670.

Sarvestani, A.S. and E. Jabbari. 2006. Modeling and experimental investigation of rheological properties of injectable poly (lactide ethylene oxide fumarate)/hydroxyapatite nanocomposites. Biomacromolecules 7(5): 1573-1580.

Sarvestani, A.S., X. He and E. Jabbari. 2007. Viscoelastic characterization and modeling of gelation kinetics of injectable *in situ* cross-linkable poly (lactide-co-ethylene oxide-co-fumarate) hydrogels. Biomacromolecules 8(2): 406-415.

Schoenfeld, C.M., G.J. Conard and E.P. Lautenschlager. 1979. Monomer release from methacrylate bone cements during simulated *in vivo* polymerization. Journal of Biomedical Materials Research 13: 135-147.

Seeherman, H.J., K. Azari, S. Bidic, L. Rogers, X.J. Li, J.O. Hollinger and J.M. Wozney. 2006. rhBMP-2 delivered in a calcium phosphate cement accelerates bridging of critical-sized defects in rabbit radii. Journal of Bone and Joint Surgery 88(7): 1553-1565.

SEER Training Modules, 2019. U.S. National Institutes of Health. https://training.seer.cancer.gov/anatomy/skeletal/tissue.html.

Sheikh, Z., C. Sima and M. Glogauer. 2015. Bone replacement materials and techniques used for achieving vertical alveolar bone augmentation. Materials 8: 2953-2993.

Simeonov, M.S., A.A. Apostolov and E.D. Vassileva. 2016. *In situ* calcium phosphate deposition in hydrogels of poly (acrylic acid)-polyacrylamide interpenetrating polymer networks. RSC Advances 6(20): 16274-16284.

Skovrlj, B., J. Gilligan, H.S. Cutler and S.A. Qureshi. 2015. Minimally invasive procedures on the lumbar spine. World Journal of Clinical Cases 3(1): 1-9.

Stankus, J.J., D.O. Freytes, S.F. Badylak and W.R. Wagner. 2008. Hybrid nanofibrous scaffolds from electrospinning of a synthetic biodegradable elastomer and urinary bladder matrix. Journal of Biomaterials Science, Polymer Edition 19(5): 635-652.

Strange, D.G.T., K. Tonsomboon and M.L. Oyen. 2014. Mechanical behaviour of electrospun fibre-reinforced hydrogels. Journal of Materials Science: Materials in Medicine 25(3): 681-690.

Tan, H., and K.G. Marra. 2010. Injectable, biodegradable hydrogels for tissue engineering applications. Materials 3: 1746-1767.

Tan, R., X. Niu, S. Gan and Q. Feng. 2009. Preparation and characterization of an injectable composite. Journal of Materials Science: Materials in Medicine 20(6): 1245-1253.

Tan, R., Q. Feng, Z. She, M. Wang, H. Jin, J. Li and X. Yu. 2010. *In vitro* and *in vivo* degradation of an injectable bone repair composite. Polymer Degradation and Stability 95(9): 1736-1742.

Tanner, K.E. 2017. Hard tissue applications of biocomposites. pp. 37-58. *In*: L. Ambrosio [ed.]. Biomedical Composites (Second Edition). Woodhead Publishing: Cambridge, UK.

Taton, T.A. 2001. Boning up on biology. Nature 412(6846): 491-492.

Temenoff, J.S. and A.G. Mikos. 2000. Injectable biodegradable materials for orthopedic tissue engineering. Biomaterials 21: 2405-2412.

Thangprasert, A., C. Tansakul, N. Thuaksubun and J. Meesane. 2019. Mimicked hybrid hydrogel based on gelatin/PVA for tissue engineering in subchondral bone interface for osteoarthritis surgery. Materials & Design 183: 108113.

Thorvaldsson, A., J. Silva-Correia, J.M. Oliveira, R.L. Reis, P. Gatenholm and P. Walkenström. 2012. Development of nanofiber-reinforced hydrogel scaffolds for nucleus pulposus regeneration by a combination of electrospinning and spraying technique. Journal of Applied Polymer Science 128(2): 1158-1163.

Tian, H., Z. Tang, X. Zhuang, X. Chen and X. Jing. 2012. Biodegradable synthetic polymers: preparation, functionalization and biomedical application. Progress in Polymer Science 37: 237-280.

Timmer, M.D., C.G. Ambrose and A.G. Mikos. 2003a. Evaluation of thermal- and photo-cross-linked biodegradable poly (propylene fumarate)-based networks. Journal of Biomedical Materials Research Part A 66: 811-818.

Timmer, M.D., C.G. Ambrose and A.G. Mikos. 2003b. *In vitro* degradation of polymeric networks of poly (propylene fumarate) and the cross-linking macromere poly (propylene fumarate)-diacrylate. Biomaterials 24: 571-577.

Tommasi, G., S. Perni and P. Prokopovich. 2016. An injectable hydrogel as bone graft material with added antimicrobial properties. Tissue Engineering Part A 22(11-12): 862-872.

Tonsomboon, K. and M.L. Oyen. 2013. Composite electrospun gelatin fiber-alginate gel scaffolds for mechanically robust tissue engineered cornea. Journal of the Mechanical Behavior of Biomedical Materials 21: 185-194.

Vo, T.N., F.M. Kasper and A.G. Mikos. 2012. Strategies for controlled delivery of growth factors and cells for bone regeneration. Advanced Drug Delivery Reviews 64(12): 1292-1309.

Wang, H., S. Xu, S. Zhou, W. Xu, M. Leary, P. Choong, M. Qian, M. Brandt and Y.M. Xie. 2016. Topological design and additive manufacturing of porous metals for bone scaffolds and orthopaedic implants: a review. Biomaterials 83: 127-141.

Wang, X., J. Ye and H. Wang. 2006. Effects of additives on the rheological properties and injectability of a calcium phosphate bone substitute material. Journal of Biomedical Materials Research Part B: Applied Biomaterials 78(2): 259-264.

Watanabe, J., M. Kashii, M. Hirao, K. Oka, K. Sugamoto, H. Yoshikawa and M. Akashi. 2007. Quick-forming hydroxyapatite/agarose gel composites induce bone regeneration. Journal of Biomedical Materials Research Part A 83A(3): 845-852.

Weiss, P. and A. Fatimi. 2010. Injectable composites for bone repair. pp. 255-275. *In*: L. Ambrosio [ed.]. Biomedical Composites. Woodhead Publishing India Private Limited. New Delhi-India.

Whitaker, M.J., R.A. Quirk, S.M. Howdle and K.M. Shakesheff. 2001. Growth factor release from tissue engineering scaffolds. Journal of Pharmacy and Pharmacology 53: 1427-1437.

Williams, C.G. and J.H. Elisseeff. 2005. Injectable systems for cartilage tissue engineering. pp. 169-188. *In*: P.X. Ma and J. Elisseeff [eds]. Scaffolding in Tissue Engineering. CRC Press, Taylor & Francis Group, Boca Raton-Florida.

Winkler, T., F.A. Sass, G.N. Duda and K. Schmidt-Bleek. 2018. A review of biomaterials in bone defect healing, remaining shortcomings and future opportunities for bone tissue engineering. Bone & Joint Research 7(3): 232-243.

Wu, G., C. Feng, J. Quan, Z. Wang, W. Wei, S. Zang, S. Kang, G. Hui, X. Chen and Q. Wang. 2018. *In situ* controlled release of stromal cell-derived factor-1α and antimiR-138 for on-demand cranial bone regeneration. Carbohydrate Polymers 182: 215-224.

Wu, S., X. Liu, K.W.K. Yeung, C. Liu and X. Yang. 2014. Biomimetic porous scaffolds for bone tissue engineering. Materials Science and Engineering: R: Reports 80: 1-36.

Xavier, J.R., T. Thakur, P. Desai, M.K. Jaiswal, N. Sears, E. Cosgriff-Hernandez, R. Kaunas and A.K. Gaharwar. 2015. Bioactive nanoengineered hydrogels for bone tissue engineering: a growth-factor-free approach. ACS Nano 9(3): 3109-3118.

Yan, S., W. Wang, X. Li, J. Ren, W. Yun, K. Zhang, G. Li and J. Yin. 2018. Preparation of mussel-inspired injectable hydrogels based on dually functionalized alginate with improved adhesive, self-healing, and mechanical properties. Journal of Materials Chemistry B 6: 6377-6390.

Yang, S., K. Leong, Z. Du and C. Chua. 2001. The design of scaffolds for use in tissue engineering. Part I: Traditional factors. Tissue Engineering 7: 679-689.

Yang, X., E. Bakaic, T. Hoare and E.D. Cranston. 2013. Injectable polysaccharide hydrogels reinforced with cellulose nanocrystals: morphology, rheology, degradation, and cytotoxicity. Biomacromolecules 14(12): 4447-4455.

Yoshizawa, T., Y. Shin-Ya, H. Kyung-Jin and T. Kajiuchi. 2004. pH- and temperature-sensitive permeation through polyelectrolyte complex films composed of chitosan and polyalkyleneoxide–maleic acid copolymer. Journal of Membrane Science 241(2): 347-354.

Zhang, H., S. Guo, S. Fu and Y. Zhao. 2017. A near-infrared light-responsive hybrid hydrogel based on UCST triblock copolymer and gold nanorods. Polymers 9(12): 238.

Zhao, D., Q. Tang, Q. Zhou, K. Peng, H. Yang and X. Zhang. 2018. A photo-degradable injectable self-healing hydrogel based on star poly (ethylene glycol)-b-polypeptide as potential pharmaceuticals delivery carrier. Soft Matter 14: 7420-7428.

Zhao, X. 2010. Injectable Degradable Composite Materials for Bone Repair and Drug Delivery. Ph.D. Thesis, University College London, UK.

8

Bio-Ceramics for Tissue Engineering

Hasan Zuhudi Abdullah*, Te Chuan Lee, Maizlinda Izwana Idris and Mohamad Ali Selimin

8.1 INTRODUCTION

Biomaterials are very important in biomedical applications for replacement, construction and repairing hard tissue and soft tissue purposes. Biomaterials can be classified into biometal, bioceramic, biopolymer and biocomposite. Bioceramics have got more attention for bone reconstruction and as an implant especially for hard tissue (bone). The properties of bioceramics were altered depending the specific application in the human body. It can be in various form and structure such as porous, dense and combination of them. In this chapter, metal oxide ceramic (gel oxidation of titanium, i.e. TiO_2), glass ceramic (Bioactive glass) and ceramic (hydroxyapatite) were discussed in association with bioactive properties and reaction with the natural bone. *In vitro* testing (simulated body fluid (SBF) and cultured cell (osteoblast)) were performed to study the bioactive properties and prediction of in vivo reaction of bioceramics. The preparation, mechanism and biological reactions are investigated and analysed to get the information for potential use in biomedical applications. The analysed results from the *in vitro* testing show the suitability of bioceramics (bioactive) for substituting or repairing hard tissue (bone).

8.2 OVERVIEW OF BIO-CERAMICS FOR TISSUE ENGINEERING

Bio-ceramics for biomedical applications can be categorised into three types, i.e. bioinert, bioactive and bioresorbable. It is classified according to the material's tissue response (Table 8.1). Bio-ceramics

TABLE 8.1 Types of bio-ceramics based on tissue response (Hench 1998)

Types of Material	Biological Properties of Materials	Tissue Response	Example of Bio-ceramics
Bioinert	Biologically inactive	A fibrous tissue of variable thickness forms	Alumina, zirconia, titania (TiO_2)
Bioactive	Biologically active	An interfacial bond forms, between the tissue and material	Hydroxyapatite, bioactive glasses
Bioresorbable	Dissolves in surrounding environment	The surrounding tissue replaces it	Tri-calcium phosphate

Department of Materials Engineering and Design, Faculty of Mechanical and Manufacturing Engineering, Universiti Tun Hussein Onn Malaysia, 86400 Parit Raja, Batu Pahat, Johor, Malaysia.

* Corresponding author: hasan@uthm.edu.my

are used for the repair and reconstruction of damaged parts of the body, including replacements for hips, knees, teeth, and fractured bones as shown in Fig. 8.1 (Liu et al. 2004). Bio-ceramics may come from metal oxides such as alumina from aluminium and titania from titanium.

The three types of bio-ceramics that are relevant to improve bioactivity with hard tissue (bone) will be discussed in this chapter.

1. Metal oxide ceramics – Titanium dioxide (TiO_2)
2. Glass ceramics – Bioactive glass
3. Ceramics – Hydroxyapatite

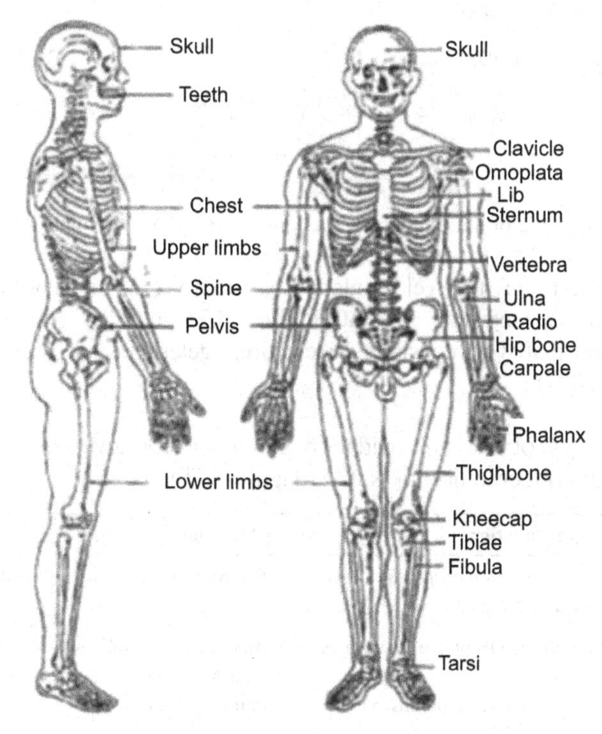

FIGURE 8.1 Schematic diagram of hard tissues in human body (Liu et al. 2004). *(Reproduced with permission of Elsevier).*

8.3 METAL OXIDE CERAMICS – TITANIUM DIOXIDE (TIO_2)

Gel oxidation is a surface modification of titanium (Ti) by using thermochemical method to prepare bioactive ceramic (TiO_2) (Kim et al. 1996). The bioactive ceramic layer formed on the Ti substrate is important to improve tissue compatibility and natural bonding between the implant and the living bone (Han and Xu 2004). Gel oxidation can be explained as a two-steps process:

1. Gelation: Sodium titanate formed on Ti substrate after treating with NaOH aqueous in glass bottle (24 hours, at 60°C) as illustrated in Fig. 8.2 (Kim et al. 1996, Abdullah 2010). Various concentrations of NaOH were used to vary the thickness of sodium titanate hydrogel.
2. Oxidation: Treated Ti oxidised by heat treatment to produce titanium dioxide (TiO_2) and stabilise the sodium titanate. Amorphous sodium titanate was produced at low temperature (<600°C) or a mixture of crystalline sodium titanate and titanium dioxide produced at high temperature (≥600°C) due to dehydration and densification of the gel (Kim et al. 1996, Jonášová et al. 2004).

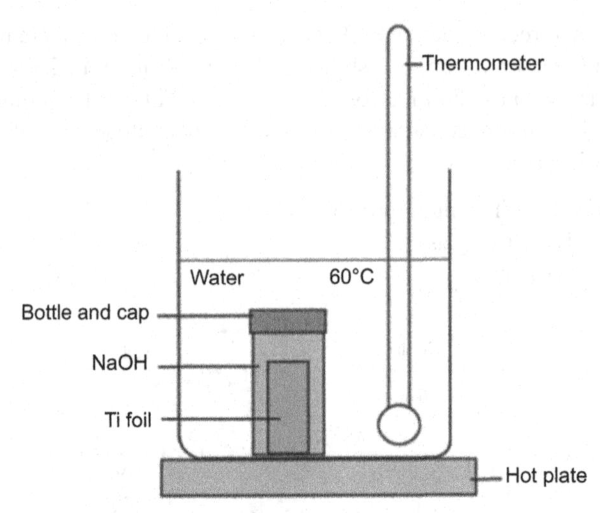

FIGURE 8.2 Schematic of the gelation of titanium sample in NaOH solution (Abdullah 2010).

A modified schematic of the gel oxidation sequence for low concentration of NaOH (Fig. 8.3a) and high concentration of NaOH (Fig. 8.3b) can be explained in five stages (Table 8.2). Parameters that were considered were time, gelation, and oxidation (modified from (Abdullah and Sorell 2007)).

TABLE 8.2 Five stages of the gel oxidation sequence for low concentration of NaOH (Figure 8.3a) and high concentration of NaOH (Figure 8.3b)

Stage	Explanation	Materials/Structure	References
1	• Passive layer of TiO_2 was present spontaneously on the Ti substrate due to humidity	• Passive anatase	Abdullah and Sorrell 2012
2	• Corrosive attack of the NaOH on the Ti substrate by the hydroxyl groups • Dissolution to form an expanded porous network	• Amorphous (a) sodium titanate hydrogel • Porous anatase	Liu et al. 2004, Kim et al. 1996
3	• Low temperatures (400°C) oxidation • Corrosion of the protective anatase layer • Slow oxidation of the underlying titanium substrate to form anatase	• Porous anatase • Amorphous (a) sodium titanate	Abdullah and Sorrell 2012
4	• Intermediate temperatures (600°C) oxidation • Decomposition of the sodium titanate to volatilise Na_2O and resultant formation of anatase • Densification of the TiO_2 to the point of near-coherence (but not completely) • Hindrance of further corrosion of the Ti substrate owing to the highly (but not fully) dense anatase diffusion barrier	• Porous anatase • Rutile • Crystalline sodium titanate	Abdullah and Sorrell 2012
5	• High temperatures (800°C) oxidation • Rapid oxidation of the underlying titanium substrate • Phase transformation from anatase to rutile • Densification of the rutile • Films adhered strongly to the Ti substrates	• Porous rutile • Dense rutile • Crystalline sodium titanate	Abdullah and Sorrell 2012, Kim et al. 1997, Jaeggi et al. 2006

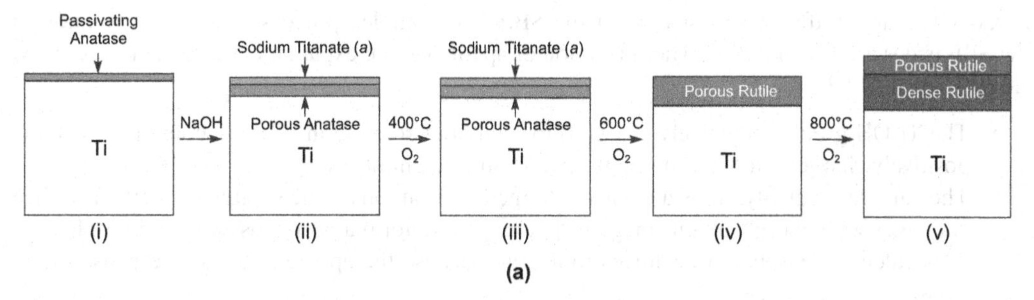

FIGURE 8.3(a) Schematic of gelation and oxidation processes for low concentration of NaOH; a = amorphous; c = crystalline (Abdullah 2010).

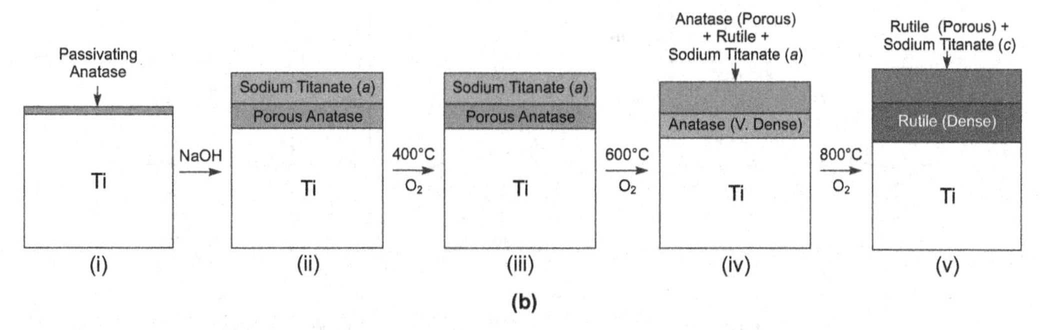

FIGURE 8.3(b) Schematic of gelation and oxidation processes for high concentration of NaOH; a = amorphous; c = crystalline) (Abdullah 2010).

The example application of gel oxidised titanium in biomedical is shown in Fig. 8.4. Rough and porous modified Ti surface will bond with living bone without using cement for fixing.

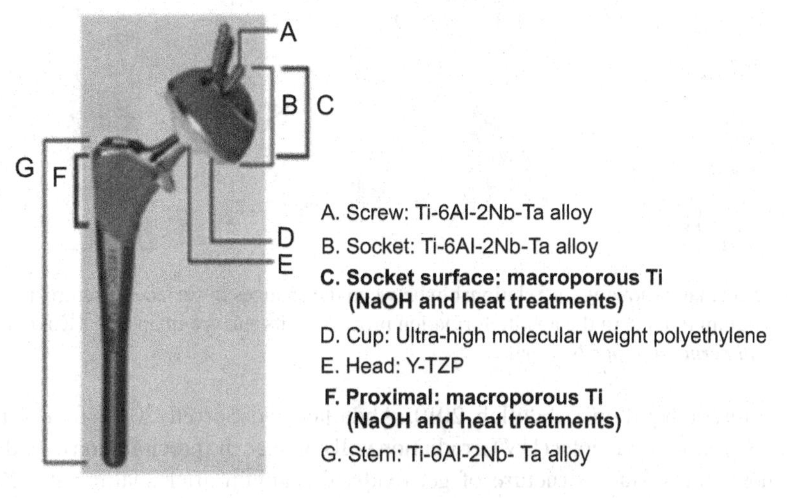

FIGURE 8.4 Bioactive titanium metal in a clinical hip joint system (Kokubo et al. 2003). *(Reproduced with permission of Elsevier).*

8.3.1 The Effects of Immersing Gel-Oxidised Titanium in Simulated Body Fluid (SBF)

The mechanism of sodium titanate (gel oxidation) reaction in simulated body fluid (SBF) was explained by Kokubo et al. (2003). It was found that the surface of sodium titanate was highly negatively charged after it was soaked in the SBF. The complex process of apatite formation in the SBF is described in Fig. 8.5. The formation of apatite can be explained in a sequence as below (Kokubo et al. 2003):

- The Ti-OH groups (negatively charged) formed after soaking in SBF are combined with the positively charged Ca^{2+} ions in the fluid to form calcium titanate.
- The surface gradually gains a positive charge (calcium ions). The positively charged surface combines with phosphate ions (negatively charged) to form amorphous calcium phosphate.
- This calcium phosphate transforms to apatite because the apatite is the stable phase (Tung 1998).

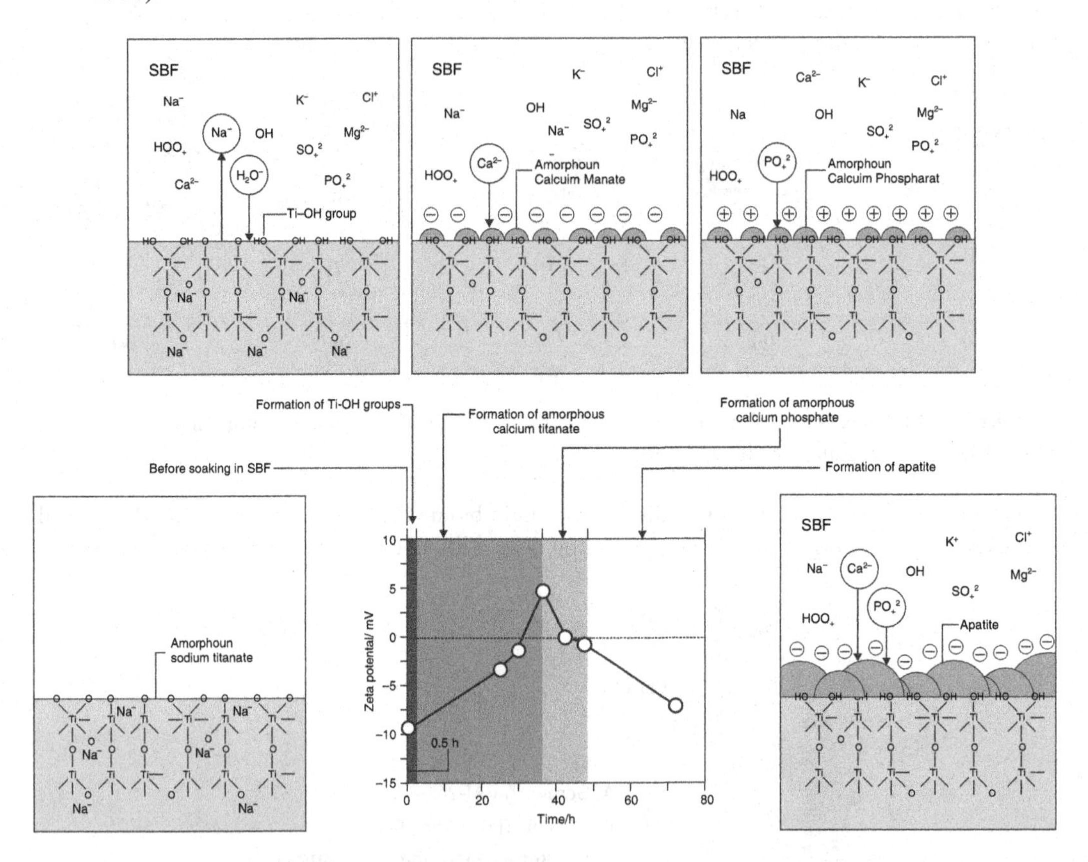

FIGURE 8.5 Schematic showing the relationship between the changes in surface structure and the potential of amorphous sodium titanate in the apatite formation process on its surface in an SBF (Kokubo et al. 2003). *(Reproduced with permission of Elsevier).*

Abdullah's research group (Abdullah 2010, Abdullah and Sorrell 2012) found that soaking oxidised Ti in SBF with ultra violet (UV) irradiation will enhance the precipitation of hydroxyapatite. Figure 8.6 shows the surface structure of gel oxidised titanium after soaking in SBF with and without UV irradiation for 24 hours. Soaking in SBF without UV appears to have thickened and smoothed the struts of the microstructures network of the samples as treated and oxidised at 400°C and 600°C, which suggests the precipitation of a thin layer of hydroxyapatite.

Soaking in SBF with UV irradiation appears to have had a significant effect on enhancing the precipitation of hydroxyapatite. It can be seen on the sample oxidised at 400°C, with precipitation of hydroxypatite and consequent densification. The sample oxidised at 600°C also has the effect of UV irradiation with the increased roundness and smoothness, meaning that hydroxyapatite precipitation was enhanced slightly.

According to Shozui et al. (2008), when TiO_2 is irradiated with UV light, electron-hole pairs are generated along with numerous Ti-OH groups on the TiO_2 surface. Ti-OH groups play the role of nucleation sites for apatite formation (Kokubo et al. 2003). The reaction occurs when UV-irradiation is conducted on TiO_2 surface in SBF. The mechanism of the reaction can be described as follows:

a. UV irradiation of TiO_2 in SBF solution.
b. Electron-hole pairs are generated on the surface. Boehm suggested equations (Kokubo et al. 2003, Cangiani 2003):

$$Ti^{4+}\text{-O-}Ti^{4+} + h\nu \rightarrow Ti^{4+}\text{-O-Ti }(h^+ + e^-) \rightarrow 2\ Ti^{3+} + Vo + \bullet O \quad (1)$$

$$\bullet O + H_2O \rightarrow 2\bullet OH \quad (2)$$

$$Ti^{3+} + Vo + H_2O \rightarrow Ti^{4+}\text{-OH} + H\bullet \quad (3)$$

$$Ti^{3+} + \bullet OH \rightarrow Ti^{4+}\text{-OH} \quad (4)$$

$$\bullet OH + H\bullet \rightarrow H_2O \quad (5)$$

where, h^+: hole, Vo : oxygen vacancy, •OH : free radical species.

c. Attachment of •OH radicals will enhance the negative charge on the surface and help in formation of Ti-OH group and also improve the hydrophilicity of the surface.

The samples oxidised at 400°C showed increased apatite formations and the samples oxidised at 600°C showed apatite formation on the surfaces after irradiation with UV light for 1 day (Fig. 8.6f, 8.6i). The samples oxidised at 400°C and 600°C were more able to form apatite on the surface as compared to those oxidised at higher temperatures. A previous study by Wang et al. (2003) also reported that titanium thermally oxidised at 400°C and 500°C showed greater ability for apatite formation than those thermally oxidised at 700°C and 800°C. This was due to lower crystallinity in the oxide layer (Fujishima et al. 2001).

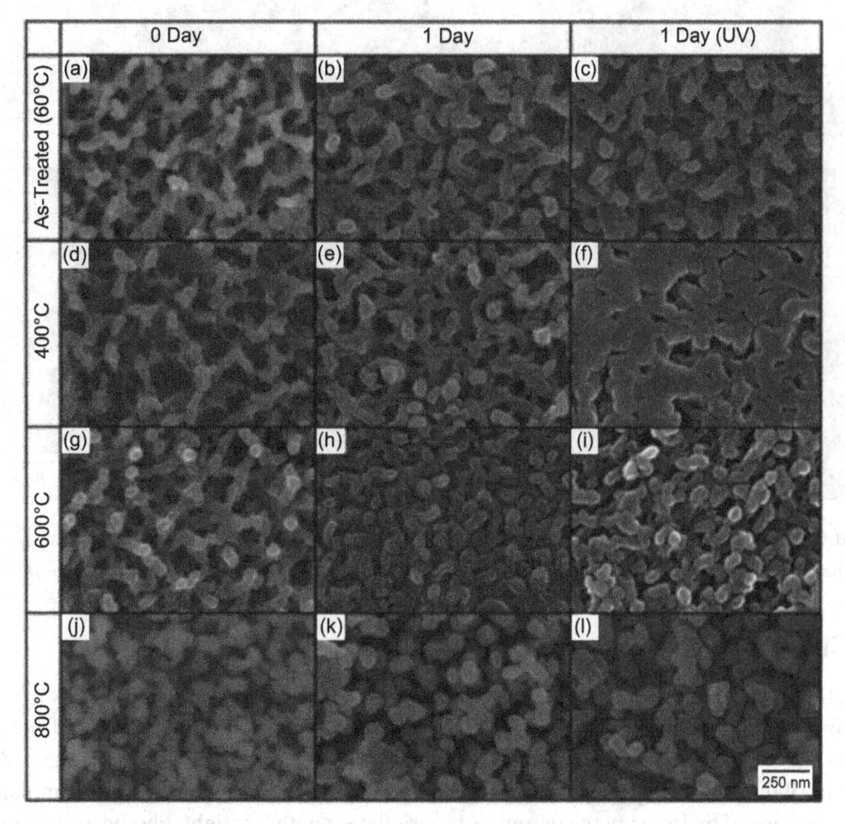

FIGURE 8.6 FESEM images of Ti substrates treated with NaOH and oxidised at 400°C, 600°C and 800°C, followed by soaking in SBF without and with UV irradiation for 24 h (Abdullah 2010).

8.3.2 Culturing Saos-2 Cell on Gel-Oxidised Titanium

The oxidised titanium underwent *in vitro* test by using Saos-2 cell culture to investigate the function of cultured osteoblasts on the oxidised titanium. For comparison, the Ti substrate and oxidised Ti were used as shown in Figure 8.7. The cell morphology on titanium and the gel-oxidised surface after 19 h of cell culture were analysed by using FESEM at low and high magnification. It can be seen that the osteoblast cells had grown on the surfaces of titanium and the gel-oxidised titanium oxide.

The FESEM images showed that osteoblasts have grown on the gel-oxidised film but in different shape compared to the titanium sample. The gel-oxidised surface showed good cell attachment and penetration in the majority of the cells, as shown in Fig. 8.7b. Here, the cells penetrated into the porous network structure and established good bonding. The appearance of this cell growth was indicated by the lack of raised edges, good adhesion and pore penetration. Hence, it was concluded that gel-oxidised Ti can provide a superior implant owing to good ceramic-cell adhesion (Abdullah 2010). Kokubo et al. (2003) also reported that the porous structure of gel-oxidised titanium greatly encouraged bone ingrowth after implantation into rabbit tibia. At an early stage in the implantation period, a hydroxyapatite layer was formed on the gel-oxidised surfaces and this enhanced bonding with the bone.

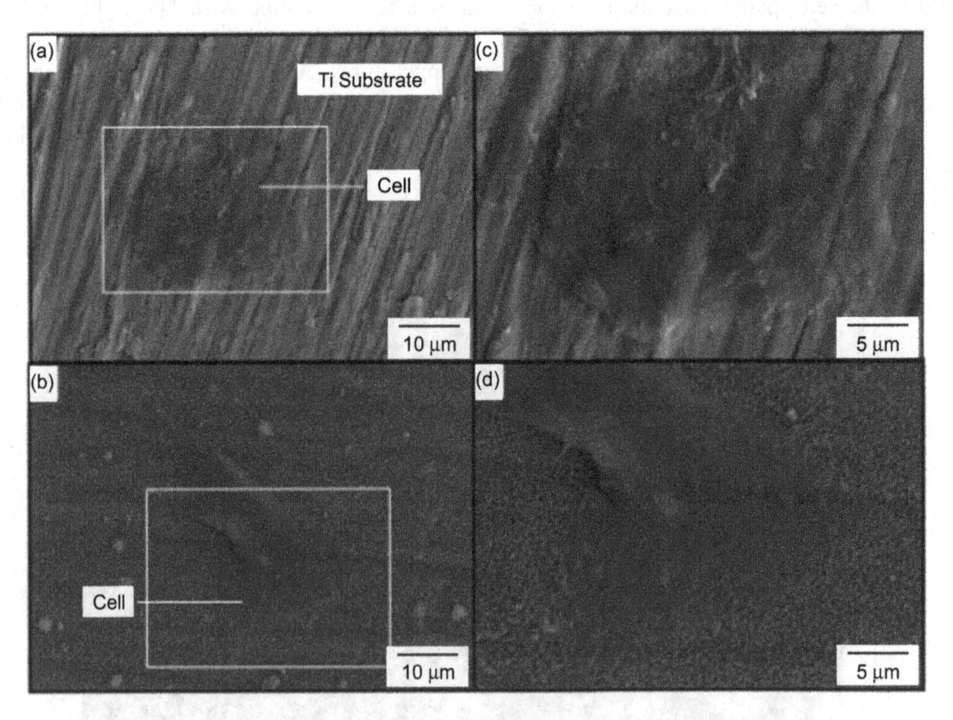

FIGURE 8.7 FESEM images of cell attachment on Ti surface at low (a, b) and high (c, d) magnifications for (a) Ti and (b) Ti treated with 5.0 M NaOH (60°C, 24 h) followed by oxidation at 400°C (Abdullah 2010).

8.3.3 Summary

Gel oxidation is a thermochemical method to produce porous but well adhered layers of titania on titanium metal. The oxidation of these layers has changed the structure and phases of the titanium surface (anatase, rutile, and residual sodium titanate). The different titania polymorphs and sodium titanate contribute to the degree of hydroxyapatite precipitation during soaking in simulated body fluid. Precipitation of hydroxyapatite increases by irradiation with UV light. Photocatalytic properties of titania enhanced the surface reaction resulting in precipitation with more hydroxyapatite on the gel-oxidised Ti. The gel-oxidised surface showed good cell attachment and penetration in the majority of the cells. The appearance of this cell growth indicated that gel-oxidised Ti can provide a superior implant owing to good ceramic-cell adhesion.

8.4 GLASS CERAMICS – BIOACTIVE GLASSES (BG)

There are several types of bioactive glasses (BG) and these include silicate based-glasses, phosphate based-glasses and borate based – glasses (Jones 2015, Rahaman et al. 2011). Miguez-Pacheco et al. (2015) listed (Table 8.3) the formulations of several bioactive glasses (BG) which have been recently studied and investigated thoroughly (Jones 2015, Rahaman et al. 2011, Nandi et al. 2011, Hench 1998, Brink et al. 1997, Brown et al. 2008, Lindfors et al. 2010, Clare 2004, Detsch et al. 2014, Filgueiras et al. 1993). It can be seen that bioactive silicate-based glasses have been found to be a revolutionary material after trials, showing effective and successful adherence and bonding to human bone and tissue. Hench et al. (Hench et al. 1971, Hench 2006, 2015) designed Bioglass 45S5 with 45 wt. % silica (SiO_2), 24.5 wt. % calcium oxide (CaO), 24.5 wt. % sodium oxide (Na_2O) and 6 wt. % phosphorus pentoxide (P_2O_5) as shown in Fig. 8.8. These chemical combinations are categorized as Class A bioactivity owing to their outstanding osteoconductive and osteoproductive characteristics (Hench 2015). This is due to the ions released from BG reacted positively in biological solutions such as in simulated body fluid (SBF) and thus induced the formation of carbonated hydroxyapatite (HCA) layers on the surface of glass (Rezwan et al. 2006, Hoppe et al. 2011).

TABLE 8.3 Bioactive glasses formulations investigated by previous researchers (Jones 2015, Rahaman et al. 2011, Nandi et al. 2011, Hench 1998, Brink et al. 1997, Brown et al. 2008, Lindfors et al. 2010, Clare 2004, Detsch et al. 2014, Filgueiras et al. 1993). *(Reproduced with permission of Elsevier)*

Bioactive Glasses	Composition (wt. %)							
	Na_2O	CaO	SiO_2	P_2O_5	K_2O	MgO	B_2O_3	CaF_2
45S5	24.50	24.50	45.00	6.00				
58S	0.00	32.60	58.20	9.20				
S53P4	23.00	20.00	53.00	4.00		0.00		
52S4.6	21.61	21.64	50.76	5.99				
S53P4	23.00	20.00	53.00	4.00				
46S6	24.00	24.00	46.00	6.00				
13-93	6.00	20.00	53.00	4.00	12.00		0.00	
45S5F	24.50	12.25	45.00	6.00	0.00	12.25	0.00	
13-93B3	5.50	18.50	0.00	3.70	11.10	4.60	56.60	0.00

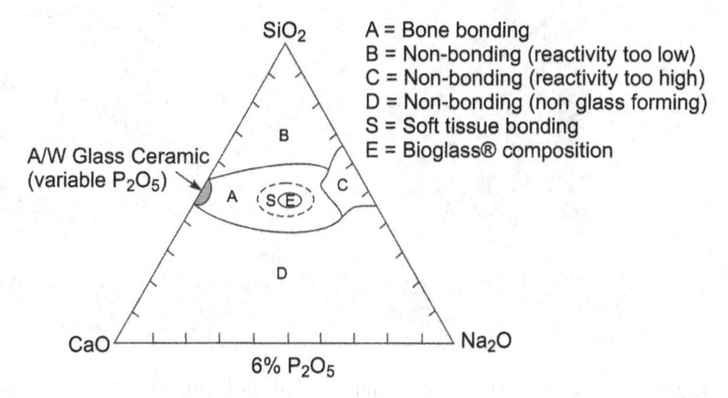

FIGURE 8.8 3D chemical composition diagram of $CaO-SiO_2-P_2O_5$ for bone-bonding (Hench 2006). *(Reproduced with permission of Springer).*

8.4.1 Mechanism of Bioactive Glasses: Ion Dissolution

Hoppe et al. (2011) proposed the schematic diagram of the biological response to ionic dissolution products of bioactive glasses (Fig. 8.9). The layer formed is biocompatible and promotes chemical bonding not only for regeneration and repair of hard tissues but for soft tissues as well (Miguez-Pacheco et al. 2015). Thus, it is seen that bioactive glasses are capable of stimulating vascularization through the secretion of angiogenic growth factors such as vascular endothelial growth factor (VEGF) from either human or animal fibroblast cells (Day et al. 2004, Day 2005) and in activating the differentiation and proliferation of osteoblast cells (Roether et al. 2002 Xynos et al. 2000, Fathi and Doostmohammadi 2009). Figure 8.10 shows micrographs of osteoblasts which have successfully been cultured and seeded on Bioglass 45S5 disk (substrate) (Xynos et al. 2000). It revealed that the filipodia and fiber-like structure occurred after 2 days (Figure 8.10a) of seeding on the Bioglass 45S5 (substrate) and later attached the substrate with several lamellipodia after being cultured for 6 days (Fig. 8.10b).

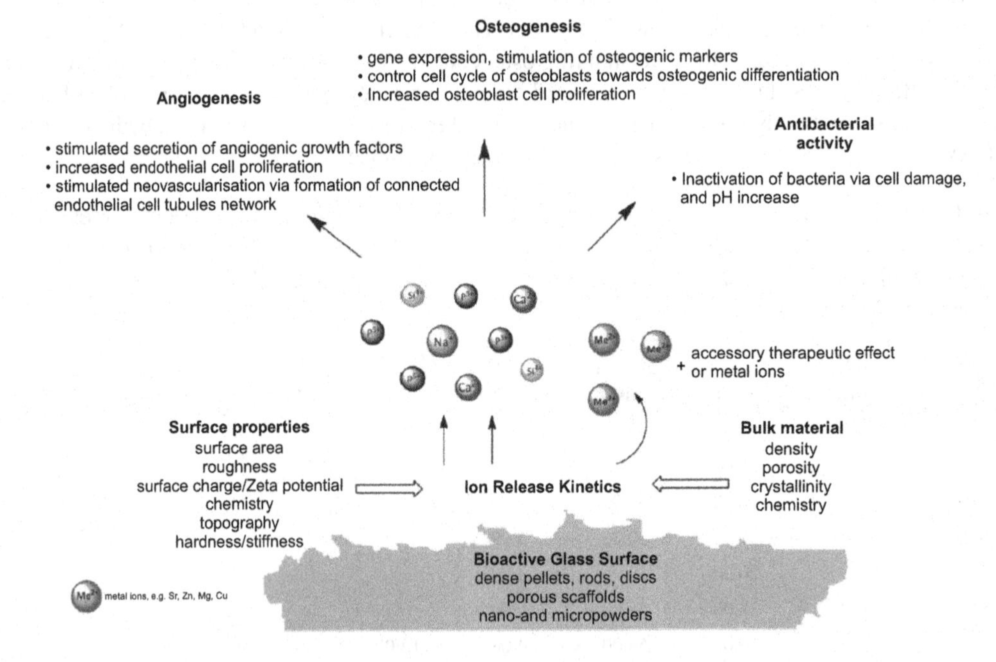

FIGURE 8.9 Biological responses to ionic dissolution products of bioactive glasses (Hoppe et al. 2011). *(Reproduced with permission of Elsevier).*

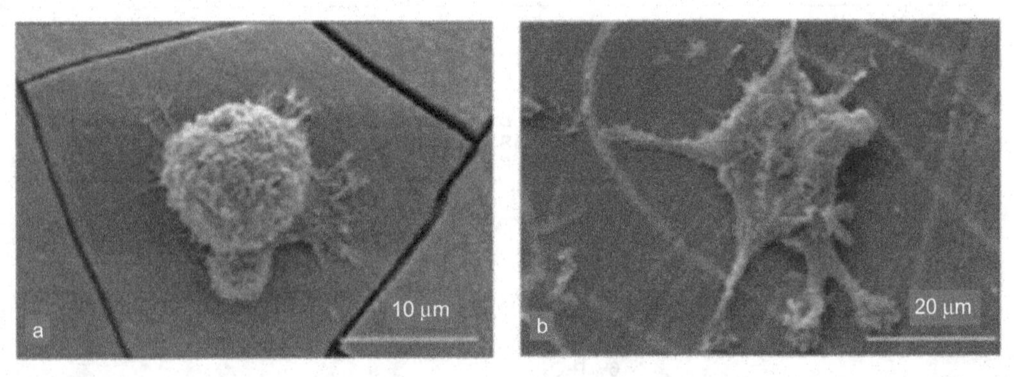

FIGURE 8.10 Micrographs of osteoblasts cultured and seeded on Bioglass 45S5: (a) 2 days and (b) 6 days. *(Reproduced with permission of Springer)* (Xynos et al. 2000).

8.4.2 Production of Bioactive Glasses

Earlier production of bioactive glasses involved the conventional method of melting the oxide compounds at high temperature (>1300°C) in a platinum crucible, followed by quenching in a graphite mould or in water; the quenching route used would result in differences in the microstructures (Jones 2015, Vichery and Nedelec 2016). Nevertheless, this melt-derived process only allows for the production of micron-sized, low porosity and dense bioactive glasses. These are denser in comparison with higher surface area mesoporous particles (Sepulveda et al. 2001) obtained via chemical approaches like the sol-gel method (Hench and West 1990, Li et al. 1991). The sol-gel method involves the hydrolysis and polycondensation reactions of a silicate precursor - tetraethyl orthosilicate (TEOS) with water either in acidic or basic solution, leading to the formation of inorganic solution, followed with agglomeration to form a 3D network or spherical nanoparticles (Stöber et al. 1968) at room temperature. Then, heat treatment is conducted at lower temperatures of 600°C-700°C to obtain calcined nanoporous bioactive glass powders (Jones 2015, Vichery and Nedelec 2016). Erol-Taygun et al. (2013) extensively reviewed the fabrication methods for nanosized bioactive glasses for medical applications from dentistry, orthopedic coatings, drug delivery, and bone tissue engineering scaffolds. The specific nanosurface morphology produced was found to be very significant since it resulted in improved contact with host tissue and bone, until the lower intrinsic properties such as brittleness and stiffness behavior hindered the application of bioactive glasses when involved with the mechanical loading. Therefore, a composite system was introduced by incorporation of bioactive glasses into the polymeric network and this allowed for an increase in the strength, flexibility and increased applicability for tissue scaffolding and regenerative medicine (Boccaccini et al. 2010).

8.4.3 Summary

The main contribution of bioactive glasses arises from the dissolution of their ions which causes stimulation of growth factors, control of the degradation rate and increased formation of apatite. The size of bioactive glasses ranging from micron to nano scale also plays a very significant role in providing a large surface area that allows the acceleration of their function and also increases the surface roughness and bioactivity of the system.

8.5 CERAMICS – HYDROXYAPATITE

8.5.1 Introduction to Hydroxyapatite (HAp)

Natural or synthetic biomaterials are anticipated to function accordingly like natural human bone in a bio-environment. Broken or fractured bone due to several factors such as accident, ageing, sports related injuries and war means there is always demand for hard tissues (bone replacement). Mainly, there are two types of applications of biomaterials: load- and non-load bearing application. Load-bearing applications involve biomaterials that require good mechanical and physical properties to stand the flexural or compression strength. Non-load bearing applications are more concerned with the bioactive surface to form bonding (osseointegration) with human bone/host. Both focuses on the functionality of replaced broken or malfunctioning bone.

Bone consists of pore structures that allow blood flow through it, which is similar to hydroxyapatite (HAp). The porous ceramic of HAp has the pore size of 100 μm, which allows the formation of apatite (Ravaglioli and Krajewski 1992). Even though the pore size of HAp is almost similar to natural bone, there is a difference in the pore structure of synthetic HAp (Lavernia and Scheonung 1991). Bone can be classified into three basic components such as non-organic, organic and water. Non-organic phase component takes up 69% of the bone weight, which is mostly HAp (Bose and Tarafder 2012). Other ceramic materials which exist in bone include dicalcium phosphate, octacalcium phosphate amorphous and tricalcium phosphate. Hence, the ceramic system of calcium phosphate is ideal for this system as it could naturally exist in natural bone.

Calcium phosphate (CaP) has been used widely in biomedical applications due to its capability in mimicking the properties of natural bone (Table 8.4). It has been used in the formation of artificial

TABLE 8.4 Different phases of calcium phosphate (Fernandez et al. 1999, Putlyaey and Safronova 2006, Basu and Nath 2009)

Name/Acronym/ Chemical Formula	Descriptions
Monocalcium phosphate monohydrate/MCPM/ $Ca(H_2PO_4)_2 \cdot H_2O$	• Ca/P ratio: 0.5 • Solubility: very high • Can be used for biomedical application • Precipitate from calcium carbonate and phosphoric acid at ambient temperature
Calcium phosphate dehydrate/DCP/ $CaHPO_4 \cdot 2H_2O$	• Ca/P ratio: 1.0 • Solubility: very high • Acts as precursor of more stable phase HAp • Easy to deposit onto metallic material
Anhydrous calcium phosphate/ADCP/$CaHPO_4$	• Ca/P ratio: 1.0 • Solubility: very high and lower than DCP • Triclinic crystalline structure • Stable at high temperature
Amorphous calcium phosphate/ACP/ $(Ca, Mg)_x(PO_4, Y')_y$	• Ca/P ratio: 1.15-1.67 • Solubility: high • Rapid release of ions gives negative impacts to the cell attachment or proliferation in short term and long-term performance
Octacalcium phosphate/ OCP/$Ca_8H_2(PO_4)_6 \cdot 5H_2O$	• Ca/P ratio: 1.33 • Solubility: very poor • Triclinic crystalline structure • Most stable at a physiological pH and temperature
Tricalcium phosphate/TCP/ $(Ca_3(PO_4)_2)$	• Ca/P ratio: 1.5 • Solubility: less stable but higher than HAp • Present in two phases: α-TCP and β-TCP • α-TCP forms when sintering temperature is up to 1250°C • β-TCP forms when the sintering temperature is between 900°C-1100°C • β-TCP has low interfacial energy
Calcium-deficient hydroxyapatite/CHDA/ $Ca_{10-x}(HPO_4)_x(PO_4)_{6-x}(OH)_{2-x}$ $(o < x < 1)$	• Ca/P ratio: 1.5-1.67 • Solubility: variable but higher than HAp • Similar to HAp in terms of composition and structure • Thermal stability is lower than HAp • Higher specific surface area and superior seeding efficiency
Biphasic calcium phosphate/ BCP/ -	• Ca/P ratio: 1.5-1.67 • Solubility: variable • Combination between HAp and TCP • It combines the exclusive properties of calcium phosphorous compounds
Hydroxyapatite/HAp/ $(Ca_5(PO_4)_3OH)$	• Ca/P ratio: 1.67 • Solubility: most stable but least soluble biomaterial • Natural mineral and chemical composition is nearer to human bone • Major mineral component of human bone and teeth • Provide nucleating sites for the apatite precipitation • Able to improve adhesion, proliferation, differentiation and increase alkaline phosphate activity of primary human osteoblasts cells
Tetracalcium phosphate/ TTCP/$Ca_4(PO_4)_2O$	• Ca/P ratio: 2.0 • Solubility : low • Formed by solid state reactions such as $CaHPO_4 + CaCO_3$ at high temperature • Very reactive and can rapidly convert to apatite in a wet atmosphere

bone (bone-graft) or as a bioactive coating on other biomaterials especially those made from bioinert metals for orthopaedic as well as orthodontic applications (Legeros et al. 2009). CaP can be observed naturally in biological systems. CaP has been synthesized and used to manufacture various forms of implants, as well as for solid and porous coating on other implants. It is necessary to mention that different phases of CaP exhibit different solubility coefficients and these are dependent on temperature, pH, and environmental composition (Shadanbaz and Dias 2012).

The most used calcium phosphate in implant fabrication is hydroxyapatite. The most interesting property is the bonding of hydroxyapatite to the bone to form indistinguishable unions (Legeros et al. 2009). However, poor mechanical properties (in particular, fatigue properties) mean that hydroxyapatite cannot be used in bulk form for load bearing applications such as orthopaedic (Sasikumar and Vijayaraghavan, 2006, Chang et al. 2010, Belluci et al. 2015). Therefore, most of HAp studies were done to enhance the strength of HAp materials for load-bearing applications. Among the advantages of CaP in mimicking natural bone are: (i) it encourages bone regeneration, (ii) similar composition and crystalline structure to human bone, (iii) biocompatible and bioactive implant material, (iv) enables newly formed bone cells to gradually substitute and integrate with the host bone, and (v) the pore size of CaP implant can be controlled (Nascimento et al. 2007). HAp and TCP are the type of ceramics that have been widely used as a bone substitute due to the ability in mimicking the human bone (Lavernia and Scheonung 1991). The nanoscales crystal size HAp and biomimetic micro-/nano- structured morphology has greater bioactivity that can promote greater adhesion, proliferation and provide better nucleation sites for osteoblasts.

8.5.2 Synthesis of Hydroxyapatite

There are countless numbers of phosphate compounds as well as highly complex systems of calcium phosphate systems. This is further complicated by the sensitive stability of phosphates to minor changes in terms of composition, pH and reaction conditions (e.g. temperature). It is important to acknowledge the fact that the purity and particle characteristics of the final synthesised powder can affect the bioactivity, mechanical and biological dissolution properties (Cox 2012). These characteristics could determine the medical application of the material which makes it imperative in developing a synthesis method that enables the control of crystal morphology, chemical composition, crystallinity, particle size distribution and agglomeration. Different synthesis techniques had been applied to obtain different CaP phases. Table 8.5 summarizes the synthesis methods of HAp from different studies.

TABLE 8.5 Synthesis methods of HAp

Method	References
1. Mechanochemical • Carried out at room temperature with great simplicity, although high temperatures can be locally reached in the colliding region of balls, powders and vial walls • The properties of the product depend on the method employed, the type of vial and the energy of the mill • Products with different morphology, stoichiometry, and level of crystallinity can be obtained depending on the technique and the materials used	Toriyama et al. 1996, Salas et al. 2004, Choi and Kumta 2007
2. Combustion Preparation • Known as self-propagating combustion synthesis (SPCS) • Simple and quick energy saving synthesis option for HAp • Very high local temperature of mixture solutions between calcium and phosphate sources, fuel and oxidiser that cause the formation of solid calcium phosphate powder • Parameters can vary, resulting in different CaP phases and/or particle morphologies, and calcination is required to remove residues and to form crystalline HAp • Can be completed in less than 20 minutes	Tas 2000, Sasikumar and Vijayaraghavan 2006, Bovand et al. 2012

Table 8.5 contd...

Table 8.5 contd.

3. Hydrothermal • Temperature of solvent (water) can reach above boiling point as the autogenous pressure within the sealed vessel has exceeded the ambient pressure • Properties, namely crystal structure nucleation, growth, and ageing can be regulated • Yields approaching 100% have relatively low-cost reagents and short reaction times • Scalability is limited to the size of the reaction vessel	Monmaturapoj and Yatongchai 2010, Hui et al. 2010, Ramesh et al. 2012
4. Sol-gel • 'Sol', a dispersion of solid particles, otherwise known as colloids in liquid • 'Gel', a diphasic system consisting of a solid and interstitial liquid phase • Drying process is needed to remove the liquid phase that leads to a significant amount of shrinkage and densification • A material specific sintering protocol is employed in practice, and this step can be time-consuming • Homogeneous molecular mixing, low processing temperatures (<400°C), and the ability to generate nano-sized particles • High cost of the reactants • Very limited scalability due to the sensitivity of the process	Sopyan 2003, Nayak 2010, Salimi and Anuar 2013
5. Precipitation from aqueous solutions • Simple method, ready availability, use of relatively inexpensive raw materials, and scalability (manufacturing) • Reactions require fine-tuning to optimise morphology and to minimise crystal growth • Time consuming; the high-temperature heat treatment may be needed to maximise the percentage of crystalline phase • The final composition is dependent on the solution's pH, concentration and temperature. Main disadvantage of all the solution methods proposed so far is the presence of metastable phases in the final product • Post-formation precipitated powders are typically calcined at 400-600°C to refine the crystal structure. In some cases, fully crystallised HAp is not formed until sintered up to 1200°C	Kwon et al. 2003, Tadic et al. 2004, Epple et al. 2010, Nath et al. 2010, Pham et al. 2013
6. Solid state • Relies on the solid diffusion of ions amongst powder raw materials; thus, it requires relatively inefficient high-temperature processing (< 1250°C) to initiate the reaction • Comparatively simple method where CaP sources are mixed with additives, a binder, and an organic vehicle to form slurry before milling • The slurry is then dried before being formed into pellets using hot or cold press at pressures up to 135 MPa • Pellets are then sintered up to 1250°C to crystallise. Re-crushing sintered sample is suggested, before further compaction and sintering at 1250°C to refine the crystal structure and reduce grain (7-8 μm) and crystal size (50-70 nm)	Pramanik et al. 2009, Safronova et al. 2009
7. Emulsion / Micro-emulsion • Heterogeneous mixtures of at least one immiscible liquid that dispersed in another in the form of droplets • Often described as either water-in-oil (W/O) or oil-in-water (O/W) • A reaction that takes place when two different droplets containing the reactants collide with each other • Micro-emulsion techniques have been reported to reduce particle agglomeration of HAp • Granules at 50-2000 μm can be formed with porosity up to 58.5% and no substantial trace of impurities up to 1250°C • HAp particles produced via micro-emulsion routes with spherical morphology and small particle sizes (>22 nm) can be used in plasma spraying	Bose and Tarafder 2012
8. Natural resources • Natural resources of HAp usually will undergo calcination to remove impurities and residues • Various methods can be used to synthesise high crystalline HAp from natural resources • Some examples of naturally derived HAp are from bovine bone, eggshells, natural coral, human tooth enamel, cuttlefish bone, oyster, nacre, sea urchin, conch, and giant clam	Salman et al. 2009, Zhang et al. 2010, Dahlan et al. 2012, Shavandi et al. 2015

Different applications of HAp are required for different physical and chemical properties; thus, it is concluded that the optimum characteristics for each application should be decided before selecting a synthesis technique (Rajendran et al. 2014). This information is used when making an informed selection of the most appropriate technique before being further developed. Furthermore, the ease of scalability and cost of each process are other factors that have to be considered as these factors could fluctuate, especially when large quantities are required.

8.5.3 Bone Mimetic Performance

Currently, the mimicking of bone focuses on the function of implant surface and its properties (mechanical and physical properties). Various studies have been reported on the performance of HAp both on the properties and also on the bioactivity of the material (*in vitro* and *in vivo*). Besides manipulating the mixing ratio, adding additives or binders into HAp, most ceramic products are dependent on the temperature (heat treatment) to enhance the brittle nature of ceramics to have better mechanical and physical properties. Table 8.6 shows some recent studies on improving HAp performance in relation to the additive materials.

TABLE 8.6 Summary of previous finding on HAp performance

Objective(s)	Finding(s)	References
To optimise the electrophoretic deposition (EPD) and suspension parameters for producing PEEK-HAp	Adhesion strength and *in vitro* bioactivity of the coatings were dependent on the PEEK and HAp relative contents. Increasing the amount of HAp improved the bioactivity while decreasing the adhesion strength of the coatings	Bastan et al. 2018
To investigate the viability of producing biogenic HAp from bio-waste animal bones	Heating the bovine and caprine bones at selected temperatures yielded porous HAp body, having hardness values that are comparable with human cortical bone. However, the sintered galline bone sample showed higher porosity levels and low hardness	Ramesh et al. 2018
To improve the mechanical strength and osseointegration with an added advantage of antibacterial activity	The synthesized HAp+ZrO_2 (rice husk) composite could offer more mechanical stability to the implants than the pristine HAp	Prema et al. 2018
To identify biomimetic synthesis reaction for the production of HAp/natural rubber (NR) composites	Cell proliferation positively influenced by HAp content at shorter culture times. The good dispersion of HAp in the composites is expected to improve NR bioactivity and direct its use towards bone applications	Dick and dos Santo 2017
To investigate the effect of aluminium substitution on the biocompatibility of HAp under the physiochemical conditions	Al-HAp nanoparticles are biocompatible on cell lines L929 and do not have toxic effects. The results of these studies confirmed the biocompatibility of Al-HAp and demonstrated the suitability for biomedical applications	Kolekar et al. 2016

8.5.4 Summary

It is proven that HAp is undoubtedly among the most promising material in the development of bioactive ceramics in the near future. HAp was produced through different methods from synthetic to natural resources to suit with the desired properties for specific applications of hard tissue. Continuous efforts to explore, develop and extend the application of HAp as a biomaterial, especially in mimicking human bone properties will undoubtedly continue to increase subsequently.

REFERENCES

Abdullah, H.Z. and C.C. Sorrell. 2007. Preparation and characterisation of TiO_2 thick film by gel oxidation. Materials Science Forum 561-565: 2167-2170.

Abdullah, H.Z. 2010. Titanium Surface Modification by Oxidation for Biomedical Application. Ph.D. Thesis, University of New South Wales, Sydney, Australia.

Abdullah, H.Z. and C.C. Sorrell. 2012. Gel oxidation of titanium and effect of UV irradiation on precipitation of hydroxyapatite from simulated body fluid. Advanced Materials Research 488-489: 1229-1237.

Bastan, F.E., M.A. Ur Rehman, Y.Y. Avcu, E. Avcu, F. Ostel and A.R. Boccaccini. 2018. Electrophoretic co-deposition of PEEK-hydroxyapatite composite coatings for biomedical applications. Colloids and Surfaces B: Biointerfaces 160: 176-182.

Basu, B. and S. Nath. 2009. Fundamentals of biomaterials and biocompatibility. pp. 3-18. *In*: B. Basu, D.S. Katti and A. Kumar [eds.]. Advance Biomaterials: Fundamentals, Processing and Applications. John Wiley & Sons, New Jersey.

Bellucci, D., A. Sola and V. Cannillo. 2015. Hydroxyapatite and tricalcium phosphate composites with bioactive glass as second phase: state of the art and current applications. Journal of Biomedical Materials Research A 104: 1030-1056.

Boccaccini, A.R., M. Erol, W.J. Stark, D. Mohn, Z. Hong and J.F. Mano. 2010. Polymer/bioactive glass nanocomposites for biomedical applications: a review. Composites Science and Technology 70: 1764-1776.

Bose, S. and S. Tarafder. 2012. Calcium phosphate ceramic systems in growth factor and drug delivery for bone tissue engineering: a review. Acta Biomaterialia 8: 1401-1421.

Bovand, N., S. Rasouli, M.-R. Mohammadi and D. Bovand. 2012. Rapid synthesis of hydroxyapatite nanopowders by a microwave-assisted combustion method. Journal of Ceramic Processing Research 13: 221-225.

Brink, M., T. Turunen, R.P. Happonen and A. Yli-Urpo. 1997. Compositional dependence of bioactivity of glasses in the system Na_2O-K_2O-MgO-CaO-B_2O_3-P_2O_5-SiO_2. Journal of Biomedical Materials Research: An Official Journal of The Society for Biomaterials and The Japanese Society for Biomaterials 37: 114-121.

Brown, R.F., D.E. Day, T.E. Day, S. Jung, M.N. Rahaman and Q. Fu. 2008. Growth and differentiation of osteoblastic cells on 13-93 bioactive glass fibers and scaffolds. Acta Biomaterialia 4: 387-396.

Cangiani, G. 2003. AB-initio Study of the Properties of TiO_2 Rutile and Anatase Polytypes. Ph.D. Thesis. Universite de Trieste, Italy.

Chang, Q., D.L. Chen, H.Q. Ru, X.Y. Yue, L. Yu and C.P. Zhang. 2010. Toughening mechanisms in iron-containing hydroxyapatite/titanium composites. Biomaterials 31: 1493-1501.

Choi, D. and P.N. Kumta. 2007. Mechano-chemical synthesis and characterization of nanostructured β-TCP powder. Materials Science and Engineering C 27: 377-381.

Clare, A. 2004. Biotechnological applications of inorganic glasses. pp. 145-161. *In*: M. Shi [ed.]. Biomaterials and Tissue Engineering. Springer, Berlin, Germany.

Cox. S. 2012. Synthesis method of hydroxyapatite. Ceramics 2: 1-10.

Dahlan, K., S.U. Dewi, A. Nurlaila and D. Soejoko. 2012. Synthesis and characterization of calcium phosphate/chitosan composites. International Journal of Basic and Applied Science 12: 50-57.

Day, R.M., A.R. Boccaccini, S. Shurey, J.A. Roether, A. Forbes, L.L. Hench and S.M. Gabe 2004. Assessment of polyglycolic acid mesh and bioactive glass for soft-tissue engineering scaffolds. Biomaterials 25: 5857-5866.

Day, R.M. 2005. Bioactive glass stimulates the secretion of angiogenic growth factors and angiogenesis *in vitro*. Tissue Engineering 11: 768-777.

Detsch, R., P. Stoor, A. Grünewald, J.A. Roether, N.C. Lindfors and A.R. Boccaccini. 2014. Increase in VEGF secretion from human fibroblast cells by bioactive glass S53P4 to stimulate angiogenesis in bone. Journal of Biomedical Materials Research A 102: 4055-4061.

Dick, T.A. and L.A. dos Santos. 2017. *In situ* synthesis and characterization of hydroxyapatite/natural rubber composites for biomedical applications. Materials Science and Engineering C 77: 874-882.

Epple, M., K. Ganesan, R. Heumann, J. Klesing, A. Kovtun, S. Neumann and V. Sokolova 2010. Application of calcium phosphate nanoparticles in biomedicine. Journal of Materials Chemistry 20:18-23.

Erol-Taygun, M., K. Zheng and A.R. Boccaccini. 2013. Nanoscale bioactive glasses in medical applications. International Journal of Applied Glass Science 4: 136-148.

Fathi, M.H., and A. Doostmohammadi. 2009. Bioactive glass nanopowder and bioglass coating for biocompatibility improvement of metallic implant. Journal of Materials Processing Technology 209: 1385-1391.

Fernandez, E., F.J. Gil, M.P. Ginebra, F.C.M. Driessens and J.A. Planell. 1999. Calcium phosphate bone cements for clinical applications. Journal of Materials Science: Materials in Medicine 10: 169-176.

Filgueiras, M.R.T., G. La Torre and L.L. Hench. 1993. Solution effects on the surface reactions of three bioactive glass compositions. Journal of Biomedical Materials Research 27: 1485-1493.

Fujishima, A., K. Hashimoto and T. Watanabe. 2001. TiO_2 Photocatalysis: Fundamentals and Applications. BKC INC, Tokyo.

Han, Y. and K. Xu. 2004. Photoexcited formation of bone apatite-like coatings on micro-arc oxidized titanium. Journal of Biomedical Materials Research A 71: 608-614.

Hench, L.L., R.J. Splinter, W.C. Allen and T.K. Greenlee. 1971. Bonding mechanisms at the interface of ceramic prosthetic materials. Journal of Biomedical Materials Research 5: 117-141.

Hench, L.L. and J.K. West. 1990. The sol-gel process. Chemical Reviews 90: 33-72.

Hench, L.L. 1998. An introduction to materials in medicine: bioceramics. Journal of the American Ceramic Society 81: 1705-1728.

Hench, L.L. 2006. The story of Bioglass®. Journal of Materials Science: Materials in Medicine 17: 967-978.

Hench, L.L. 2015. Opening paper 2015-some comments on bioglass: four eras of discovery and development. Biomedical Glasses 1: 1-11.

Hoppe, A., N.S. Güldal and A.R. Boccaccini. 2011. A review of the biological response to ionic dissolution products from bioactive glasses and glass-ceramics. Biomaterials 32: 2757-2774.

Hui, P., S.L. Meena, G. Singh, R.D. Agarawal and S. Prakash. 2010. Synthesis of hydroxyapatite bio-ceramic powder by hydrothermal method. Journal of Minerals and Material Characterization and Engineering 9: 683-692.

Jaeggi, C., P. Kern, J. Michler, J. Patscheider, J. Tharian and F. Munnik. 2006. Film formation and characterization of anodic oxides on titanium for biomedical applications. Surface and Interface Analysis 38: 182-185.

Jones, J.R. 2015. Reprint of: review of bioactive glass: from hench to hybrids. Acta Materialia 23: S53-S82.

Jonášová, L., F.A. Müller, A. Helebrant, J. Strnad and P. Greil. 2004. Biomimetic apatite formation on chemically treated titanium. Biomaterials 25: 1187-1194.

Kim, H.M., F. Miyaji, T. Kokubo and T. Nakamura. 1996. Preparation of bioactive Ti and its alloys via simple chemical surface treatment. Journal of Biomedical Materials Research 32: 409-417.

Kim, H.M., F. Miyaji and T. Kokubo. 1997. Effect of heat treatment on apatite-forming ability of Ti metal induced by alkali treatment. Journal of Materials Science: Materials in Medicine 8: 341-347.

Kokubo, T., H.M. Kim and M. Kawashita. 2003. Novel bioactive materials with different mechanical properties. Biomaterials 2: 2161-2175.

Kolekar, T.V., N.D. Thorat, H.M. Yadav, V.T. Magalad, M.A. Shinde, S.S. Bandgar, J.H. Kim and G.L. Agawane. 2016. Nanocrystalline hydroxyapatite doped with aluminium: a potential carrier for biomedical applications. Ceramics International 42: 5304-5311.

Kwon, S.-H., Y.-K. Jun, S.-H. Hong and H.-E. Kim. 2003. Synthesis and dissolution behaviour of β-TCP and HA/β-TCP composite powders. Journal of the European Ceramic Society 23: 1039-1045.

Lavernia, C. and J.M. Schoenung. 1991. Calcium phosphate ceramics as bone substitutes. Ceramic Bulletin 70: 95-100.

Legeros, R.Z., A. Ito, K. Ishikawa, T. Sakae and P.L. John. 2009. Fundamental of hydroxyapatite and related calcium phosphates. pp. 19-52. *In*: B. Basu, D.S. Katti and A. Kumar [eds.]. Advance Biomaterials: Fundamentals, Processing and Applications. John Wiley & Sons, New Jersey.

Li, R., A.E. Clark and L.L. Hench. 1991. An investigation of bioactive glass powders by sol-gel processing. J Journal of Applied Biomaterials 2: 231-239.

Lindfors, N.C., P. Hyvönen, M. Nyyssönen, M. Kirjavainen, J. Kankare, E. Gullichsen and J. Salo. 2010. Bioactive glass S53P4 as bone graft substitute in treatment of osteomyelitis. Bone 47: 212-218.

Liu, X., P.K. Chu and C. Ding. 2004. Surface modification of titanium, titanium alloys, and related materials for biomedical applications. Materials Science and Engineering R: Reports 47: 49-121.

Miguez-Pacheco, V., L.L. Hench and A.R. Boccaccini. 2015. Bioactive glasses beyond bone and teeth: emerging applications in contact with soft tissues. Acta Biomaterialia 13: 1-15.

Monmaturapoj, N. and C. Yatongchai. 2010. Effect of sintering on microstructure and properties of hydroxyapatite produced by different synthesizing methods. Journal of Metals, Materials and Minerals 20: 53-61.

Nandi, S.K., B. Kundu and S. Datta. 2011. Development and applications of varieties of bioactive glass compositions in dental surgery, third generation tissue engineering, orthopaedic surgery and as drug delivery system. In Biomaterials Applications for Nanomedicine. InTech.

Nascimento, C., J.P.M. Issa, R.R. Oliveira, M.M. Iyomasa, S. Siessere and S.C.H Regalo. 2007. Biomaterials applied to the bone healing process. International Journal of Morphology 25: 839-846.

Nath, S., K. Biswas, K. Wang, R.K. Bordia and B. Basu. 2010. Sintering, phase stability, and properties of calcium phosphate-mullite composites. Journal of the American Ceramic Society 93: 1639-1649.

Nayak, A.K. 2010. Hydroxyapatite synthesis methodologies: an overview. International Journal of ChemTech Research 2: 903-907.

Pham, T.T.T., T.P. Nguyen, T.N. Pham, T.P. Vu, D.L. Tran, H. Thai and T.M.T. Dinh. 2013. Impact of physical and chemical parameters on the hydroxyapatite nanopowder synthesized by chemical precipitation method. Advances in Natural Sciences: Nanoscience and Nanotechnology 4: 1-9.

Pramanik, N., S. Mohapatra, P. Bhargava and P. Pramanik. 2009. Chemical synthesis and characterization of hydroxyapatite (HAp)-poly (ethylene co vinyl alcohol) (EVA) nanocomposite using a phosphoric acid coupling agent for orthopedic applications. Materials Science and Engineering C 29: 228-236.

Prema, D., S. Gnanavel, S. Anuraj and C. Gopalakrishnan. 2018. Synthesis and characterization of different chemical combination of hydroxyapatite for biomedical application. Materials Today: Proceedings 5: 8868-8874.

Putlyaev, V.I., and T.V. Safronova. 2006. A new generation of calcium phosphate biomaterials: the role of phase and chemical compositions. Glass and Ceramics 63: 30-33.

Rahamana, M.N., D.E. Delbert, B.S. Bal, Q. Fu, S.B. Jung, L.F. Bonewald and A.P. Tomsia. 2011. Bioactive glass in tissue engineering. Acta Biomaterialia 7: 2355-2373.

Rajendran, A., R.C. Barik, D. Natarajan, M.S. Kiran and D.K. Pattanayak. 2014. Synthesis, phase stability of hydroxyapatite-silver composite with antimicrobial activity and cytocompatibility. Ceramics International 40: 10831-10838.

Ramesh, S., C.Y. Tan, R. Tolouei, M. Amiriyan, J. Purbolaksono, I. Sopyan and W.D. Teng. 2012. Sintering behaviour of hydroxyapatite prepared from different routes. Materials & Design 34: 148-154.

Ramesh, S., Z.Z. Loo, C.Y. Tan, W.J. Kelvin Chew, Y.C. Ching, F. Tarlochan, H. Chandran, S. Krishnasamy, L.T. Bang and A.A.D. Sarhan. 2018. Characterization of biogenic hydroxyapatite derived from animal bones for biomedical applications. Ceramics International 44: 10525-10530.

Ravaglioli, A. and A. Krajewski. 1992. Bioceramics: Materials, Properties, Application. 1st ed. Springer, Netherlands.

Rezwan, K., Q.Z. Chen, J.J. Blaker and A.R. Boccaccini. 2006. Biodegradable and bioactive porous polymer/inorganic composite scaffolds for bone tissue engineering. Biomaterials 27: 3413-3431.

Roether, J.A., J.E. Gough, A.R. Boccaccini, L.L. Hench, V. Maquet and R. Jérôme. 2002. Novel bioresorbable and bioactive composites based on bioactive glass and polylactide foams for bone tissue engineering. Journal of Materials Science: Materials in Medicine 13: 1207-1214.

Safronova, T.V., V.I. Putlyaev, M.A. Shekhirev, Y.D. Tretyakov, A.V. Kuznetsov and A.V. Belyakov. 2009. Densification additives for hydroxyapatite ceramics. Journal of the European Ceramic Society 29: 1925-1932.

Salas, J., Z. Benzo and G. Gonzalez. 2004. Synthesis of hydroxyapatite by mechanochemical transformation. Revista Latinoamericana de Metalurgia y Materiales 24: 12-16.

Salimi, M.N and A. Anuar. 2013. Characterizations of biocompatible and bioactive hydroxyapatite particles. Procedia Engineering 53: 192-196.

Salman, S., O. Gunduz, S. Yilmza, M.L. Ovecoglu, R.L. Snyder, S. Agathopoulos and F.N. Oktar. 2009. Sintering effect on mechanical properties of composites of natural hydroxyapatites and titanium. Ceramics International 35: 2965-2971.

Sasikumar, S and R. Vijayaraghavan. 2006. Low temperature synthesis of nanocrystalline hydroxyapatite from egg shells by combustion method. Trends in Biomaterials and Artificial Organs 19: 70-73.

Sepulveda, P., J.R. Jones and L.L. Hench. 2001. Characterization of melt-derived 45S5 and sol-gel–derived 58S bioactive glasses. Journal of Biomedical Materials Research: An Official Journal of The Society for Biomaterials, The Japanese Society for Biomaterials, and The Australian Society for Biomaterials and the Korean Society for Biomaterials 58: 734-740.

Shadanbaz, S. and G.J. Dias. 2012. Calcium phosphate coatings on magnesium alloys for biomedical applications: a review. Acta Biomaterialia 8: 20-30.

Shavandi, A., A.E.-D.A. Bekhit, Z. Sun and A. Ali. 2015. A review of synthesis methods, properties and use of hydroxyapatite as a substitute of bone. Journal of Biomimetic, Biomaterials and Biomedical Engineering 25: 98-117.

Shozui, T., K. Tsuru, S. Hayakawa and A. Osaka. 2008. Enhancement of *in vitro* apatite-forming ability of thermally oxidized titanium surfaces by ultraviolet irradiation. Journal of Ceramics Society Japan 116: 530-535.

Sopyan, I. 2003. Preparation of hydroxyapatite powders for medical applications via sol-gel technique. Indonesia Journal of Materials Science 4: 46-51.

Stöber, W., A. Fink and E. Bohn. 1968. Controlled growth of monodisperse silica spheres in the micron size range. Journal of Colloid and Interface Science 26: 62-69.

Tadic, D., F. Beckmann, K. Schwarz and M. Epple. 2004. A novel method to produce hydroxyapatite objects with interconnecting porosity that avoids sintering. Biomaterials 25: 3335-3340.

Tas, A.C. 2000. Combustion synthesis of calcium phosphate bioceramic powders. Journal of the European Ceramic Society 20: 2389-2394.

Toriyama, M., A. Ravagliolo, A. Krajewski, G. Celotti and A. Piancastelli. 1996. Synthesis of hydroxyapatite-based powders by mechano-chemical method and their sintering. Journal of the European Ceramic Society 16: 429-436.

Tung, M.S. 1998. Calcium Phosphate in Biological and Industrial System. Kluwer Academic Publishers, Dordrecht, Netherlands.

Vichery, C. and J.M. Nedelec. 2016. Bioactive glass nanoparticles: from synthesis to materials design for biomedical applications. Materials 9: 1-17.

Wang, X.X., W. Yan, S. Hayakawa, K. Tsuru and A. Osaka. 2003. Apatite deposition on thermally and anodically oxidized titanium surfaces in a simulated body fluid. Biomaterials 24: 4631-4637.

Xynos, I.D., M.V.J. Hukkanen, J.J. Batten, L.D. Buttery, L.L. Hench and J.M. Polak. 2000. Bioglass® 45S5 stimulates osteoblast turnover and enhances bone formation *in vitro*: implications and applications for bone tissue engineering. Calcified Tissue International 67: 321-329.

Zhang, X., Y. Ding, S. Wang, J. Xu and Y. Feng. 2010. Sintering behaviour and kinetic evaluation of hydroxyapatite bio-ceramics from bovine bone. Ceramic – Silikaty 54: 248-252.

9

Stimulus-Receptive Conductive Polymers for Tissue Engineering

Naznin Sultana

9.1 INTRODUCTION

Traditionally, all carbon-based polymers such as plastics, rubbers, etc. have long been considered insulating materials (Inzelt et al. 2000), offering significant resistance to electrical conduction. Hence, electrical conduction in polymers would have been regarded as absurd (Inzelt et al. 2000). Indeed, polymers are widely used by the electronic industry as inactive packaging and insulators. For instance, a layer of plastic is coated on the outside of metal cables to insulate them (Inzelt and György 2008). However, in the last few decades, the invention of conductive polyacetylene in the 1970s has spurred an opposite trend as a new class of polymer called conducting polymers (Hesketh and Misra 2012), which simply means that plastics are able to conduct electricity. This peaked in 2000 when Alan J. Heeger, Alan G. MacDiarmid, and Hideki Shirakawa were awarded the Nobel Prize for their discovery and the development of conductive polymers. With advances in the stability of the materials and improved control of their properties, a wide range of promising applications in conducting polymers has been investigated (Hesketh and Misra 2012), although it is still in the infancy stage. For the preparation and characterization of conducting polymers, electrochemistry techniques play a significant role because they are suitable for controlled synthesis of these compounds by tuning to a well-defined oxidation state (Inzelt et al. 2000). Nevertheless, much research is still necessary to reach a comprehensive understanding of all related processes of conducting polymers.

Recently, conducting polymers have become promising biomaterials which have exhibited excellent properties in biomedical applications (Green et al. 2008, Guimard et al. 2007, Bendrea et al. 2011, Higgins et al. 2012, Svennersten et al. 2011, Sayyar et al. 2013, Otero et al. 2012. The characteristics that make these polymers attractive include: (i) the fact that they show a wide range of electrical conductivity which can be accomplished by altering its doping levels while maintaining its mechanical flexibility and high thermal stability (Hesketh and Misra 2012), and (ii) the fact that they can be modified to become a biodegradable and biocompatible (Oh et al. 2013) polymer, which is very useful for tissue engineering applications (Fig. 9.1).

Therefore, it is intended to discuss some of the available topics on the most commonly used conductive polymers, their mechanisms, properties, processing and tissue engineering applications. It is expected that despite the incompleteness and subjectivity, this paper will be useful for other

Medical Academy, Prairie View A&M University, TX 77446, USA.
E-mail: nasultana@pvamu.edu

researchers who are interested in developing conducting polymer-based scaffolds for tissue engineering applications.

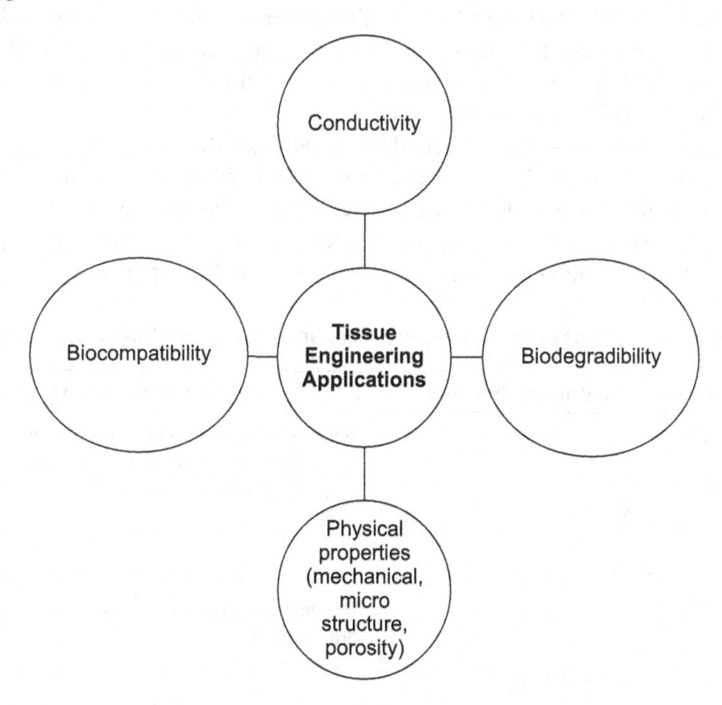

FIGURE 9.1 Conductive polymers together with other required properties for tissue regeneration.

9.2 Conductive Polymers: A Brief History

A conductive polymer is a very long chain of organic and inorganic molecules that can conduct electricity. The oldest conductive polymer is Polyaniline, which was discovered by Letheby in 1862, using the method of anodic oxidation of aniline in sulfuric acid (Inzelt et al. 2000). However, only partly conductive material was obtained (Heeger et al. 2000).

The discovery and evolution of a variety of conductive polymers were at their peak in the 1970s. In 1974, Shirakawa and co-workers fabricated polyacetylene silver films by Ziegler-Natta Polymerization, but these were not electrically conductive, despite showing a metallic appearance (Heeger et al. 2000). Meanwhile, in 1975, Alan MacDiarmid and Alan Heeger began to collaborate in order to investigate the metallic properties of covalent inorganic polymers, poly (sulfur nitride) (SN)x to oxygen, air and water vapor (Mikulski et al. 1976). However, they shifted their focus to polyacetylene after a meeting with Hideki Shirakawa (Zhou et al. 2010). Shirakawa enhanced the polymerization of polyacetylene (Chiang et al. 1977), while MacDiarmid wanted to use iodine treatment to modify the polyacetylene (Chiang et al. 1978a).

Although Shirakawa and co-workers had previously found that the infrared transmission of polyacetylene silver films was reduced while treating with bromine or chlorine (Ito et al. 1974), MacDiarmid and Heeger successfully obtained an increase of seven orders of magnitude over the polyacetylene for iodine-modified trans-polyacetylene (Chiang et al. 1978a). These underlying phenomena using agents, i.e. iodine (halogen) were named "doping" (Chiang et al. 1977, Chiang et al. 1978a), which is essential for conductive polymers. Shirakawa was able to regulate the ratio of cis/trans double bonds using Ziegler Natta catalyst (Shirakawa et al. 1978). The isomers of trans-polyacetylene were able to undergo efficient doping (defect-free) and the overall results were great (Shirakawa et al. 1978). The further discovery of the conductivity of cis-polyacetylene doped with AsF5 resulted in an increase in conductivity of 10^{11} (Shirakawa et al. 1977).

The publication of the paper entitled "Synthesis of electrically conducting organic polymers: Halogen derivatives of polyacetylene (CH)x" on May 16, 1977 by Shirawaka, MacDiarmid and Heeger (Chiang et al. 1978b), followed by two other papers elaborating further and deeper about the same topic, have revealed the "plastic electronics" field. Therefore, Alan J. Heeger, Alan G MacDiarmid and Hideki Shirakawa were awarded the Nobel Prize for chemistry in 2000 for the discovery and study of conducting polymers.

Although polyacetylene was the first conducting polymer discovered by doping with Iodine (halogen), it was unstable in air, easily oxidized by oxygen and sensitive to humidity (Shirakawa 2001). Hence, the search for a suitable conductive polymer has increased extensively since the early 1980s, i.e. polypyrrole (PPy), Polyaniline (PANI) and Polythiophenes. Table 9.1 shows some commonly used conductive polymers and their potential biomedical applications.

TABLE 9.1 Commonly used conductive polymers and their potential biomedical applications

Names, Abbreviations and Primary Structures	Potential Biomedical Applications
1. Polypyrrole (PPy) 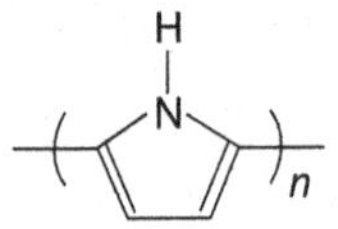	Biosensors, drug delivery systems, biomaterial in neural tissue regeneration, neural probes, nerve guidance channels, blood conduits
2. Polyaniline (PANI) 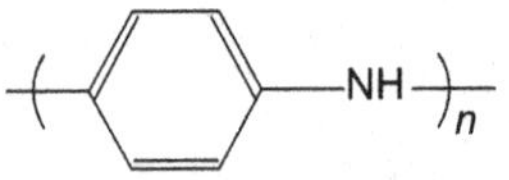	Biosensors, neural probes, tissue engineering, controlled drug delivery
3. Poly(3,4-ethylenedioxythiophene) (PEDOT) 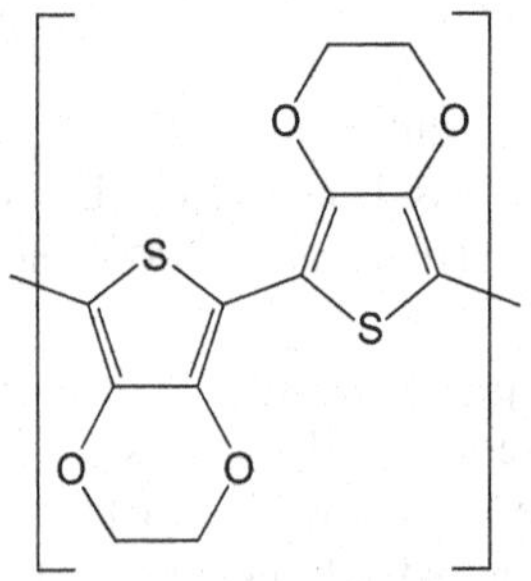	Nanofibers for tissue engineering, neural electrode, nerve grafts and heart muscle patches
4. Poly (3,4-ethylenedioxythiophene) (PEDOT): Polystyrene Sulfonate (PSS) 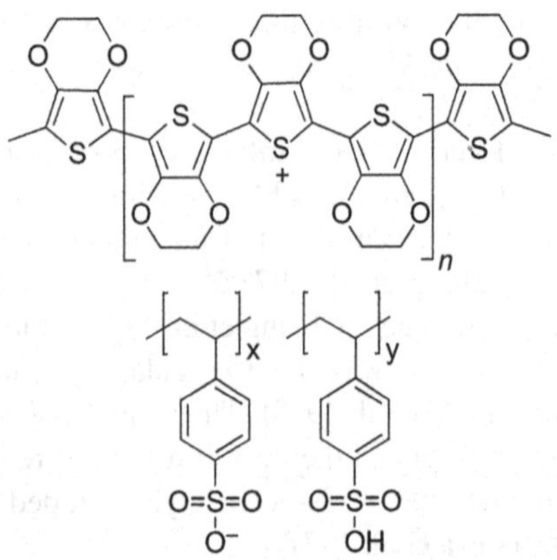	Bone and neural tissue engineering, drug delivery

9.3 MECHANISMS AND PROPERTIES OF CONDUCTIVE POLYMER

9.3.1 Conducting Mechanism: Conductivity and Doping

The polymers in their original state are commonly known as electrical insulators or are shown in very low conductivities because most organic polymers do not consist of mobile charge carriers or "free electrons" that can move from one atom to another under an applied electric field. Therefore, it is important to have free moving electrons jumping within and between the chains of a polymer in order to enable it to conduct electricity.

Hence, for essential mechanisms, doping is the key element for the conductive polymer. Doping is also recognized as a charge transfer reaction, which creates active sites (polarons) of the polymer, removing an electron from the valence band (*p*-doping) or adding an electron to the conduction band (*n*-doping) through redox reaction to enable the carriers (electron and holes) to move from one site to another (Zhou et al. 2010, Ghasemi-Mobarakeh et al. 2009, Bredas and Street 1985). Where an electron is missing, it is regarded as a hole. A new hole is created and allows the charge to migrate over a long distance when such a hole is filled by a jumping in electron from the neighboring position. This movement of charge is responsible for electrical conductivity. Simply, the doping process generates charge carriers producing a number of mobile carriers in a polymer to contribute to the conductivity. The charge carrier, in the form of extra electrons or "holes", should be injected into the material, which is also known as 'dopant' (Pelto et al. 2010). Conductivity increases very rapidly as dopant is added.

Conductivity in PPy is precisely dependent on *p*-type bipolaron conduction (Ateh et al. 2006a). For instance, the originated PPy is neutral, but it is positively charged when oxidized (Ateh et al. 2006a). In order to maintain its conductivity, some counter ions should diffuse into the polymers while charging and diffuse out throughout the neutralization process (Ateh et al. 2006a). However, over-oxidation of PPy above the standard oxidative potential leads to the loss of conductivity. This process is known as de-doping (Shiigi et al. 2002).

9.3.2 Properties of Conductive Polymers: Biodegradability

The main drawback of conductive polymers is their characteristic incapability to degrade for *in vivo* applications (Asplund et al. 2009), which may elicit an inflammatory response, and the requirement for a second surgical removal (Guo et al. 2013). Hence, it is highly desirable to synthesize materials with both biodegradable and electrical conductive properties and remains a challenge (Asplund et al. 2009). Basically, researchers have found three ways to address the drawbacks. The most common attempt is blending the conductive polymer together with a suitable biodegradable polymer (Zelikin et al. 2001). For example, PPy nanoparticle–polylactide (PLA) (Shi et al. 2008a), PPy–PLLA (Guimard et al. 2009) and PPy-thiophene polymer blending (Meng et al. 2008) have been successfully prepared. For PANI, in order to make it biocompatible, blending the aniline pentamer with the polymer of polyglycolide (Shi et al. 2008b), polyphosphazene (Rivers et al. 2002) and hydroxyl-capped poly (L-lactide) (Ding et al. 2007) was carried out. The biodegradable properties of PEDOT.PSS can be enhanced by adding Mangnesium (Mg) (Zhang et al. 2010). The linkage of biodegradable polymer can be conjoined using enzymes *in vivo*. Thus, these blending polymers have the distinctive properties of being both electroactive and biodegradable (Huang et al. 2015). Polymer blending or composite combines the advantages of both polymers (Huang et al. 2008), but it is still unable to solve the problem of eliminating the conductive polymer *in vivo* after the degradable polymer has broken down. However, this problem can be minimized by selecting the optimum ratio of the two polymers to control its degradation rate (Guimard et al. 2007). The second attempt is to change the conformation of the conductive polymer itself. The ionizable (butyric acid) or hydrolysable (butyric ester) side groups were added to the structure of PPy, altering its conformation, which was shown to enhance its degradation rate (Sebaa et al. 2013, Wang et al. 2004, Lakard et al. 2009). As the third solution to this problem, small chains of conductive polymer enable it to undergo gradual erosion because of the small size (Lakard et al. 2009). PPy was discovered to degrade using this solution (Lakard et al. 2009).

9.3.3 Biocompatibility

The ability to support the appropriate cellular growth and provide good cellular response is necessary for many biomedical applications (Guiseppi-Elie 2010), especially to those which need to be implanted in the human body, avoiding any inflammatory response from the human immune system. Therefore, the biocompatibility of conductive polymers is the key property, and if insufficient, can be enhanced through several ways, including bond with biocompatible molecules, segments, or side chains onto the conductive polymer (Zelikin et al. 2002). Studies of the application of conductive polymers in the biomedical field were greatly explored in the 1980s due to their biocompatible characteristic with many biological molecules.

Polypyrrole (PPy), the conductive synthetic polymer, possesses good *in vivo* and *in vitro* biocompatibility. Many researchers have demonstrated that PPy can support the growth of large variety of cell types including endothelial cells (Huang et al. 2007, Williams and Doherty 1994, Collier et al. 2000), fibroblasts (Zhang et al. 2001, Wang et al. 2003), keratinocytes (Wang et al. 2003), bone cells (Collier et al. 2000), neural cells (Zhang et al. 2001, Wang et al. 2003, Ateh et al. 2006b), and mesenchymal stem cells (Ateh et al. 2006b). Iodine-doped PPy and PPy-polyethylene glycol (PEO) were fabricated by Olayo et al. and these polymers were proven to be biocompatible with trans-sectioned spinal cord tissue after implantation, demonstrating the suitability of this conducting polymer for the repair of spinal cord damage (Richardson et al. 2007). The composite of PPy-PLGA meshes was also verified to be biocompatible with PC-12 cells.

The good biocompatibility of PPy was also exhibited in an animal model. PPy was shown by Wong and colleagues to consist of good biocompatibility with rat peripheral nerve tissue, making it a promising substrate for connecting the peripheral nerve gap. In addition, PPy has also been proven to be cytocompatible to mouse fibroblast (L929) and neuroblastoma cells with only a minimal tissue inflammatory response after 4 weeks *in situ* by Williams and Doherty. Under a constant current or voltage, PPy has been reported to enhance the axonal extensions of nerve cells.

Besides, the investigation of PANI for biomedical engineering applications has recently increased after PPy. There are more evidences showing the ability of PANI and PANI variants in supporting the cell growth (Alikacem et al. 1999). Different oxidation states of PANI have shown no distinguishing characteristic results from tissue incompatibility after implantation in order to investigate the *in vivo* tissue response to PANI by Wang's research team (George et al. 2005). Furthermore, in the vicinity of the implants, there was no significant inflammation at the implant site and no indication of anomaly of the muscle; also, the adipose tissues were observed (Guimard et al. 2007). Both the emeraldine base and emeraldine salt form of PANI were investigated by Bidez et al. to be cytocompatible to the H9c2 cardiac myoblasts and also to further enhance the cellular adhesion and proliferation. In general, the good biocompatible property of PANI *in vivo* has spurred much attention in tissue-engineering applications.

Poly (3,4-ethylenedioxythiophene) (PEDOT), the most effective polythiophene derivative, is a biocompatible conducting polymer which is recently being employed in biomedical applications, especially in nerve tissue engineering. PEDOT has shown its biocompatible properties which exhibit very low intrinsic cytotoxicity and inflammatory response upon implantation with wide range of cell types, including epithelial cells, NIH3T3 and L929 fibroblasts (Zhang et al. 2007), and neural (Castano et al. 2004) and neuroblastoma cells (Olayo et al. 2008). PEDOT-coated PET nanofibers fabricated by Bolin et al. showed excellent human neuroblastoma SH-SY5Y cellular adhesion and proliferation (Mattioli-Belmonte et al. 2003), while PEDOT-coated platinum electrodes exhibited non-cytotoxic effects and no significant differences in immunological response in cortical tissue in comparison to pure platinum controls.

9.4 CONDUCTIVE POLYMER IN TISSUE ENGINEERING APPLICATION

Tissue engineering, also known as regenerative medicine, is an interdisciplinary field involving principles of engineering and life sciences (Wang et al. 1999) to solve the limitation of organ and tissue transplantation issues (Bidez et al. 2006) as well as the immune repulsion between receptors and donors (Karagkiozaki et al. 2013). Hence, tissue engineering has been emphasized for tissue regeneration and repair by integrating scaffolds, cells and bioactive molecules (Kim et al. 2007). In the discipline of tissue engineering, cells are first cultured on a scaffold to generate natural tissues *in vitro*, and the newly generated tissues are then transferred and implanted into the defective part of the patients *in vivo*. The main goal in tissue engineering is to mimic the native human tissue environment as much as possible. Hence, a successful tissue engineering implant is highly dependent on the design of a porous scaffold. Ideal scaffolds should be biodegradable to support the replacement of new tissues and biocompatible to avoid any inflammation or immune reactions as well as providing a suitable mechanical property to support the growth of new tissues.

Conductive biomimetic materials applied in nerve, cardiac and bone tissue engineering provide a promising alternative to scaffold fabrication. This is due to the importance of electrical stimulation in maintaining bioelectricity hemostasis such as signaling of the nervous system; therefore, electrical stimulation plays an integral role in stimulating cell proliferation and differentiation, especially for fibroblasts, neurons, and osteoblasts. Conductive polymer provides advantages in efficiently delivering electrical signals from an external source to the seeded cells. The desired conductive polymer properties for tissue engineering applications include conductivity, reversible oxidation, redox stability, biocompatibility, non-toxic and hydrophilicity (Guimard et al. 2007).

9.5 PROCESSABILITY OF CONDUCTIVE POLYMERS TO FABRICATE SCAFFOLDS AND AS COATING MATERIAL

Electrospinning has emerged as one of the promising and attractive fiber-forming techniques using the application of electrostatic force to synthesize fibers with controlled diameters ranging from nanometers to sub-micrometer from a variety of polymers (Bolin et al. 2009). During electrospinning, the polymer solution will be induced as a charge solution when high voltage is applied. Once the electrical forces in the solution are balanced by surface tension, a Taylor cone (Bolin et al. 2009) will be formed; if the electrostatic force is able to overcome the surface tension of the charged polymer solutions, the charged fiber jets are ejected from the tip of the Taylor cone and accelerate towards the collector. The morphology and fiber diameter dimensions can be controlled by altering parameters of electrospinning, including solution parameters (viscosity and conductivity of polymer solution, solvent selection), processing parameters (voltage, flow rate, distance between spinnerets and collectors) and ambient parameters (temperature, humidity).

Despite having many benefits in processing polymers, pure conductive polymers are unable to directly electrospin due to high conductivity. During electrospinning, the electrical instability caused by the highly charged conductive polymer solution disrupts the formation of a Taylor cone, leading to repulsion of the polymer chain before entanglement to form fibers at collectors (Taylor 1969). Furthermore, pure conductive polymers are typically rigid and brittle because of the tight coil-like conformation in the polymer chain backbone, causing mechanical instability (Low et al. 2015). For instance, literature has found that pure PEDOT nanofibers are difficult to form by electrospinning due to the rigid main chains, thus causing poor entanglement and interactions between polymer chains result (Rutledge and Fridrikh 2007). However, researchers have found that PPy is able to form nanofibers by electrospinning but it requires the organic solvent soluble polypyrrole, [(PPy3) C(DEHS)K]x(Hohman et al. 2001). Hence, in order to enhance the processability of electrospinning, conductive polymers are usually incorporated with non-conducting host polymers (Xia et al. 1995, Huang et al. 2015) via blending or coating. Meanwhile, the morphology of fibers and the mechanical properties were able to be controlled and improved. The PLA/PHBV-based electrospun membrane was coated with 30% PEDOT:PSS which rendered the membrane conductive and more hydrophilic. Human skin fibroblasts (HSF) were cultured on the membrane and it was observed that the cell viability was significantly higher than uncoated counterparts (Fig. 9.2). PEDOT:PSS were successfully incorporated with Chitosan and hydroxyapatite to fabricate three-dimensional scaffolds using a freeze-drying technique. A model drug, Tetracycline.HCL, was loaded in the scaffolds and the antibacterial activity was tested using the zone inhibition method. Cell viability of HSF was significantly higher in PEDOT:PSS-incorporated scaffold (Fig. 9.3).

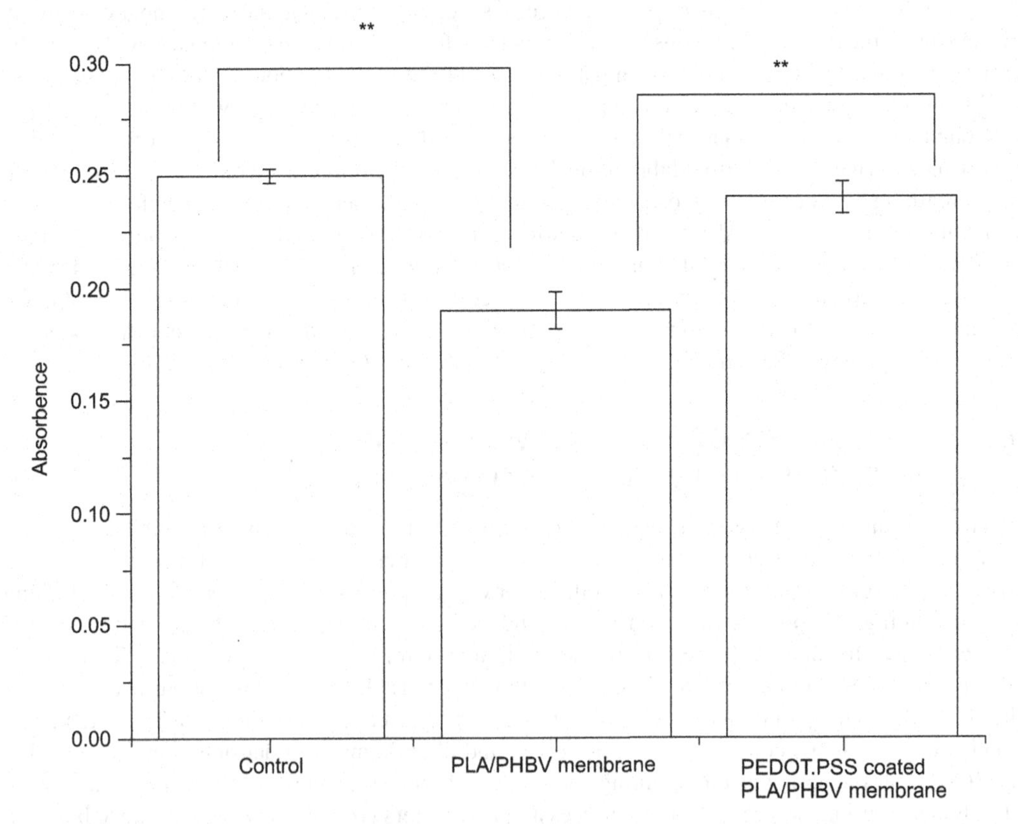

FIGURE 9.2 Cell viability of HSF cultured on PEDOT:PSS coated and un-coated scaffold. The values are mean ± standard deviation (number of samples = 3) (**$p < 0.001$).

Thus, polymer-based electrospun nanofibers and freeze-dried scaffolds carrying the electrical properties open up an opportunity to improve the properties of scaffolds using existing technologies for tissue engineering applications.

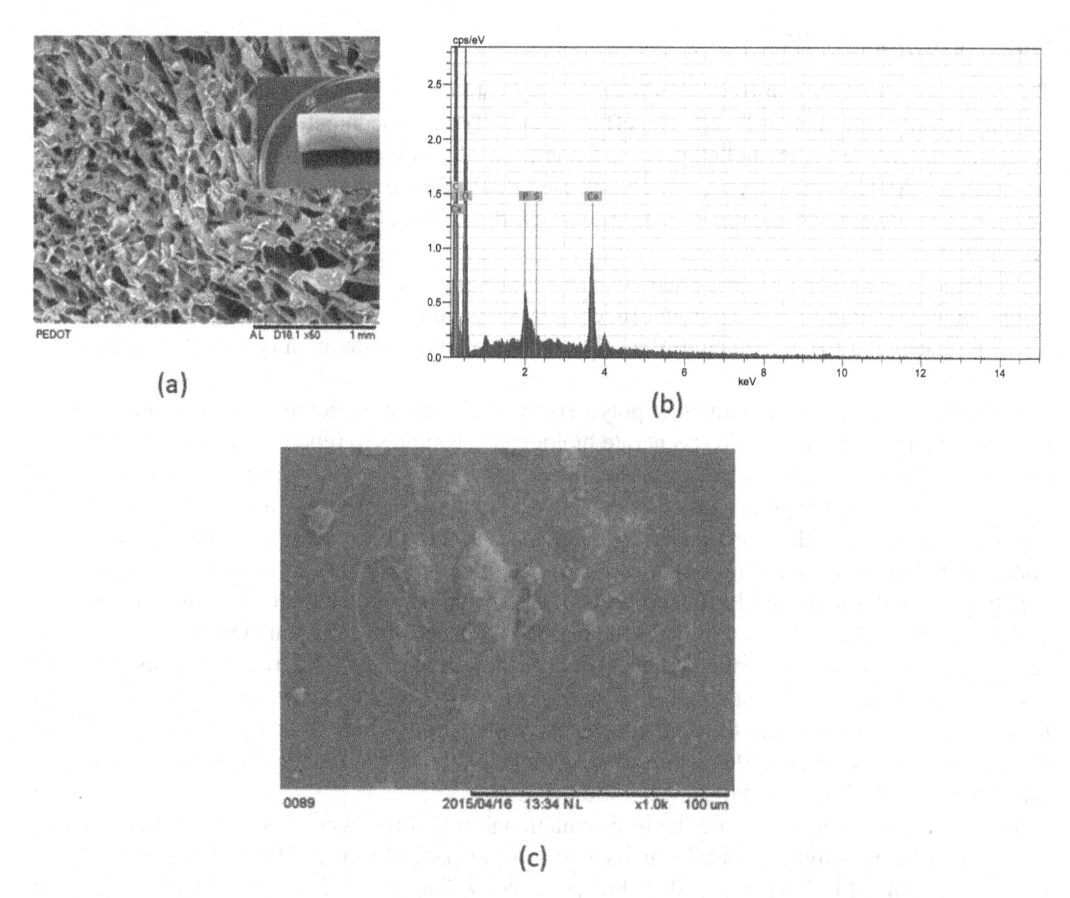

FIGURE 9.3 (a) General appearance and scanning electron micrograph of chitosan/HA/PEDOT: PSS scaffold using freeze-drying technique, (b) EDX spectrum of the scaffold confirming the presence of sulphur (S) due to PEDOT:PSS, (c) scanning electron micrograph of the HSF cells attached on the surface of chitosan/HA/PEDOT:PSS scaffold.

9.6 RECENT STUDY OF THE CONDUCTIVE SCAFFOLDS FABRICATED USING PPy, PANI AND PEDOT FOR TISSUE ENGINEERING APPLICATION

9.6.1 Polypyrrole (PPy)

Over the last decade, PPy has gained much interest as a potential conductive biomaterial in tissue engineering scaffolds (Chronakis et al. 2006, Subramanian et al. 2012, Zhao et al. 2015). However, material composed solely of PPy is unsuitable for tissue engineering applications, especially for nerve regeneration, due to a number of limitations (Macagnano et al. 2011). The poor mechanical property of PPy marks the weakness and brittleness of the polymers while the low solubility of PPy in most solvents makes it difficult to process into desirable structures for tissue engineering scaffolds. Furthermore, PPy is also a non-degradable conductive polymer, which is difficult to degrade after implantation into the human body. Therefore, to overcome these limitations, different approaches have been attempted. Overall, PPy is a potential material for conductive scaffolds in future studies involving electrical stimulation, especially in nerve tissue engineering. The limitations of PPy can be improved by blending with other polymers.

9.6.2 Polyaniline (PANI)

The rapid growth of PPy in tissue engineering applications has spurred the development of other conductive polymers such as PANI. The effort to identify the biocompatibility of PANI *in vivo* has generated interest in being applied as a biomimetic conductive scaffold for the tissue engineering application. PANI has the advantages of being used in scaffold to transmit the electrical stimulation which can modulate and accelerate activities of cardiac tissue, nerve tissue, skeletal muscle and bone tissue.

Unlike PPy, PANI can be processed alone for tissue engineering scaffold. There have been few studies demonstrating the behavior and function of cells on pure PANI scaffold. Pure PANI films were fabricated and found to support the cell attachment and proliferation of PC-12 cells (Li et al. 2006).

PANI has been combined with other polymers to form conductive composites or blended to gain the desired and key properties to regenerate biological substitutes to repair or replace the damaged and lost tissues for cardiac, skeletal muscle, nerve and bone tissue engineering in order to mimic as closely as possible the properties of native extracellular matrix (ECM). In view of cardiac tissue engineering, generally, the human heart is made up of electrically conductive tissue. The sinoatrial node, located in the right atrium, sends electrical impulses throughout the rest of myocardium via the atrioventricular node and Purkinje fibers. The propagation of electrical stimulation causes the contraction of cardiac cells, and initiates the heartbeat. Hence, it is important to regenerate cardiac cells in a conductive scaffold which is able to deliver these electrical stimulations to the seeded cells. Some studies have focused on conductive scaffold fabrication using composite or blended PANI for cardio tissue engineering. Cardiomyocytes were shown to adhere to and proliferate on PANI/PLGA nanofibers (Li et al. 2006), while another study exhibited that PANI/gelatin also successfully supported cell attachment and the proliferation of H9c2.

In addition, neurons, which are the basic functional unit of the nervous system, are electrically excitable cells, transmitting neural impulses at a rapid pace (Li et al. 2006, Wang et al. 2008, Hsiao et al. 2013, McKeon et al. 2010, Jun et al. 2009, Feng et al. 2013). Electrical stimulation has been recognized to be an effective element in enhancing nerve regeneration such as neurite growth and nerve regeneration. Thus, an electrically conductive scaffold acts as an effective and excellent scaffold in delivering electrical impulses to cultured neurons. The neural cells stimulated on nanofibrous PANI-PCL/Gelatin scaffold were demonstrated to enhance proliferation and neurite outgrowth (Li et al. 2006). Similarly, the density of viable neurons was reported to experience a greater than ten-fold increase after culturing on the composite of blended PANI and polypropylene (PANI-PP). Nanofibers of PANI/PLLA were proved to enhance the viability and proliferation of the cultured nerve stem cells significantly.

9.6.3 PEDOT:PSS

Poly(3,4-ethylenedioxythiophene) (PEDOT) is used in different fields, including biomedicine and tissue engineering due to its properties of high electrical conductivity and chemical stability. It was reported that epithelial cells presented significant activity on the surface of PEDOT electrodeposited on stainless steel electrodes and no cytoxicity was detected due to this polymer. In addition, the electrochemical characteristic of PEDOT covered with cells was determined in different biological media using cyclic voltammetry experiments. A significant increase in the electroactivity of this material is observed when it is covered with a cellular monolayer. Hence, the overall results showed that PEDOT is biocompatible and electrocompatible with Hep-2 cells (Jun et al. 2009, Feng et al. 2013, Qazi et al. 2014, Del Valle et al. 2007, Liu et al. 2009).

The doping process introduces charge carriers into the polymers and makes it conductive. Dopants can be categorized based on their molecular size. An example of a small dopant is Cl^-, while sodium polystrenesulfonate, PSS, is an example of large dopant. The difference between large and small dopants is that large dopants will affect the material properties more, and cause it to be more integrated into the polymer, allowing the polymer to have greater electrochemical stability (Liu et al. 2009). PSS is an example of dopant that is commonly used with PEDOT to improve the biocompatibility of PEDOT. Other examples of dopants which have been used with PEDOT are heparin, hyaluronic acid (HA), nerve growth factor (NGF) and fibrinogen.

9.7 CONCLUSIONS

Stimulus-receptive conductive polymers can provide smart surfaces for tissue engineering scaffolds. Scaffold-based tissue engineering still relies on the initiation of innovative approaches to serve as temporary support on wound site. Smart conductive polymers trigger the signaling process and thus permit biological responses. Conductive polymers have vast flexibility. Here, we have concentrated on the application of stimulus-receptive polymers in the fabrication of tissue engineering scaffolds, either alone or as a composite with non-conductive biodegradable or biocompatible polymers which can serve as a template. The physical and chemical properties were also discussed. However, the properties of stimulus-receptive polymer based scaffolds should be carefully modulated in order to allow greater efficacy. The overall behavior of smart polymer-based scaffolds under biological conditions, especially inside the human body, is still challenging.

9.8 ACKNOWLEDGEMENT

Authors would like to acknowledge Lor Huai Chong, Chang Hui Chung, Mohd. Izzat Hassan and Alireza Lari for their contributions to prepare the chapter.

REFERENCES

Alikacem, N., Y. Marois, Z. Zhang, B. Jakubiec, R.W. Roy, M. King and R. Guidoin. 1999. Tissue reactions to polypyrrole-coated polyesters: a magnetic resonance relaxometry study. Artificial Organs 23(10): 910-919.

Asplund, M., E. Thaning, J. Lundberg, A.C. Sandberg-Nordqvist, B. Kostyszyn, O. Inganäs and H. von Holst. 2009. Toxicity evaluation of PEDOT/biomolecular composites intended for neural communication electrodes. Biomedical Materials 4(4): 1-12.

Ateh, D.D., H.A. Navsaria and P. Vadgama. 2006a. Polypyrrole-based conducting polymers and interactions with biological tissues. Journal of The Royal Society Interface 3(11): 741-752.

Ateh, D.D., P. Vadgama and H.A. Navsaria. 2006b. Culture of human keratinocytes on polypyrrole-based conducting polymers. Tissue Engineering 12(4): 645-655.

Bendrea, A.-D., L. Cianga and I. Cianga. 2011. Review paper: progress in the field of conducting polymers for tissue engineering applications. Journal of Biomaterials Applications 26(1): 3-84.

Bidez, P.R., S. Li, A.G. Macdiarmid, E.C. Venancio, Y. Wei and P.I. Lelkes. 2006. Polyaniline, an electroactive polymer, supports adhesion and proliferation of cardiac myoblasts. Journal of Biomaterials Science, Polymer Edition 17(1-2): 199-212.

Bolin, M.H., K. Svennersten, X. Wang, I.S. Chronakis, A. Richter-Dahlfors, E.W.H. Jager and M. Berggren. 2009. Nano-fiber scaffold electrodes based on PEDOT for cell stimulation. Sensors and Actuators B: Chemical 142(2): 451-456.

Bredas, J.L. and G.B. Street. 1985. Polarons, bipolarons, and solitons in conducting polymers. Accounts of Chemical Research 18(10): 309-315.

Castano, H., E.A. O'Rear, P.S. McFetridge and V.I Sikavitsas. 2004. Polypyrrole Thin films formed by admicellar polymerization support the osteogenic differentiation of mesenchymal stem cells. Macromolecular Bioscience 4(8): 785-794.

Chiang, C.K., C.R. Jr. Fincher, Y.W. Park, A.J. Heeger, H. Shirakawa, E.J. Louis, S.C. Gau and A.G. MacDiarmid. 1977. Electrical Conductivity in Doped Polyacetylene. Physical Review Letters 39(17): 1098-1101.

Chiang, C.K., Y.W. Park and A.J. Heeger. 1978a. Conducting polymers: halogen doped polyacetylene. The Journal of Chemical Physics 69(11): 5098-5104.

Chiang, C.K., M.A. Druy, S.C. Gau, A.J. Heeger, E.J. Louis, A.G. MacDiarmid, Y.W. Park and H. Shirakawa. 1978b. Synthesis of highly conducting films of derivatives of polyacetylene, (CH)x. Journal of the American Chemical Society 100(3): 1013-1015.

Chronakis, I.S., S. Grapenson and A. Jakob. 2006. Conductive polypyrrole nanofibers via electrospinning: Electrical and morphological properties. Polymer 47(5): 1597-1603.

Collier, J.H., J.P. Camp, T.W. Hudson and C.E. Schmidt. 2000. Synthesis and characterization of polypyrrole–hyaluronic acid composite biomaterials for tissue engineering applications. Journal of Biomedical Materials Research 50(4): 574-584.

Del Valle, L.J., D. Aradilla, R. Oliver, F. Sepulcre, A. Gamez, E. Armelin, C. Alemán and F. Estrany. 2007. Cellular adhesion and proliferation on poly (3, 4-Ethylenedioxythiophene): benefits in the electroactivity of the conducting polymer. European Polymer Journal 43(6): 2342-2349.

Ding, C., Y. Wang and S. Zhang. 2007. Synthesis and characterization of degradable electrically conducting copolymer of aniline pentamer and polyglycolide. European Polymer Journal 43(10): 4244-4252.

Feng, Z.-Q., J. Wu, W. Cho, M.K. Leach, E.W. Franz, Y.I. Naim, Z.-Z. Gu, J.M. Corey and D.C. Martin. 2013. Highly aligned poly(3,4-ethylene dioxythiophene)(PEDOT) nano- and microscale fibers and tubes. Polymer 54(2): 702-708.

George, P.M., A.W. Lyckman, D.A. LaVan, A.H. Yuika Leung, R. Avasare, C. Testa, P.M. Alexander, R. Langer and M. Sur. 2005. Fabrication and biocompatibility of polypyrrole implants suitable for neural prosthetics. Biomaterials 26(17): 3511-3519.

Ghasemi-Mobarakeh, L., M.P. Prabhakaran, M. Morshed, M.H. Nasr-Esfahani and S. Ramakrishna. 2009. Electrical Stimulation of nerve cells using conductive nanofibrous scaffolds for nerve tissue engineering. Tissue Engineering Part A 15(11): 3605-3619.

Green, R.A., N.H. Lovell, G.G. Wallace and L.A. Poole-Warren. 2008. Conducting polymers for neural interfaces: challenges in developing an effective long-term implant. Biomaterials 29(24-25): 3393-3399.

Guimard, N.K., N. Gomez and C.E. Schmidt. 2007. Conducting polymers in biomedical engineering. Progress in Polymer Science 32(8-9): 876-921.

Guimard, N.K.E., J.L. Sessler and C.E. Schmidt. 2009. Toward a biocompatible and biodegradable copolymer incorporating electroactive oligothiophene units. Macromolecules 42(2): 502-511.

Guiseppi-Elie, A. 2010. Electroconductive hydrogels: synthesis, characterization and biomedical applications. Biomaterials 31(10): 2701-2716.

Guo, B., L. Glavas and A.-C. Albertsson. 2013. Biodegradable and electrically conducting polymers for biomedical applications. Progress in Polymer Science 38(9): 1263-1286.

Heeger, A.J., A.G. MacDiarmid and H. Shirakawa. 2000: The Nobel prize in chemistry, 2000: conductive polymers. KUNGL. Vetenskapsakademien, The Royal Swedish Academy of Sciences 1-16.

Hesketh, P.J. and D. Misra. 2012. Conducting polymers and their applications. The Electrochemical Society Interface 21: 61-62.

Higgins, M.J., P.J., Molino, Z. Yue and G.G. Wallace. 2012. Organic conducting polymer–protein interactions. Chemistry of Materials 24(5): 828-839.

Hohman, M.M. 2001. Electrospinning and electrically forced jets. I. Stability theory. Physics of Fluids 13(8): 2201-2220.

Hsiao, C.-W., M.-Y. Bai, Y. Chang, M.-F. Chung, T.-Y. Lee, C.-T. Wu, B. Maiti, Z.-X. Liao, R.-K. Li and H.-W. Sung. 2013. Electrical coupling of isolated cardiomyocyte clusters grown on aligned conductive nanofibrous meshes for their synchronized beating. Biomaterials 34(4): 1063-1072.

Huang, L., J. Hu, L. Lang, X. Wang, P. Zhang, X. Jing, X. Wang, X. Chen, P.I. Lelkes, A.G. MacDiarmid and Y. Wei. 2007. Synthesis and characterization of electroactive and biodegradable ABA block copolymer of polylactide and aniline pentamer. Biomaterials 28(10): 1741-1751.

Huang, L., X. Zhuang, J. Hu, L. Lang, P. Zhang, Y. Wang, X. Chen, Y. Wei and X. Jing. 2008. Synthesis of biodegradable and electroactive multiblock polylactide and aniline pentamer copolymer for tissue engineering applications. Biomacromolecules 9(3): 850-858.

Huang, Y.-C., T.-Y. Lo, C.-H. Chen, K.-H. Wu, C.-M. Lin and W.-T. Whang. 2015. Electrospinning of magnesium-ion linked binder-less PEDOT:PSS nanofibers for sensing organic gases. Sensors and Actuators B: Chemical 216: 603-607.

Inzelt, G., M. Pineri, J.W. Schultze and M.A. Vorotyntsev. 2000. Electron and proton conducting polymers: recent developments and prospects. Electrochimica Acta 45(15-16): 2403-2421.

Inzelt and György. 2008. Conducting Polymers. 1 ed. Monographs in Electrochemistry. Springer-Verlag Berlin Heidelberg. 269.

Ito, T., H. Shirakawa and S. Ikeda. 1974. Simultaneous polymerization and formation of polyacetylene film on the surface of concentrated soluble Ziegler-type catalyst solution. Journal of Polymer Science: Polymer Chemistry Edition 12(1): 11-20.

Jun, I., S. Jeong and H. Shin. 2009. The stimulation of myoblast differentiation by electrically conductive sub-micron fibers. Biomaterials 30(11): 2038-2047.

Karagkiozaki, V., P.G. Karagiannidis, M. Gioti, P. Kavatzikidou, D. Georgiou, E. Georgaraki and S. Logothetidis. 2013. Bioelectronics meets nanomedicine for cardiovascular implants: PEDOT-based nanocoatings for tissue regeneration. Biochimica et Biophysica Acta (BBA) - General Subjects 1830(9): 4294-4304.

Kim, D.-H., S. M. Richardson-Burns, J.L. Hendricks, C. Sequera and D.C. Martin. 2007. Effect of immobilized nerve growth factor on conductive polymers: electrical properties and cellular response. Advanced Functional Materials 17(1): 79-86.

Lakard, B., L. Ploux, K. Anselme, F. Lallemand, S. Lakard, M. Nardin and J.Y. Hihn. 2009. Effect of ultrasounds on the electrochemical synthesis of polypyrrole, application to the adhesion and growth of biological cells. Bioelectrochemistry 75(2): 148-157.

Li, M., Y. Guo, Y. Wei, A.G. MacDiarmid and P.I. Lelkes. 2006. Electrospinning polyaniline-contained gelatin nanofibers for tissue engineering applications. Biomaterials 27(13): 2705-2715.

Liu X., K.J. Gilmore, S.E. Moulton and G.G. Wallace. 2009. Electrical stimulation promotes nerve cell differentiation on polypyrrole/poly(2-methoxy-5 aniline sulfonic acid) composites. Journal of Neural Engineering 6(065002): 1-10.

Low, K., C.B. Horner, C. Li, G. Ico, W. Bosze, N.V. Myung and J. Nam. 2015. Composition-dependent sensing mechanism of electrospun conductive polymer composite nanofibers. Sensors and Actuators B: Chemical Part A 207: 235-242.

Macagnano, A., E. Zampetti, S. Pantalei, F. De Cesare, A. Bearzotti and K.C. Persaud. 2011. Nanofibrous PANI-based conductive polymers for trace gas analysis. Thin Solid Films 520(3): 978-985.

Mattioli-Belmonte, M., G. Giavaresi, G. Biagini, L. Virgili, M. Giacomini, M. Fini, F. Giantomassi, D. Natali, P. Torricelli and R. Giardino. 2003. Tailoring biomaterial compatibility: *in vivo* tissue response versus *in vitro* cell behavior. The International Journal of Artificial Organs 26(12): 1077-1085.

McKeon, K.D., A. Lewis and J.W. Freeman. 2010. Electrospun poly(D, L-lactide) and polyaniline scaffold characterization. Journal of Applied Polymer Science 115(3): 1566-1572.

Meng, S., M. Rouabhia, G. Shi and Z. Zhang. 2008. Heparin dopant increases the electrical stability, cell adhesion, and growth of conducting polypyrrole/poly(L, L-lactide) composites. Journal of Biomedical Materials Research Part A 87A(2): 332-344.

Mikulski, C.M., A.G. MacDiarmid, A.F. Garito and A.J. Heeger. 1976. Stability of polymeric sulfur nitride, (SN)x, to air, oxygen, and water vapor. Inorganic Chemistry 15(11): 2943-2945.

Oh, W.-K., O.S. Kwon and J. Jang. 2013. Conducting polymer nanomaterials for biomedical applications: cellular interfacing and biosensing. Polymer Reviews 53(3): 407-442.

Olayo, R., C. Ríos, H. Salgado-Ceballos, G.J. Cruz, J. Morales, M.G. Olayo, M. Alcaraz-Zubeldia, A.L. Alvarez, R. Mondragon, A. Morales and A. Diaz-Ruiz. 2008. Tissue spinal cord response in rats after implants of polypyrrole and polyethylene glycol obtained by plasma. Journal of Materials Science: Materials in Medicine 19(2): 817-826.

Otero, T.F., J.G. Martinez and J. Arias-Pardilla. 2012. Biomimetic electrochemistry from conducting polymers: a review: artificial muscles, smart membranes, smart drug delivery and computer/neuron interfaces. Electrochimica Acta 84: 112-128.

Pelto, J., S. Haimi, E. Puukilainen, P.G. Whitten, G.M. Spinks, M. Bahrami-Samani, M. Ritala and T. Vuorinen. 2010. Electroactivity and biocompatibility of polypyrrole-hyaluronic acid multi-walled carbon nanotube composite. Journal of Biomedical Materials Research Part A 93A(3): 1056-1067.

Qazi, T.H., R. Rai and A.R. Boccaccini. 2014. Tissue engineering of electrically responsive tissues using polyaniline based polymers: a review. Biomaterials 35(33): 9068-9086.

Richardson, R.T., B. Thompson, S. Moulton, C. Newbold, M.G. Lum, A. Cameron, G. Wallace, R. Kapsa, G. Clark and S. O'Leary. 2007. The effect of polypyrrole with incorporated neurotrophin-3 on the promotion of neurite outgrowth from auditory neurons. Biomaterials 28(3): 513-523.

Rivers, T.J., T.W. Hudson and C.E. Schmidt. 2002. Synthesis of a novel, biodegradable electrically conducting polymer for biomedical applications. Advanced Functional Materials 12(1): 33-37.

Rutledge, G.C. and S.V. Fridrikh. 2007. Formation of fibers by electrospinning. Advanced Drug Delivery Reviews 59(14): 1384-1391.

Sayyar, S., E. Murray, B.C. Thompson, S. Gambhir, D.L. Officer and G.G. Wallace. 2013. Covalently linked biocompatible graphene/polycaprolactone composites for tissue engineering. Carbon 52: 296-304.

Sebaa, M., S. Dhillon and H. Liu. 2013. Electrochemical deposition and evaluation of electrically conductive polymer coating on biodegradable magnesium implants for neural applications. Journal of Materials Science: Materials in Medicine 24(2): 307-316.

Shi, G., M. Rouabhia, S. Meng and Z. Zhang. 2008a. Electrical stimulation enhances viability of human cutaneous fibroblasts on conductive biodegradable substrates. Journal of Biomedical Materials Research Part A 84A(4): 1026-1037.

Shi, G., Z. Zhang and M. Rouabhia. 2008b. The regulation of cell functions electrically using biodegradable polypyrrole–polylactide conductors. Biomaterials 29(28): 3792-3798.

Shiigi, H., M. Kishimoto, H. Yakabe, B. Deore and T. Nagaoka. 2002. Highly selective molecularly imprinted overoxidized polypyrrole colloids: one-step preparation technique. Analytical Sciences 18(1): 41-44.

Shirakawa, H., E.J. Louis, A.G. MacDiarmid, C.K. Chiang and A.J. Heeger. 1977. Synthesis of electrically conducting organic polymers: halogen derivatives of polyacetylene, (CH)x. Journal of the Chemical Society, Chemical Communications 16: 578-580.

Shirakawa, H., T. Ito and S. Ikeda. 1978. Electrical properties of polyacetylene with various cis-trans compositions. Die Makromolekulare Chemie 179(6): 1565-1573.

Shirakawa, H. 2001. The Discovery of Polyacetylene Film: The Dawning of an Era of Conducting Polymers (Nobel Lecture). Angewandte Chemie International Edition 40(14): 2574-2580.

Subramanian, A., U. Krishnan and S. Sethuraman. 2012. Axially aligned electrically conducting biodegradable nanofibers for neural regeneration. Journal of Materials Science: Materials in Medicine 23(7): 1797-1809.

Svennersten, K., K.C. Larsson, M. Berggren and A. Richter-Dahlfors. 2011. Organic bioelectronics in nanomedicine. Biochimica et Biophysica Acta (BBA) - General Subjects 1810(3): 276-285.

Taylor, G. 1969. Electrically driven jets. Vol. 313: 453-475.

Wang, C.H., Y.Q. Dong, K. Sengothi, K.L. Tan and E.T. Kang. 1999. *In vivo* tissue response to polyaniline. Synthetic Metals 102(1-3): 1313-1314.

Wang, H.-J., L.-W. Ji, D.-F. Li and J.-Y. Wang. 2008. Characterization of nanostructure and cell compatibility of polyaniline films with different dopant acids. The Journal of Physical Chemistry B 112(9): 2671-2677.

Wang, Z., C. Roberge, Y. Wan, L.H. Dao, R. Guidoin and Z. Zhang. 2003. A biodegradable electrical bioconductor made of polypyrrole nanoparticle/poly(D,L-lactide) composite: a preliminary *in vitro* biostability study. Journal of Biomedical Materials Research Part A 66A(4): 738-746.

Wang, Z., C. Roberge, L.H. Dao, Y. Wan, G. Shi, M. Rouabhia, R. Guidoin and Z. Zhang. 2004. *In vivo* evaluation of a novel electrically conductive polypyrrole/poly(D,L-lactide) composite and polypyrrole-coated poly(D,L-lactide-co-glycolide) membranes. Journal of Biomedical Materials Research Part A 70A(1): 28-38.

Williams, R.L. and P.J. Doherty. 1994. A preliminary assessment of poly(pyrrole) in nerve guide studies. Journal of Materials Science: Materials in Medicine 5(6-7): 429-433.

Xia, Y., J.M. Wiesinger, A.G. MacDiarmid and A.J. Epstein. 1995. Camphorsulfonic acid fully doped polyaniline emeraldine salt: conformations in different solvents studied by an ultraviolet/visible/near-infrared spectroscopic method. Chemistry of Materials 7(3): 443-445.

Zelikin, A., V. Shastri, D. Lynn, J. Farhadi, I. Martin and R. Langer. 2001. Bioerodible polypyrrole. Materials Research Society Symposium Proceedings 711: 193-197.

Zelikin, A.N., D.M. Lynn, J. Farhadi, I. Martin, V. Shastri and R. Langer. 2002. Erodible conducting polymers for potential biomedical applications. Angewandte Chemie 114(1): 149-152.

Zhang, Q., Y. Yan, S. Li and T. Feng. 2010. The synthesis and characterization of a novel biodegradable and electroactive polyphosphazene for nerve regeneration. Materials Science and Engineering: C 30(1): 160-166.

Zhang, Z., R. Roy, F.J. Dugré, D. Tessier and L.H. Dao. 2001. *In vitro* biocompatibility study of electrically conductive polypyrrole-coated polyester fabrics. Journal of Biomedical Materials Research 57(1): 63-71.

Zhang, Z., M. Rouabhia, Z. Wang, C. Roberge, G. Shi, P. Roche, J. Li and L.H. Dao. 2007. Electrically Conductive Biodegradable Polymer Composite for Nerve Regeneration: Electricity-Stimulated Neurite Outgrowth and Axon Regeneration. Artificial Organs 31(1): 13-22.

Zhao, W., B. Yalcin and M. Cakmak. 2015. Dynamic assembly of electrically conductive PEDOT:PSS nanofibers in electrospinning process studied by high speed video. Synthetic Metals 203(0): 107-116.

Zhou D.D., X.T. Cui, A. Hines and R.J. Greenberg. 2010. Conducting polymers in neural stimulation applications. pp. 217-252. *In*: D. Zhou and E. Greenbaum [eds.]. Implantable Neural Prostheses 2. Springer, New York.

10

Evaluation of PCL/Chitosan/Nanohydroxyapatite/Tetracycline Composite Scaffolds for Bone Tissue Engineering

Rashid Bin Mad Jin[1], Naznin Sultana[2,*], Chin Fhong Soon[3] and Ahmad Fauzi Ismail[4]

10.1 INTRODUCTION

Tissue engineering (TE) aims to repair and restore damaged tissue function. In this field, it utilizes cells, scaffolds and growth factors (GFs) as tools to achieve potential alternative grafts (Biondi et al. 2008). Conventional methods showed major drawbacks related to organ transplantation, surgical reconstruction (to treat the loss or failure of an organ or tissues) and the lack of a suitable donor. A new field of tissue engineering approaches uses scaffolds, living cells and growth factors as synthetic grafts that are able to heal or improve damaged organs (Griffith and Naughton 2002).

Regeneration of damaged tissues in tissues engineering through the substitution of engineered tissues can help to restore the functions during regeneration and subsequent integration with the host tissues. In bone tissue engineering, the biodegradable scaffolds will play a role as a temporary template for new bones tissues. The scaffolds will degrade and be replaced by newly formed bone tissue (Griffith and Naughton 2002). Three-dimensional scaffolds with a well-defined microstructure, proper mechanical properties, interconnected pore network, and biocompatibility are among the ideal properties of the scaffolds. Other than that, the scaffolds should be able to deliver bioactive factors to help the regeneration process (Biondi et al. 2008).

Tissue engineering is an important therapeutic strategy for current and future medicine. Limitation efficacy of conventional treatment strategies for large bones, osteochondral defects, and cartilage has inspired research into the development of scaffold-based tissue engineering solutions. Tissue engineering is a promising alternative approach for treating patients suffering from this problem. In theory, porous 3D temporary scaffolds manipulate cell functions, where specifically isolated cells were grown on 3D scaffolds under controlled culture conditions (*in vitro*). Later, the construct was implanted to the desired site in the body to direct new tissue formation onto scaffolds that can degrade over time. Scaffold-based tissue engineering strategies for the repair and regeneration of skeletal tissues should meet the physical, mechanical and biological requirements of the target

[1] Universiti Teknologi Malaysia, Malaysia.
[2] Medical Academy, Prairie View A&M University, TX 77446, USA.
[3] Faculty of Electrical and Electronic Engineering, Universiti Tun Hussein Onn Malaysia, Batu Pahat, Johor, Malaysia.
[4] Advanced Membrane Technology Research Center, Universiti Teknologi Malaysia, Malaysia.
* Corresponding author: nasultana@pvamu.edu

tissue. Clinically viable scaffold designs, ease of manufacture and sterilization, reproducibility of material properties, and cost effectiveness also play important roles in the successful strategies of scaffold-based tissue engineering.

In order to repair damaged tissues, several techniques and materials have been used to produce synthetic scaffolds that can imitate the extracellular matrix of the organ that has been repaired (Gomes et al. 2002). There are four sequences in tissues engineering: first is to identify, isolate and produce sufficient cell source; then, the synthesis and manufacturing of the cell in the desired dimension and shape of suitable biocompatible carrier that can act as cell carrier; third, uniformly seeding the cells into or onto the carrier and incubating in the bioreactor, and lastly, implanting the carrier in the animal model. Vascularisation may be needed depending on the site and the structure of the animal model (Ma and Elisseeff 2005, Ehrenfreund-Kleinman et al. 2006).

Bone is a complex and highly specialized form of connective tissues pertaining to the formation of the skeleton of the body (Hussain et al. 2010). It is a physically hard and strong tissue that provides mechanical function to various types of muscle groups, load bearing, and locomotion and serves as protection to vital organs such as brain, lung, heart, bladder, and many others (Nather 2005). Bone is composed of 70% of minerals and 30% of proteins. Composition of bone consists of minerals, collagen, water, non-collagenous protein, lipid, vascular element and cellular element. Table 10.1 shows the detailed composition of the bone.

Main component of the bone matrix is 60% of calcium phosphate–hydroxyapatite $(Ca_{10}PO_4)_6(OH)_2$ and association of trace element such as carbonate, citrate, sodium, magnesium, fluoride, potassium and iron to provide rigidity and toughness to the bone (Hussain et al. 2010). Meanwhile, type I collagen fibres are made up of bundles of fibrils to resist pulling force. Combination of both collagen and hydroxyapatite forms hard and tough bone that not only provides mechanical support but also serves as a reservoir for minerals, especially calcium and phosphate (Nather 2005, Hussain et al. 2010).

Bone can be categorised into four types which are long, short, flat and irregular bone. Long bone such as femur, tibia, humerus, radius and ulna can be found in the limbs. Short bones are carpal and tarsal. Flat bone are ribs, sternum, cranium and scapula. Irregular bones are vertebrae, sacrum, coccyx, temporal, mandible, palatine, inferior nasal concha, hyoid, sphenoid, ethmoid, zygomatic, and maxilla (Buckwalter et al. 1995) and lastly, sesamoid that is patella and smaller bones found in flexor hallucis longus and peroneus longus tendons (Buckwalter et al. 1995).

There are two types of structures of adult skeleton bone which are cortical bone (compact pattern) and cancellous bone (trabecular pattern). Cortical bone or compact bone is the bone with a dense outer layer that is made up of a structure of Haversian systems or Osteon that can resist bending. It only has 10% porosity and is omnipresent in long, short and flat bones. In contrast, cancellous bone or sponge bone consists of a series of interconnecting plates of bone or a porous sponge-like pattern, which the bone element places in the direction of functional pressure according to Wolff's law. Cancellous bone can be found harbouring a large part of bone marrow, metaphysis of long bones and also the vertebral bodies (Buckwalter et al. 1995).

TABLE 10.1 Composition of bone tissue (Hussain et al. 2010)

Inorganic Phase (wt. %)	Organic Phase (wt. %)
Hydroxyapatite ~60	Collagen ~20%
Carbonate ~4	Water ~9
Citrate ~0.9	Non collagenous protein ~3:
Sodium ~0.7	Osteocalcin, osteonectin, protein,
Magnesium ~0.5	Serum protein and etc.
Other traces:	Other traces:
Cl, F, K, Sr, Pb, Zn, Cu, Fe	Polysaccharides, lipid, cytokines
	Primary bone cells:
	Osteoblast, osteocytes, osteoclast

The ideal scaffolds should possess several characteristics such as biocompatibility, mechanical properties, pore size, and bioresorbability (Bose et al. 2012).

The main materials for fabrication of scaffolds in tissues engineering application are polymers, especially biodegradable polymeric materials. These materials can be divided into two groups, that is, natural-based materials including polysaccharides (starch, alginate, and chitin/chitosan) or proteins (soy, collagen, fibrin and silk). The other one is synthetic polymer such as poly (lactic acid) (PLA), poly (glycolic acid) (PGA), poly (3-caprolactone) (PCL), and poly (hydroxyl butyrate) (PHB). Synthetic polymers have relatively good mechanical strength and their shape and degradation rate can be easily modified. Meanwhile, natural polymers possess many functional groups such as amino, carboxylic and hydroxyl groups that are available for chemical reactions such as hydrolysis, oxidation, reduction, esterification, cross-linking and many others. In addition to various properties such as pseudo plastic behaviour, gelation ability, water binding capacity, biodegradability makes them suitable for tissues engineering application (Ehrenfreund-Kleinman et al. 2006).

Polycaprolactone (PCL) is one of the polymers that is widely used in the biomaterials fields and a number of drug-delivery devices as it possesses excellent biocompatibility, long-term biodegradation (up to 3-4 years), and has the ability to form compatible blends with others polymers to improve crack resistance, affects the degradation kinetics as well as facilitates tailoring to achieve desired release profiles (Lieberman and Friedlaender 2005). Deacetylation of chitin with concentrated strong alkaline (NaOH) produces chitosan or poly-D-glucosamine that contains ionic primary amino group which endows it with polycationic, chelating and also improves its solubility in acidic aqueous media. In addition, amino and hydroxyl groups can be modified chemically to provide a high chemical versatility and can be metabolized by certain human enzymes which makes it biodegradable (Malafaya et al. 2007).

To obtain material with related benefits and improve the performance of one single polymer, polymers can be blended. Mixture of at least two macromolecular substances or polymers or copolymers where the ingredient content is above 2 wt. % are known as polymer blending. There are 3 types of polymer blending; (1) miscible polymer blending, the capability of a mixture to form a single phase over certain range of temperature, pressure and also composition (Work et al. 2004).

This chapter is focused on the fabrication and evaluation of PCL/chitosan-based scaffolds using a freeze-drying technique. Parameter setup was studied and controlled to prepare the scaffolds with the desired properties. Nano-sized HA (nHA) and tetracycline HCL were incorporated into the scaffolds to study their characteristics, release and biocompatibility with human osteoblast cells via *in vitro* studies. The scaffolds were characterized using FESEM, SEM, FTIR, TGA, and XRD. An antibacterial study using *E. coli* and *Bacillus* sp. was carried out. The *in vitro* biological evaluation of scaffolds were conducted through cell culture using osteoblast cells. Cytotoxicity, cell attachments, and cell proliferation activity were investigated.

10.2 FABRICATION OF COMPOSITE SCAFFOLDS VIA FREEZE DRYING TECHNIQUE

Freezing/freeze-drying technique involves creating a solution using polymers and solvent, quickly freezing the solution to lock liquid state structure and then remove the solvent by freeze drying. In other words, freeze drying is the process whereby solutions are completely frozen and then the frozen solvents are removed via sublimation under vacuum (Qian and Zhang 2011). This process is also known as lyophilisation. This process has been used in pharmaceutical industry to improve the stability of labile drugs, and also in tissue engineering application as it will fabricate porous materials in the process (Qian and Zhang 2011).

Freeze drying technique has certain advantages compared to other methods as it uses ice crystal as porogens. Most importantly, with changing variables during freezing, it may produce materials with appropriate pore morphologies and nanostructure (Qian and Zhang 2011). With all these benefits, freeze drying technique has been widely used in the fields of bioengineering, drug delivery, catalyst support and also separation.

Various concentrations of PCL and chitosan were prepared to identify the suitable blend ratio. In this case, PCL was weighed and dissolved in glacial acetic acid. The solution was stirred using a magnetic stirrer until it had all dissolved. Then, chitosan powder was added to the solution and stirred until the solution became a clear, viscous solution. A hand-held homogenizer (IKA Ultra-Turax, model T25, Germany) was used to homogenize the blends. Following this, the homogenized blend was molded in a capped glass tube. The glass tube was immediately placed in a freezer at –80°C and solidified overnight. The frozen emulsion was transferred to the freeze dryer vessel and lyophilized for 72 hours to remove all the solvent phase. The prepared scaffolds were transferred into a desiccator and kept until evaluation. The same technique was used to incorporate nano-hydroxyapatite (nHA) and tetracycline HCL (TCH).

10.3 EVALUATION OF SCAFFOLDS

Blending PCL and chitosan in a single phase using glacial acetic acid produced scaffolds with irregular pores and a high pore distribution. In the freeze drying process, these porous structures, formed as the ice crystals, act as porogens and were removed via sublimation process. Figure 10.1a shows the general appearances of Tetracycline HCL-incorporated scaffolds fabricated by freeze drying process. Three types of scaffolds were fabricated: PCL/CS, nHA/PCL/CS and nHA/PCL/CS/TCH.

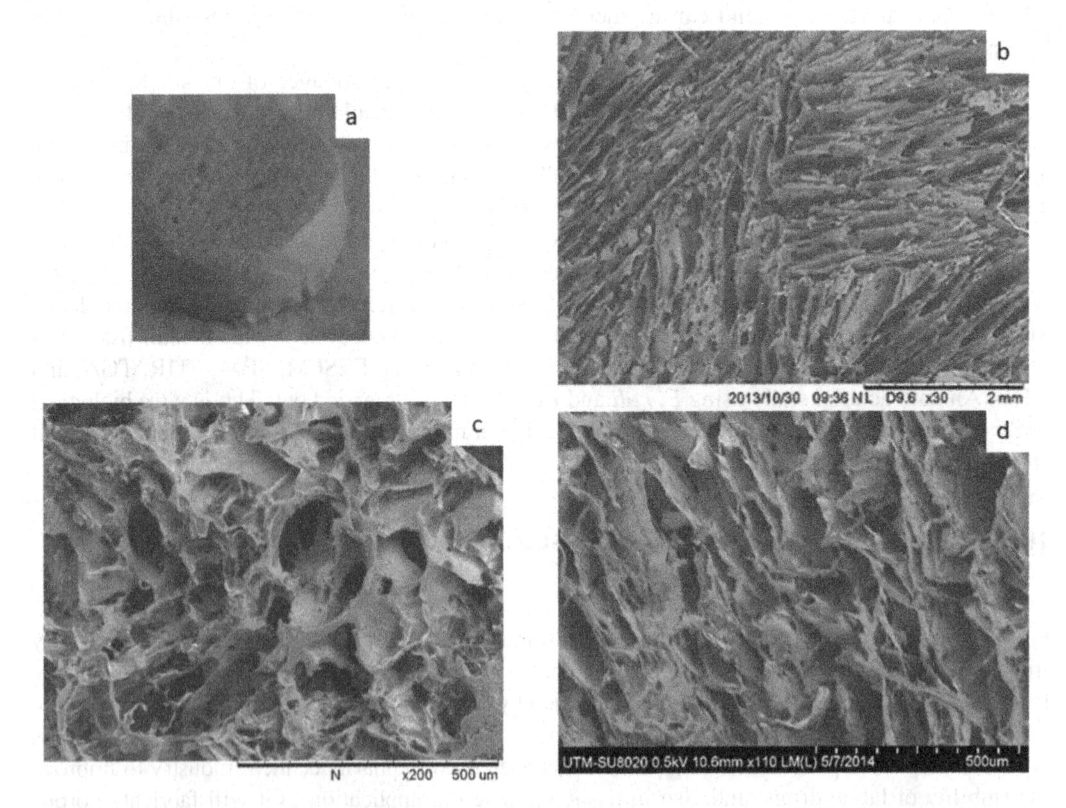

FIGURE 10.1 (a) General appearances of tetracycline HCL-incorporated scaffolds fabricated by freeze drying process. SEM micrographs of scaffolds of (b) PCL/CS, (c) nHA/PCL/CS, (d) nHA/PCL/CS/TCH.

The scaffolds were examined under SEM (Hitachi TM-3000, Japan) and field emission scanning microscopy (FESEM, Hitachi SU8000, Japan). The scaffolds were mounted on an aluminum stub and coated with gold using a sputter-coating machine. The microstructure of the scaffolds was observed at an accelerating high voltage of 15 kV. Surface morphology of the scaffolds was examined and the measurements of pore sizes were recorded. The micrographs of porous scaffolds are shown in Figure 10.1(b-d). SEM analysis of the cross-section of the scaffolds shows highly anisotropic structures which is the typical morphology formed by the freeze drying technique. SEM micrograph of the scaffolds showed in Fig. 10.1 shows PCL/CS scaffolds had slightly different pore structures compared to nHA/PCL/CS and nHA/PCL/CS/TCH scaffolds. PCL/CS scaffolds are highly anisotropic, while the addition of nHA cause the structure change to smaller pore sizes and lower porosity. However, the range of pore sizes did not change significantly. Pore size and pore morphology are dependent on different factors such as freezing temperature, solution concentration and many others (Qian and Zhang 2011). Importantly, the method used to fabricate scaffolds with a higher chance of reproducibility make it advantageous in tissue engineering applications (Bose et al. 2012).

The results of density and porosity of nHA/PCL/CS/TCH nanocomposite scaffolds were measured and compared with PCL/CS and nHA/PCL/CS scaffolds in Table 10.2. Incorporation of nHA and TCH resulted in increased density, while this decreased the porosity of the scaffolds.

TABLE 10.2 The porosity and density measurement of the scaffolds

Scaffold	Porosity%	Density (g/cm^3)
PCL/CS	85 ± 1.05	0.0773 ± 0.009
nHA/PCL/CS	75 ± 0.96	0.085 ± 0.003
nHA/PCL/CS/TCH	74.5 ± 0.46	0.080 ± 0.025

The surface hydrophobicity was measured by measuring contact angle through the water spread of droplets on the surface of scaffolds. Hydrophilic properties are very important characteristics for tissue engineering implants, as one of the factors controlling *in vitro* cell adhesion, migration, intracellular signaling as well as phenotype maintenance (Ma et al. 2007). The results showed that the contact angles were within the range of 75.0° to 65.0° for the scaffolds prepared, which indicated that surface of the scaffolds were hydrophilic. Chitosan scaffolds showed a contact angle of zero ($\theta = 0$), as it quickly formed a flat puddle and was absorbed, indicating the hydrophilic properties of chitosan. However, after blending with PCL, the contact angles increased. However, the contact angle of PCL/CS had a higher contact angle compared to nHA/PCL/CS and nHA/PCL/CS/TCH, respectively. The results also showed that the wettability of the composite scaffold of nHA/PCL/CS increased with the addition of nHA as well as the addition of nHA and tetracycline HCl. The reason for this is that the incorporation nHA increases the water contact angle because the surface morphology becomes rougher. The results showed that all of the scaffolds were hydrophilic because of the amino groups present on the surfaces (Jin et al. 2015). Chitosan is more hydrophilic than PCL, which is hydrophobic, so the blending of these polymers changes the wettability of the composite scaffold. The cationic characteristic of chitosan plays an important role in bioadhesion, absorption enhancement and biological activities such as antimicrobial, anti-inflammatory and antitumor effects. The wettability and swelling behavior of the scaffolds are important parameters to study as they are the key factors to govern cell response. As the scaffolds or implants come into contact with cell culture conditions, protein adsorption to the surface occurs, providing signals to the cell through cell adhesion receptors (integrins) which mediate cell adhesion (Chang and Wang 2006). If the materials are highly hydrophobic, this will inhibit protein adsorption by expelling any protein molecules and causing cell adhesion to become more difficult or absent.

The compressive mechanical properties of the scaffolds were tested with an Instron mechanical tester (Instron 5848, USA) with a crosshead speed of 0.5 mm/min and load 1 kN. From each polymer concentration sample, cylindrical specimens (diameter of 1.2 cm and height of 0.2 cm) were prepared. Three specimens from each scaffold were cut from different locations (top, bottom, and middle). Compression testing was performed in the longitudinal directions. The averages and standard deviations were calculated. Two sample independent t-tests were performed to determine the statistical significance ($p<0.05$) of the different mechanical properties.

It was found that both the compressive modulus and yield strength increased with the addition of nHA. Table 10.3 shows that the same processing conditions led to a varied compressive modulus from 4.0 MPa to 12.5 MPa, and a yield strength that varied from 0.48 MPa and 0.75 MPa for the PCL/CS scaffold and nHA/PCL/CS scaffold, respectively. From the statistical analysis, it was found that both the compressive modulus and the compressive yield strength of the PCL/chitosan scaffolds were significantly different to those of nHA/PCL/chitosan scaffolds ($p^*<0.05$). These data demonstrate the positive effects of the addition of HA in enhancing the mechanical properties of scaffolds. It was also found that the compressive modulus and yield strength of the HA/PCL/CS scaffold were not significantly different from HA/PCL/CS/TCH ($p<0.05$) scaffold. These results show that the incorporation of TCH did not change the mechanical strength of the scaffolds.

Mechanical properties of the scaffold should match the host bone properties and have proper load transferred to support the harboring of new bone tissues and support and also the mechanical integrity of the implanted area. In bones in our body, there is a large variation in mechanical properties and geometry between cancellous and cortical bones. Cancellous bones have a Young's modulus between 0.1 and 2 GPa, while cortical bones are between 15 and 20 GPa. Also, the compression strength for cortical bone is 100 to 200 Mpa and that of cancellous bone is 2 to 20 MPa (Bose et al. 2012).

TABLE 10.3 Compressive yield strength and compressive modulus of scaffolds

Scaffolds	Compressive Yield Strength (MPa)	Compressive Modulus (MPa)
PCL/CS	0.48 ± 0.05	4.0 ± 0.2
nHA/ PCL/CS	0.75 ± 0.06	12.5 ± 0.5
nHA/ PCL/CS/TCH	0.75 ± 0.08	12.1 ± 0.1

Chitosan is widely used as an antimicrobial agent, either alone or blended with other natural polymers because it has high biodegradability, is non-toxic and has antimicrobial properties (Kim et al. 2008). Bacterial contamination during operative procedures, such as open procedures, always poses contamination risks; however, the presence of biomaterials increases the risk of infection due to their susceptibility to bacterial colonization (Miola et al. 2013). In fact, the pores or porous structure of the scaffold are more likely to be suitable places for bacterial colonization (Miola et al. 2013). Antibacterial analyses of PCL/CS, nHA/PCL/CS and nHA/PCL/CS/TCH composite scaffolds using *E. coli* (gram-negative bacteria) and *Bacillus cereus* (gram-positive bacteria) were conducted.

Antibacterial analysis of the PCL/CS and nHA/PCL/CS scaffolds using *E. coli* (gram-negative bacteria) and *B. cereus* (gram-positive bacteria), as shown in Fig. 10.2 (a and b) and Fig. 10.3 (a and b), show negative results for antibacterial properties. It has been documented that chitosan shows antibacterial activities against *E. coli, Salmonella typhimurium, B. cereus, Staphylococcus aureus* and many others (Fujita et al. 2004). As for the mechanism of antibacterial of chitosan, this is still not fully explained, but it has been suggested that interactions between polycationic chitosan and negatively charged cell walls cause cell permeability and the leakage of intracellular electrolytes and proteins (Fujita et al. 2004).

However, the addition of tetracycline HCl in nHA/PCL/CS/TCH scaffolds toward *E. coli* and *B. cereus,* as shown in Fig. 10.2c and Fig. 10.3c, led to a clear inhibition zone. Based on calculation, nHA/PCL/CS/TCH showed a diameter of the inhibition zone of 2.8 ± 0.06 cm for *B. cereus* and 1.8 ± 0.15 cm for *E. coli*. According to the definition of the inhibition zone, 10 mm can be defined as light antibacterial action, 11 to 15 mm is medium activity and greater than 16 mm is high antibacterial activity. TCH acts as an inhibitor of protein synthesis for *B. cereus* or gram-positive bacteria, as tetracycline prevents the association of aminoacyl-tRNA by the bacterial ribosome, and the results inhibit bacterial protein synthesis, leading to the death of bacteria. For gram-negative bacteria, tetracycline moves through membranes via porin channels and moves across cytoplasmic membranes. Then, tetracycline molecules bind reversibly with the prokaryotic 30S ribosomal subunit, leading to the cessation of protein synthesis.

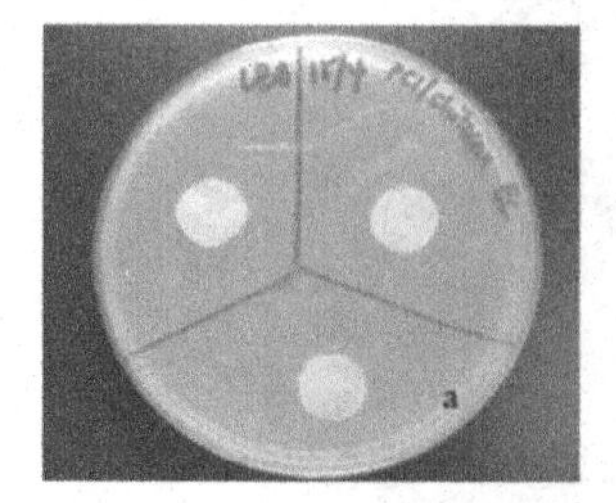

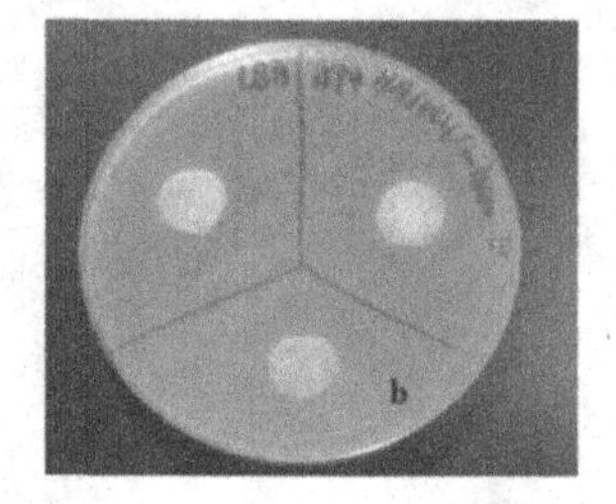

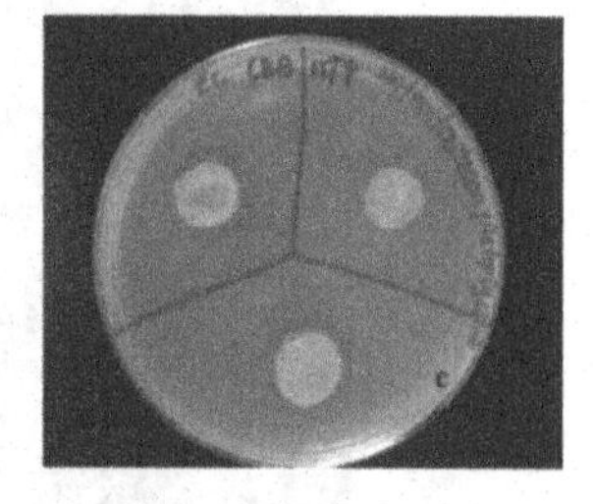

FIGURE 10.2 Antibacterial properties of the (a) PCL/CS, (b) nHA/PCL/CS and (c) nHA/PCL/CS/TCH composite scaffolds using *E. coli* from gram-negative bacteria. The diameter of the clear inhibition zone was measured after a 24-hour incubation period.

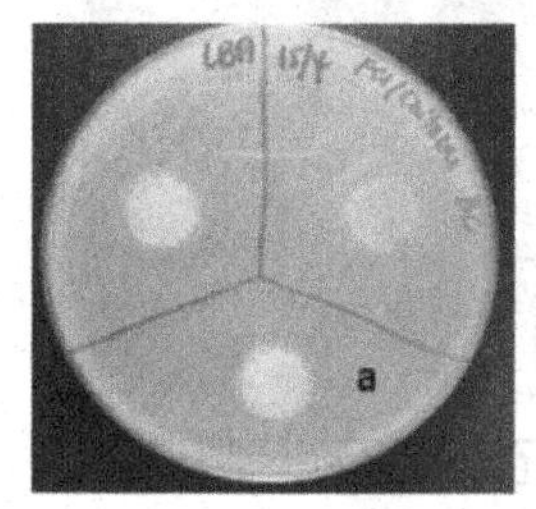

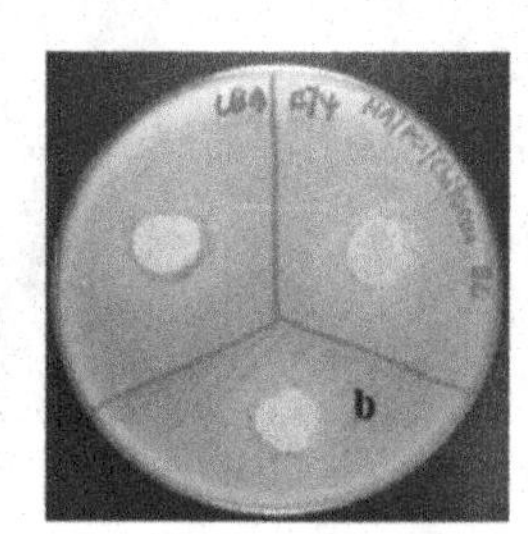

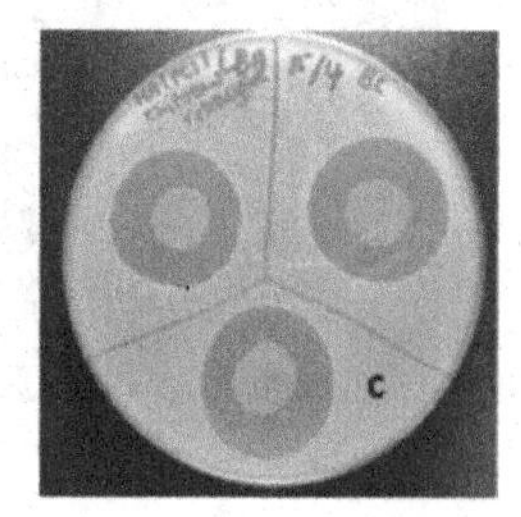

FIGURE 10.3 Antibacterial properties of the (a) PCL/CS, (b) nHA/PCL/CS and (c) nHA/PCL/CS/TCH composite scaffolds using *B. cereus* from gram-positive bacteria. The image was taken after a 24-hour incubation period.

The osteogenic potential of the composites can be evaluated by submerging the scaffolds in simulated body fluid (SBF) to form the apatite layer (Marc and Jacques 2009). This method has become the "gold standard" to ascertain whether materials are bioactive or not since the invention of SBF by Kokubo et al. in 2003 (Wu et al. 2014). The ion concentration of SBF is the same as in human blood plasma, which is suitable for the *in vitro* study of the biomineralization of scaffolds. Bioactivity of the scaffolds is said to be a performance indicator for the biomaterial for biomineralization in an *in vitro* and *in vivo* study. Biomineralization is a process for the deposition or growth of bone-minerals, like crystals such as hydroxyapatite, apatite and CaP compounds on the scaffolds. In the matrix of organisms, biomineralization induces the formation of the bone mineral-like skeletal structure during development (Wu et al. 2014).

In the bioactivity assay of the scaffold, sterile SBF solution was used as a medium. All of the sample scaffold samples were prepared as cylinders with a diameter of 1.2 cm and a height of 0.2 cm. The samples were placed in 50 ml centrifuge tubes containing 20 ml of SBF solution sealed with parafilm. The centrifuge tube was kept in a shaking water bath at 37°C and shaken at 40 rpm to ensure that fluid flowed through the entire scaffold. The SBF was renewed every 7 days. After 3 weeks of immersion, the samples were taken out, rinsed with distilled water several times and dried. The samples were observed under SEM to investigate the deposition of apatite layer on the samples of scaffolds.

The results were positive for the formation of an appetite layer on the surface of scaffolds, as shown in Fig. 10.4. Figure 10.4a shows the apatite nucleation on the surface of the scaffolds with different distribution throughout the sample. EDX analysis shows that the element of the apatite layer is Ca/P: 1.8. However, nHA/PCL/chitosan scaffolds showed the increased formation of apatite layer compared to PCL/chitosan. This may be due to nHA increasing the properties of the scaffolds, which favors the formation of apatite layer. The success of forming the apatite layer through immersion in SBF showed that the scaffolds prepared had osteogenic potential and would show a good performance during application both *in vitro* and *in vivo* studies.

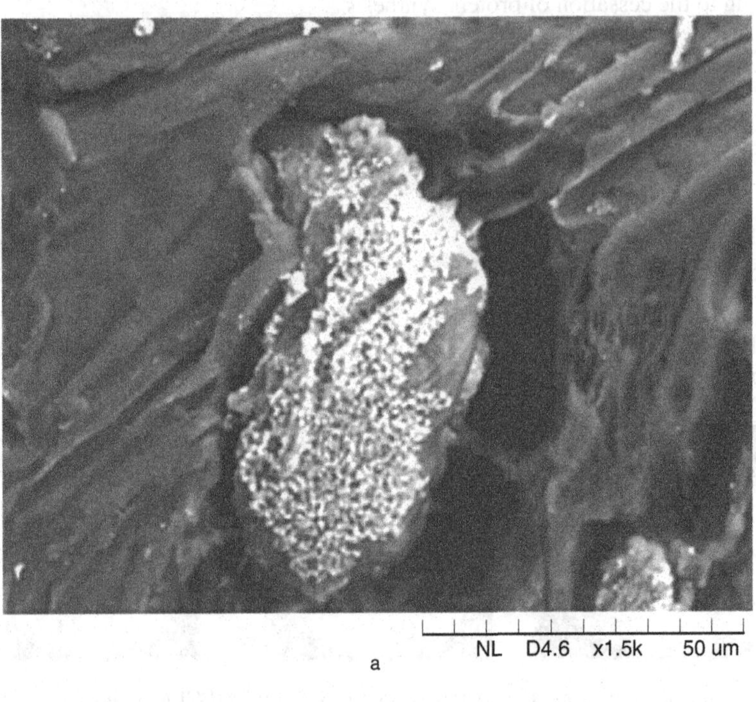

a

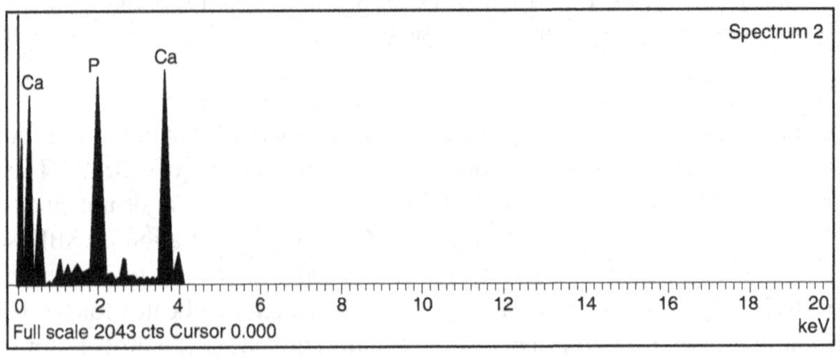

b

Element	Weight %	Weight % σ	Atomic %
Phosphorus	35.102	0.441	41.173
Calcium	64.898	0.441	58.827

c

FIGURE 10.4 The (a) SEM micrograph (b) EDX spectrum and (c) elemental component percentage of Ca/P deposited on PCL/CS scaffolds after 3 weeks' immersion.

10.4 *IN VITRO* CELL CULTURE STUDY

In this chapter, PCL/chitosan-based scaffolds were fabricated for bone tissue engineering applications. Normal human osteoblast cell studies will provide insight into the performance of the scaffolds as an implant for bone regeneration. The osteoblast is a fully differentiated cell that is responsible for the synthesis and mineralization of bone during bone formation and remodeling. It can produce many cell products such as the enzymes alkaline phosphate and collagenase, growth factors, hormones and many others (Velasco et al. 2015). VEGF growth factors were used as their "enhanced vascularization" provides abundant osteoprogenitor cells to the defect site along with a direct stimulating effect on the osteoblast migration and differentiation, leading to higher bone deposition" (Bose et al. 2012).

Cell morphology can be viewed under SEM or FESEM. Cells that were attached and grown on the scaffolds surface were fixed using glutaraldehyde at 4°C followed by dehydration with a serial of graded ethanol. Cells maintained their morphology on the scaffolds. Figure 10.5 shows SEM micrograph of a single osteoblast cell, anchoring on the surface of the scaffolds. The circle indicates the body or nucleus of an osteoblast cell with long conical protrusions or dendritic extensions.

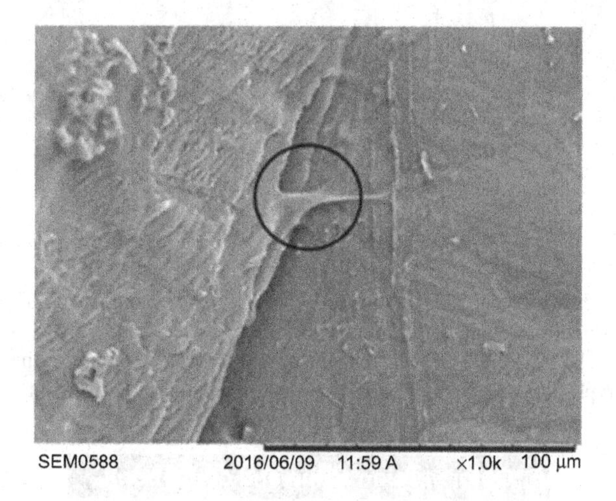

FIGURE 10.5 SEM micrograph of a single osteoblast cell, anchoring between two pores of the scaffolds.

For application as implants, it is crucial to produce scaffolds that are biocompatible and release non-toxic by-products that can be excreted by the body. A colorimetric method, such as the MTT tetrazolium assay or MTT assay, has been widely used to determine the number of cells required for adhesion and proliferation studies. MTT or the chemical formula of 3-[4,5-dimethythiazol-2-yl]-2,5-diphenyltetrazolium bromide assays utilize cells demonstrating metabolic activity. The tetrazolium salt that is water soluble will be converted to an insoluble purple formazan crystal by cleavage of the tetrazolium ring by succinate dehydrogenase within the mitochondria. Thus, the quantity of formazan formed is directly proportional to the number of viable cells.

Indirect MTT assays were performed to study the cytotoxicity of extraction medium from the scaffold specimens. The cell viability of the treatment was compared to control cells, with Triton-X 100 acting as a positive control, showing greater toxicity than the control. Figure 10.6 shows positive results toward cell viability. The results showed that cell viability was above 80% compared to the control cells. From Figure 10.6, it can be seen that PCL/CS scaffolds showed the lowest cell viability compared to the rest of the scaffolds. In addition, the incorporation of nHA boosts the rate of cell viability after 72 hours of incubation. Moreover, tetracycline-incorporated scaffolds also show the positive growth of NHOst cells. This shows that the release of TCH was not fatal with regard to cell growth.

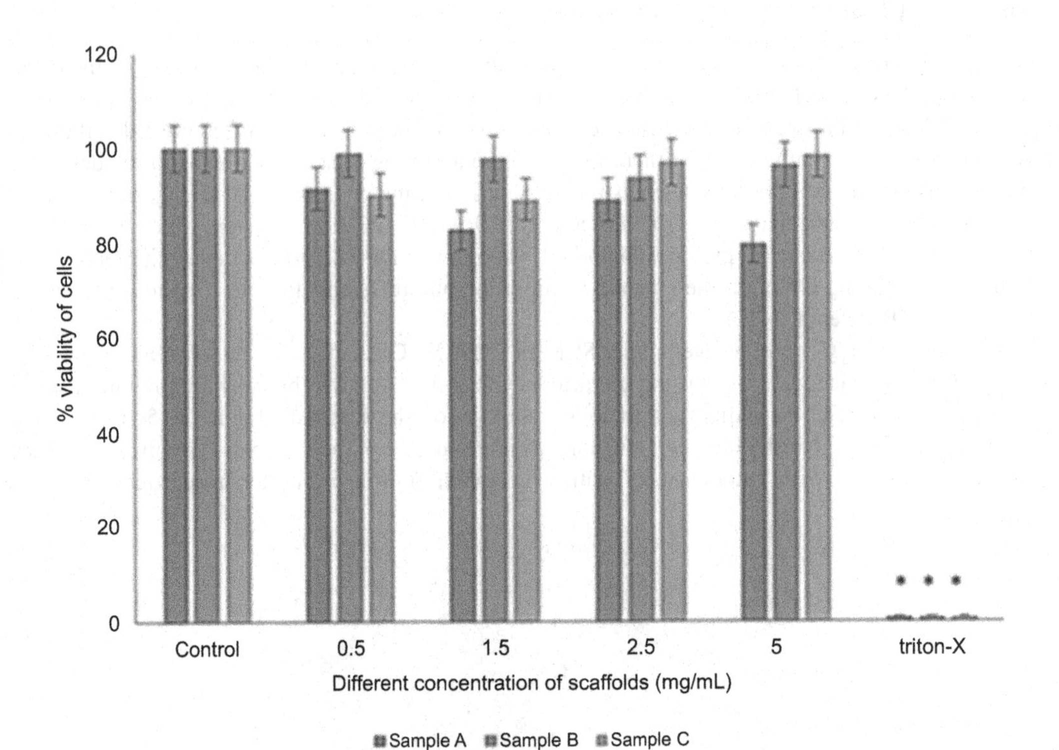

FIGURE 10.6 Cell viability of seeded NHOst cells (2×10^5/cm^2) using the MTT assay. The assay was performed after 72 h of treatment with different concentrations of scaffolds for each sample: (a) PCL/CS, (b) nHA/PCL/CS, and (c) nHa/PCL/CS/TCH scaffolds and controls (cell without treatment). The treatment was carried out with three replicates ($n = 5$). The paired t-test was statistically significant at $P < 0.05$ (*) compared to control cells.

Color version at the end of the book

Figure 10.7 illustrates the cell proliferation of NHOst cells seeded directly onto the scaffolds; sample (a) PCL/CS, sample (b) nHA/PCL/CS, and sample (c) nHA/PCL/CS/ TCH scaffolds. Control cells have no scaffolds for the comparison of cell growth without stress. All of the prepared scaffolds show increasing trends for cell growth. On the 3rd day, PCL/CS scaffolds showed 26.9% cell viability, nHa/PCL/CS scaffolds showed 26.1% and nHA/PCL/CS/TCH scaffolds showed 41.7% cell viability. The amount of cell viability increased on the 5th day, with PCL/CS scaffolds showing 29.6%, nHa/PCL/CS scaffolds showing 33.2% and nHA/PCL/CS/TCH scaffolds showing 54.4% cell viability. On the 7th day, PCL/CS scaffolds showed 41.7%, nHa/PCL/CS scaffolds showed 54.4% and nHA/PCL/CS/TCH scaffolds showed 63.3% cell viability.

Tetracycline HCL is one of the broad-spectrum antibiotics that acts against both gram-positive and gram-negative bacteria (Eliopoulos et al. 2003). The study of nHA/PCL/CS/TCH with NHOst cells showed that the incorporation of tetracycline HCL is not toxic towards cells. This may mean that the release of tetracycline is suitable and does not inhibit NHOst cell growth. In fact, the growth of NHOst cells was higher compared to that with the other two scaffolds. Tetracycline may show advantages due to its ability to inhibit bacterial growth as well as to enhance the growth of osteoblast cells. The vitality of osteoblasts is important to determine the bone formation. Thus, bone is formed by the mineralization of an organic matrix (largely collagen), through the nucleation and growth of minerals (HAp) regulated by osteoblasts (Harada and Rodan 2003, Chien et al. 2000).

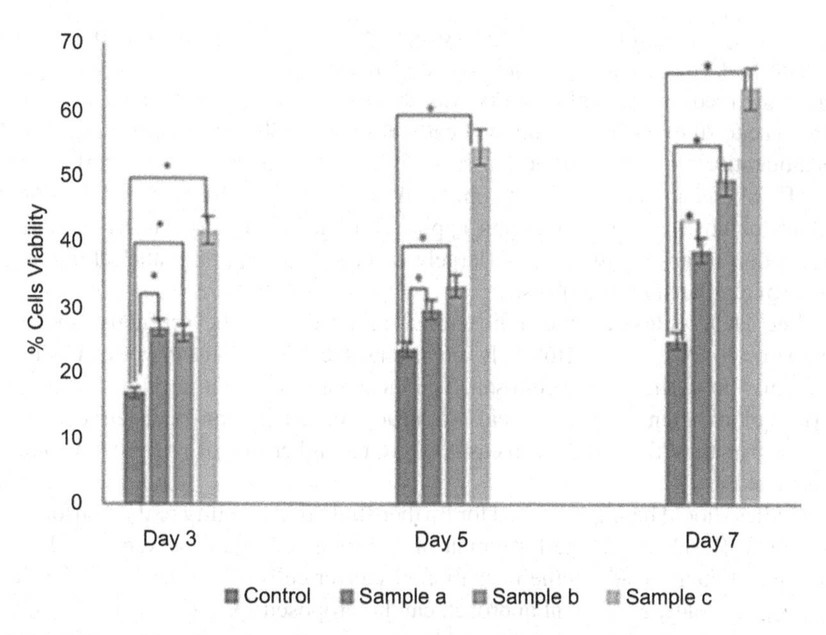

FIGURE 10.7 Cell proliferation of NHOst cells seeded directly onto the scaffolds; sample (a) PCL/CS, sample (b) nHA/PCL/CS, and sample (c) nHA/PCL/CS/TCH scaffolds and a control (cell without scaffolds). The MTT assay was performed on days 3 to 7. The treatment was carried out with 3 replicates ($n = 5$). The paired t-test analysis was statistically significant at $P < 0.05$ (*) compared to control cells.

Color version at the end of the book

Based on the results of *in vitro* studies using normal osteoblast cells (NHOst), it was shown that the design of 3D scaffolds was able to maintain its structural integrity during *in vitro* growth studies. Positive cell attachment and proliferation assays indicated that the architecture of scaffolds allowed cell attachment and subsequent migration into and through the matrix, and allowed the mass transfer of nutrients and metabolites. Moreover, apatite nucleation formation proved that the scaffolds provided a suitable environment for the cells to grow and proliferate in, as well as enabling their differentiation. Thus, the ability to produce apatite nucleation makes the scaffolds suitable for enhanced bone healing processes.

10.5 SUMMARY AND RECOMMENDATIONS

This chapter described the successful fabrication of PCL/chitosan scaffolds using a freeze-drying technique without complex chemical modifications. This technique is also capable of fabricating scaffolds with the incorporation of nHA and tetracycline HCL to produce nHA/PCL/CS and nHA/PCL/chitosan/TCH composite scaffolds. Blends of PCL/CS had suitable pore sizes that allowed the diffusion of essential nutrients and oxygen for cell survivability. However, its pore size and porosity decreased with the incorporation of nHA and drugs. Other than that, blending chitosan and PCL showed good wettability and hydrophilicity of the scaffolds as the hydrophilic properties of chitosan improved the hydrophobicity of the PCL on the polymers. The addition of nHA increased the compressive yield strength and modulus of the nHA/PCL/chitosan scaffolds compared to PCL/chitosan scaffolds.

In vitro cell culture studies used normal human osteoblast cells to study the biocompatibility of the scaffolds. FESEM/SEM analysis confirmed that NHOst cells were able to adhere and proliferate on the surface of the scaffolds, especially on the pore of the scaffolds. Indirect MTT assays were performed to study the cytotoxicity of the extraction medium from the scaffold specimens. The results showed that none of the scaffolds released toxic ions or by-products during the 24 hour immersion in the medium, as cell viability was above 80% compared to the control. Direct MTT assays were done to study the proliferation of cells, which also showed positive results in the growth of cells as a quantitative study showed increasing trends. The incorporation of nHA increased the secretion of ECM and apatite nucleation compared to PCL/CS scaffolds after 7 days of incubation. As a scaffold for bone tissue engineering applications, it is very crucial, as bone is formed by the mineralization of an organic matrix (largely collagen), through the nucleation and growth of minerals (HAp) regulated by osteoblasts.

Tetracycline HCL enhanced the antibacterial properties of the scaffolds, showing positive results compared to PCL/CS scaffolds. It seems that blending chitosan with PCL had inhibited the antimicrobial properties of the chitosan. The incorporation of tetracycline helped to eliminate any bacterial colonization within the scaffold structure. Other than that, tetracycline HCL also influenced the growth of the NHOst cells as it showed a higher rate of cell growth compared to the two other scaffolds.

Animal studies should be implemented for further study of the ability of the scaffold to withstand the real situation. Besides that, the formation of new bone can be observed by CT scans. If any contamination or toxicity leads to the formation of cancer cells or the death of native cells due to the presence of scaffolds, a different approach can be proposed for the preparation, incorporation, sterilization and many others to avoid the problem.

REFERENCES

Biondi, M., F. Ungaros, F. Quaglia and P.A. Netti. 2008. Controlled drug delivery in tissue engineering. Advanced Drug Delivery Reviews 60(2): 229-242.

Bose, S., M. Roy and A. Bandyopadhyay. 2012. Recent advances in bone tissue engineering scaffolds. Trends in Biotechnology 30(10): 546-554.

Buckwalter, J.A., M.J. Glimcher, R.R. Cooper and R. Recker 1995. Bone biology. The Journal of Bone and Joint Surgery 77(8): 1256-1275.

Chang, H. and Y. Wang. 2006. Cell Responses to Surface and Architecture of Tissue Engineering Scaffolds. InTech Open Access Publisher.

Chien, H. H., W.L. Lin and M.I. Cho. 2000. Down-regulation of osteoblastic cell differentiation by epidermal growth factor receptor. Calcified Tissue International 67(2): 141-150.

Ehrenfreund-Kleinman, T., Golenser, J. and A.J. Domb. 2006. Polysaccharide scaffolds for tissue engineering. pp. 27-44. *In*: P.X. Ma and J. Elisseeff [eds.]. Scaffolding in Tissue Engineering, CRC press, Taylor & Francis.

Eliopoulos, G.M. and M.C. Roberts. 2003. Tetracycline therapy: update. Clinical Infectious Diseases 36(4): 462-467.

Fujita, M., M. Kinoshita, M. Ishihara, Y. Kanatani, Y. Morimoto, M. Simizu and B. Takase. 2004. Inhibition of vascular prosthetic graft infection using a photocrosslinkable chitosan hydrogel. Journal of Surgical Research 121(1): 135-140.

Gomes, M.E., J.S. Godinho, D. Tchalamov, A.M. Cunha and R.L. Reis. 2002. Alternative tissue engineering scaffolds based on starch: processing methodologies, morphology, degradation and mechanical properties. Materials Science and Engineering: C 20(1): 19-26.

Griffith, L.G. and G. Naughton. 2002. Tissue engineering–current challenges and expanding opportunities. Science 295(5557): 1009-1014.

Harada, S. and G.A. Rodan. 2003. Control of osteoblast function and regulation of bone mass. 423(May): 349-355.

Hussain, N.S., C.M. Botelho, L. MA, M. Santos, J.V. Lobato, R.M. Pinto and J.D. Santos. 2010. Calcium Phosphate-based materials for bone regenerative medicine. Materials Science Foundations (Monograph Series) 62: 151-180.

Jin, R.M., N. Sultana, S. Baba, S. Hamdan, and A.F. Ismail. 2015. Porous PCL/chitosan and nHA/PCL/chitosan scaffolds for tissue engineering applications: fabrication and evaluation. Journal of Nanomaterials 16(1): 138.

Kim, I.Y., S.J. Seo, H.S. Moon, M.K. Yoo, I.Y. Park, B.C. Kim and C.S. Cho. 2008. Chitosan and its derivatives for tissue engineering applications. Biotechnology Advances 26(1): 1-21.

Kokubo, T., H.M. Kim and M. Kawashita. 2003. Novel bioactive materials with different mechanical properties. Biomaterials 24(13): 2161-2175.

Liao, C.J., C.F. Chen, J.H. Chen, S.F. Chiang, Y.J. Lin and K.Y. Chang. 2002. Fabrication of porous biodegradable polymer scaffolds using a solvent merging/particulate leaching method. Journal of Biomedical Materials Research: An Official Journal of The Society for Biomaterials, The Japanese Society for Biomaterials, and The Australian Society for Biomaterials and the Korean Society for Biomaterials 59(4): 676-681.

Lieberman, J.R. and G.E. Friedlaender (eds.). 2005. Bone Regeneration and Repair: Biology and Clinical Applications. Humana Press, New Jersey, USA.

Ma, P.X. and J. Elisseeff (eds.). 2005. Scaffolding in Tissue Engineering. CRC press, Boca Raton, FL, USA.

Ma, Z., Z. Mao and C. Gao. 2007. Surface modification and property analysis of biomedical polymers used for tissue engineering. Colloids and Surfaces B: Biointerfaces 60(2): 137-157.

Malafaya, P.B. G.A. Silva and R.L. Reis. 2007. Natural–origin polymers as carriers and scaffolds for biomolecules and cell delivery in tissue engineering applications. Advanced Drug Delivery Reviews 59(4): 207-233.

Marc, B. and L. Jacques. 2009. Biomaterials can bioactivity be tested *in vitro* with SBF solution? Biomaterials 30(12): 2175-2179.

Miola, M., Bistolfi, A., Carmen, M., Bianco, C., Fucale, G. and Verné, E. 2013. Antibiotic-loaded acrylic bone cements: an *in vitro* study on the release mechanism and its efficacy. Materials Science & Engineering C 33(5): 3025-3032.

Nam, Y.S., J.J. Yoon and T.G. Park. 2000. A novel fabrication method of macroporous biodegradable polymer scaffolds using gas foaming salt as a porogen additive. Journal of Biomedical Materials Research: An Official Journal of The Society for Biomaterials, The Japanese Society for Biomaterials and The Australian Society for Biomaterials and the Korean Society for Biomaterials 53(1): 1-7.

Nather, A. (ed.). 2005. Bone Grafts and Bone Substitutes: Basic Science and Clinical Applications. World Scientific, Singapore.

Nireesha, G.R., L. Divya, C. Sowmya, N. Venkateshan, M.N. Babu and V. Lavakumar. 2013. Lyophilization/freeze drying - an review. 3(4): 87-98.

Qian, L. and H. Zhang. 2011. Controlled freezing and freeze drying: a versatile route for porous and micro-/nano-structured materials. Journal of Chemical Technology and Biotechnology 86(2): 172-184.

Svetlane, S., G. Khaddiatou and L. Treena. 2015. A *in vitro* and *in vivo* evaluation of composite scaffolds for bone tissue engineering. pp. 615-636. *In*: I.V. Antoniac [ed.]. Handbook of Bioceramics and Biocomposites. Springer International Publishing Switzerland.

Velasco, M.A., C.A. Narváez-Tovar and D.A. Garzón-Alvarado. 2015. Design, materials, and mechanobiology of biodegradable scaffolds for bone tissue engineering. BioMed Research International: 2015.

Work, W.J., K. Horie, M. Hess and R.F.T. Stepto. 2004. Definition of terms related to polymer blends, composites, and multiphase polymeric materials (IUPAC Recommendations 2004). Pure and Applied Chemistry 76(11): 1985-2007.

Wu, S., X. Liu, W.K. Yeung, C. Liu and X. Yang. 2014. Biomimetic porous scaffolds for bone tissue engineering. Materials Science and Engineering: R: Reports 80: 1-36.

Index

Color Plate Section

Chapter 2

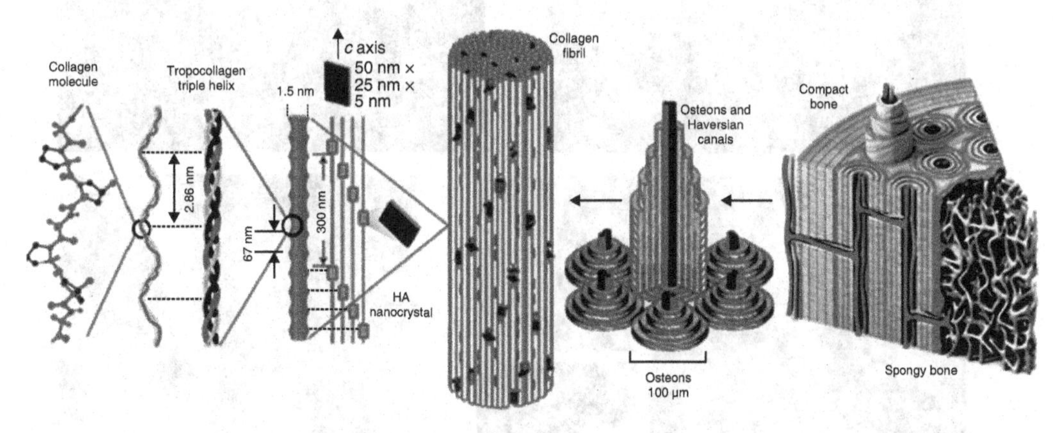

FIGURE 2.1 Hierarchical structure of natural bone, representing the sub-nanostructure of collagen molecules and tropocollagen helix, submicrostructure (collagen fibrils) and its macrostructure. Image adapted with permission from Pethig (1985).

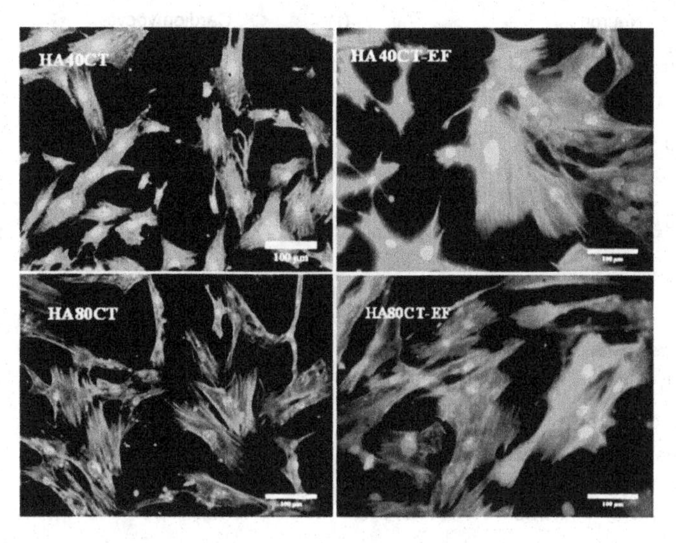

FIGURE 2.3 Change in hMSC morphology observed on conducting ceramics with external electric field stimulation. Adapted with permission from Ravikumar et al. 2017.

Chapter 6

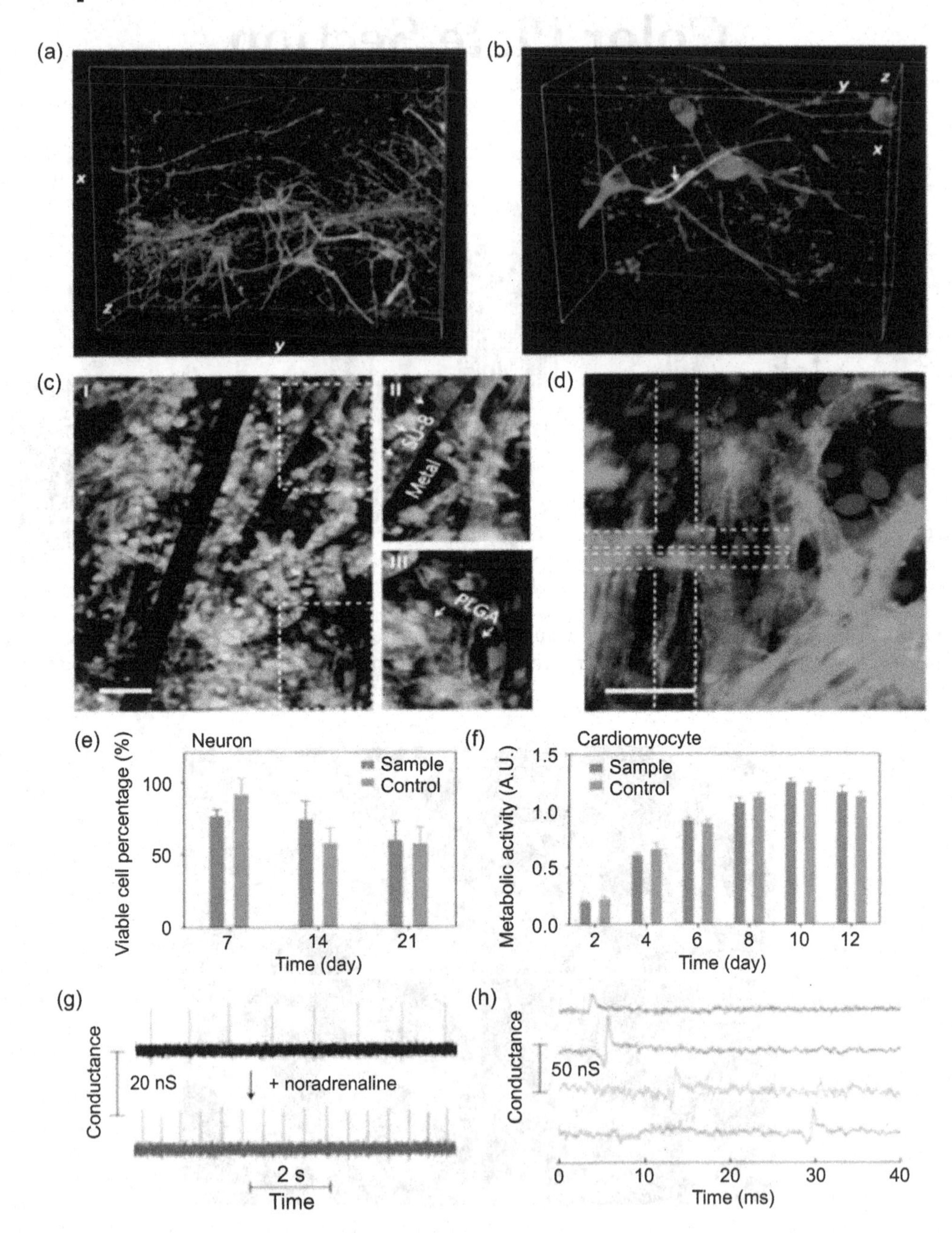

FIGURE 6.5 3D cell culture and electrical sensing from the nanoelectronic scaffold. (a) and (b) confocal images of rat hippocampal neurons after a two-week culture in nanoelectronic scaffold/matrigel matrix, (c) confocal fluorescence micrographs of a cardiac 3D culture on nanoelectronic scaffold, scale bar: 40 μm, (d) epifluorescence micrograph of the surface of the cardiac patch in Fig. 3c, scale bar: 40 μm, (e) cell viability of neurons cultured on nanoelectronic scaffolds and control samples, (f) metabolic activity assay for cardiomyocytes cultured on nanoelectronic scaffolds and control samples, (g) Electrical output from the nanoelectronic scaffold before (black) and after (blue) applying noradrenaline, (h) simultaneous electrical recording from different channels of the nanoelectronic scaffold. (Reprinted with permission from (Tian et al. 2012)).

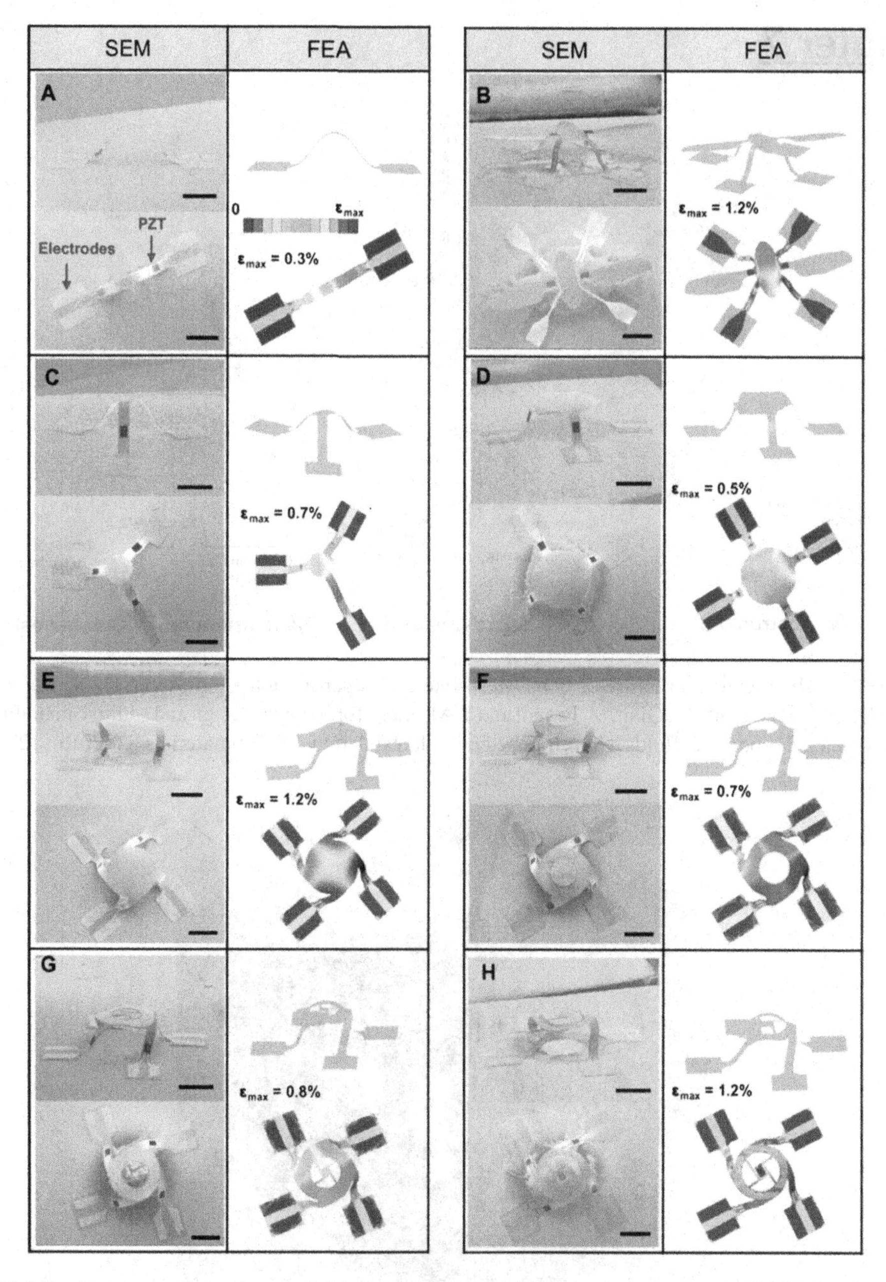

FIGURE 6.7 Demonstration of typical 3D architectures with PZT micro-actuators. (A) "Bridge" structure with two PZT micro-actuators. (B) "Fly" structure with a pair of actuators on the wings. (C) "Tilted pyramid truss" structure with three actuators. (D) "Four-leg table" structure with an actuator on each leg. (E) "Rotated table" structure with an actuator on each leg. (F) "Rotated table" with a central hole on top and four actuators. (G) "Double-floor rotated table" structure that consists of a large rotated table and a small one on the top, with four actuators. (H) "Double-floor rotated table" structure with five actuators (the additional one on top). Each panel includes a side and a top view. The yellow and blue regions correspond to the electrodes and PZT microactuators, respectively. Scale bars, 500 μm. The contour plots show results of FEA modeling for the maximum principal strain in the electrodes and PZT microactuators (Ning et al. 2018). (Reprinted with permission from (Ning et al. 2018)).

Chapter 7

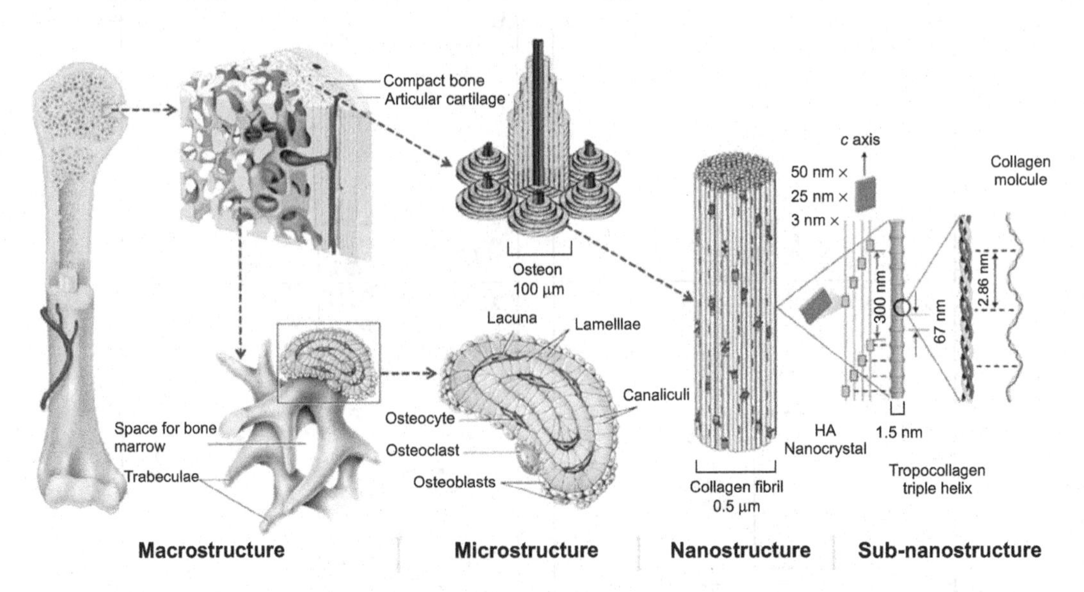

FIGURE 7.1 Hierarchical structure of bone. (Reprinted with permission from Wang, H., S. Xu, S. Zhou, W. Xu, M. Leary, P. Choong, M. Qian, M. Brandt and Y. M. Xie. "Topological design and additive manufacturing of porous metals for bone scaffolds and orthopaedic implants: A review," Biomaterials 83 (2016): 127-41).

FIGURE 7.10 Intraoperative view of a cylindrical 7 × 10 mm osseouscritical-sized defect at the distal femoral end of the rabbit. (Reprinted with permission from Huang, Z., Y. Chen, Q.-L. Feng, W. Zhao, B. Yu, J. Tian, S.-J. LI and B.-M. Lin. "*In vivo* bone regeneration with injectable chitosan/hydroxyapatite/collagen composites and mesenchymal stem cells", Front. Mater. Sci., 5(3) (2011): 301-10).

Chapter 10

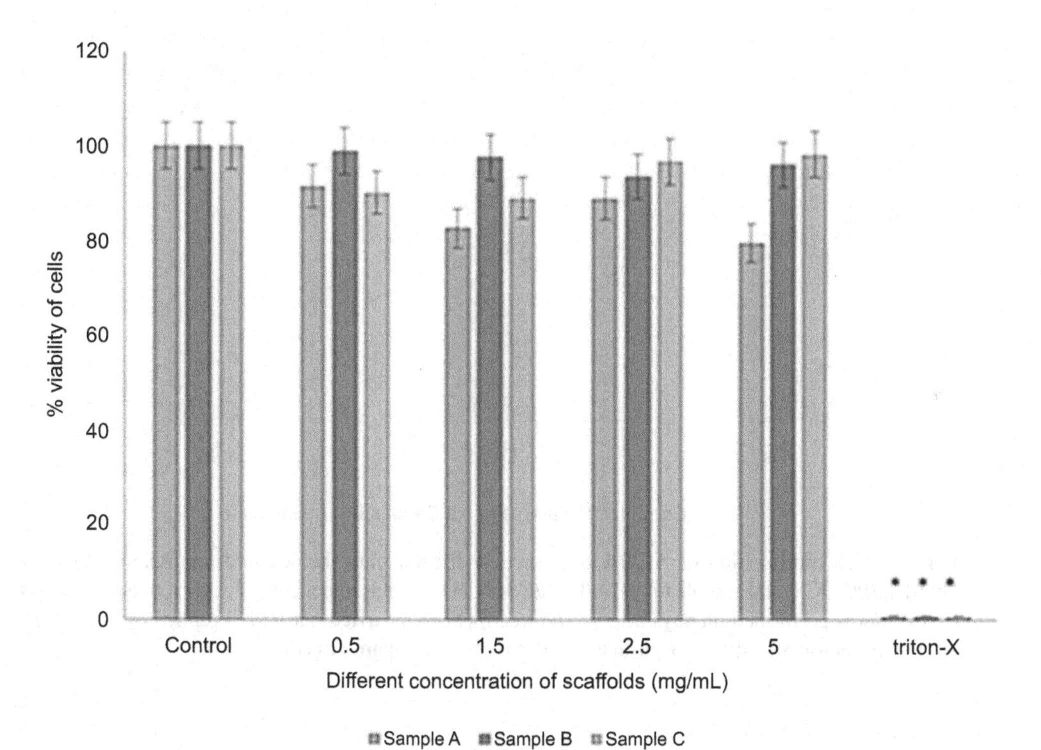

FIGURE 10.6 Cell viability of seeded NHOst cells ($2 \times 10^5/cm^2$) using the MTT assay. The assay was performed after 72 h of treatment with different concentrations of scaffolds for each sample: (a) PCL/CS, (b) nHA/PCL/CS, and (c) nHa/PCL/CS/TCH scaffolds and controls (cell without treatment). The treatment was carried out with three replicates ($n = 5$). The paired t-test was statistically significant at $P < 0.05$ (*) compared to control cells.

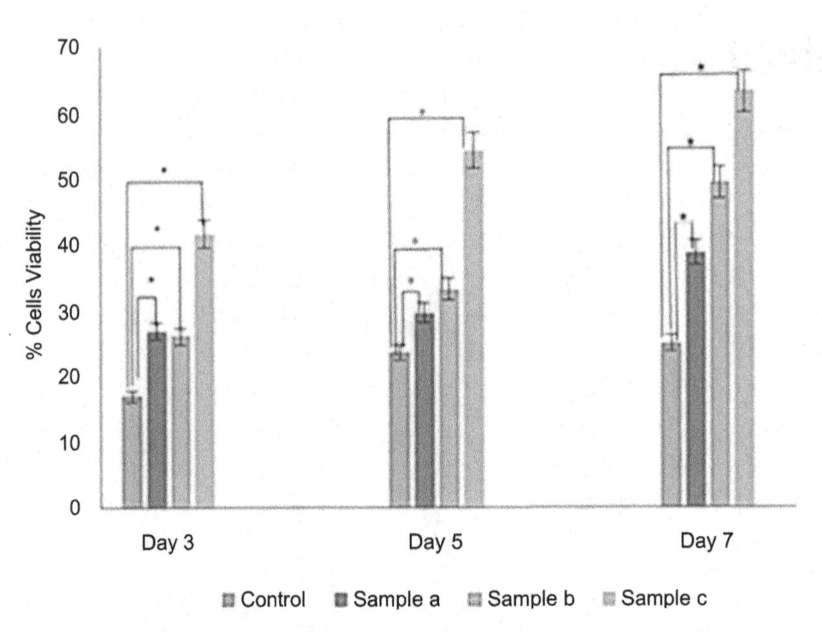

FIGURE 10.7 Cell proliferation of NHOst cells seeded directly onto the scaffolds; sample (a) PCL/CS, sample (b) nHA/PCL/CS, and sample (c) nHA/PCL/CS/TCH scaffolds and a control (cell without scaffolds). The MTT assay was performed on days 3 to 7. The treatment was carried out with 3 replicates ($n = 5$). The paired t-test analysis was statistically significant at $P < 0.05$ (*) compared to control cells.